Material Behavior Under High Stress and Ultrahigh Loading Rates

SAGAMORE ARMY MATERIALS
RESEARCH CONFERENCE PROCEEDINGS

Recent volumes in the series:

Material Behavior Under High Stress and Ultrahigh Loading Rates

Edited by
John Mescall

Army Materials and Mechanics Research Center
Watertown, Massachusetts

and
Volker Weiss

Syracuse University
Syracuse, New York

PLENUM PRESS • NEW YORK AND LONDON

Library of Congress Cataloging in Publication Data

Sagamore Army Materials Research Conference (29th: 1982: Lake Placid, N.Y.)
  Material behavior under high stress and ultrahigh loading rates.

  (Sagamore Army Materials Research Conference proceedings; 29th)
  "Proceedings of the 29th Sagamore Army Materials Research Conference...held July
19–23, 1982, at Lake Placid Club, Lake Placid, New York"—T.p. verso.
  Includes bibliographical references and index.
  1. Materials—Congresses. 2. Strains and stresses—Congresses. I. Mescall, John. II.
Weiss, Volker, 1930–   . III. Title. IV. Series.
TA401.3.S23   1982   620.1′12   83-17706
ISBN-13:978-1-4613-3789-8   e-ISBN-13:978-1-4613-3787-4
DOI: 10.1007/978-1-4613-3787-4

Proceedings of the 29th Sagamore Army Materials Conference
entitled Material Behavior Under High Stress and Ultrahigh
Loading Rates, held July 19–23, 1982, at Lake Placid Club,
Lake Placid, New York

©1983 Plenum Press, New York
Softcover reprint of the hardcover 1st edition 1983

A Division of Plenum Publishing Corporation
233 Spring Street, New York, N.Y. 10013

29TH SAGAMORE CONFERENCE COMMITTEE

Co-Chairmen

JOHN MESCALL
Army Materials and Mechanics Research Center

VOLKER WEISS
Syracuse University

PROGRAM COMMITTEE

RODNEY CLIFTON
Brown University

HARRY ROGERS
Drexel University

ROBERT ASARO
Brown University

M. F. KANNINEN
Battelle Columbus Laboratories

DONALD SHOCKEY
Stanford Research Institute International

ERIC KULA
Army Materials and Mechanics Research Center

T. NICHOLAS
Air Force Materials Laboratories

Conference Coordinators
MARY ANN HOLMQUIST and ROBERT J. SELL
Syracuse University

PREFACE

The Army Materials and Mechanics Research Center in coopera-
tion with the Materials Science Group of the Department of Chemical
Engineering and Materials Science of Syracuse University has been
conducting the Annual Sagamore Army Materials Research Conference
since 1954. The specific purpose of these conferences has been to
bring together scientists and engineers from academic institutions,
industry, and government who are uniquely qualified to explore in
depth a subject of importance to the Department of Defense, the
Army, and the scientific community.

The proceedings of this conference, entitled MATERIAL BEHAVIOR
UNDER HIGH STRESS AND ULTRAHIGH LOADING RATES, will be published
in two parts. The topics covered in the present volume include
dynamic plasticity, adiabatic shear/localized deformation, and
dynamic fracture mechanics. Papers dealing with ordnance applications, projectile launch environment, and recent work-in-progress
will appear as an AMMRC Technical Report and will have more limited
distribution in accordance with recent Army guidelines.

The Conference Chairmen are particularly grateful to the
members of the Program Committee. We wish also to acknowledge the
assistance of Mr. Charles Polley of the Army Materials and Mechanics
Research Center, Mr. Robert Sell, Ms. Helen Brown DeMascio, and
Ms. Mary Ann Holmquist of Syracuse University throughout the conference planning stages and the publication of the text.

The continued active interest in and support of these conferences by Dr. E. Wright and Col. George Sibert, Direct and Deputy
Director/Commander, respectively, of the Army Materials and Mechanics Research Center, is appreciated.

Syracuse University   John Mescall
Syracuse, New York   Volker Weiss

CONTENTS

SECTION III: DYNAMIC FRACTURE MECHANICS

DYNAMIC PLASTIC RESPONSE OF METALS UNDER PRESSURE-SHEAR IMPACT

Rodney J. Clifton,[*] Amos Gilat[**] and Chin-Ho Li[†]

*Division of Engineering
Brown University
Providence, RI 02912

**Department of Engineering Mechanics
Ohio State University
Columbus, OH 43210

†Nutech Engineers, Inc.
San Jose, CA 95119

INTRODUCTION

Two types of pressure-shear impact experiments have been developed for studying the dynamic plastic response of metals. In one type, called a symmetric impact experiment, two parallel plates of the same material are impacted with their impact faces inclined relative to the direction of approach. Dynamic plastic response of the material is inferred from the velocity-time profiles of the waves propagated through the plates. In the other type, called a high strain rate pressure-shear experiment, a thin specimen (0.2 - 0.4 mm thick) is sandwiched between two high impedance elastic plates. After several reverberations of waves through the specimen thickness the state of stress in the specimen becomes nominally homogeneous. Then, a dynamic stress-strain curve for the specimen can be obtained from a time-history of the traction at one face of the specimen and the difference between particle velocities at the two faces. These stress-time and velocity-time records can be obtained by monitoring the stress waves generated in one of the elastic plates.

Techniques for conducting pressure-shear experiments have been developed in a series of investigations beginning with the introduction of the skewed plate impact configuration (Abou-

1

Sayed, Clifton, and Hermann, 1976). The next step was the development of the transverse displacement interferometer (TDI) for monitoring the in-plane motion of the rear surface of the target plate (Kim, Clifton and Kumar, 1977). These developments were combined (Kim and Clifton, 1980) in a pilot investigation of the use of the symmetric impact experiment for studying the dynamic plastic response of 6061-T6 aluminum under pressure-shear loading. In related work, Chhabildas, Sutherland and Asay (1979) studied pressure-shear waves in 6061-T6 aluminum by using combined longitudinal and transverse waves generated by normal impact of an anisotropic elastic plate (y-cut quartz) to impose pressure-shear loading on an aluminum specimen bonded to the rear surface of the elastic plate. They monitored the longitudinal and transverse motion of the rear surface of the aluminum plate by using two VISARs (Barker and Hollenbach, 1972) with incident beams inclined relative to the normal to the plate in order to obtain two compo-nents of the particle velocity at the rear surface. Gilat (1982), using the symmetric impact experiment with a TDI and a normal velocity interferometer (NVI) to monitor simultaneously the in-plane and normal components of the velocity of the rear surface, conducted a systematic investigation of the plastic response of 6061-T6 aluminum and a commercially pure alpha-titanium. Li (1981), using the same instrumentation, conducted high strain rate pressure-shear experiments on thin specimens of commerically pure aluminum, 6061-T6 aluminum and OFHC copper. He obtained dynamic stress-strain curves at strain rates up to 3×10^5 s^{-1}. A brief discussion of his early results on commerically pure aluminum has been given by Li and Clifton (1981).

From the results of the foregoing experiments it appears that two principal conclusions can be drawn. First, for the materials listed, the flow stress for plastic straining increases markedly with increasing strain rates at strain rates greater than 10^4 s^{-1}. Second, the high strain rate response of the materials studied de-pends only weakly on the hydrostatic pressure. Experimental evi-dence for these conclusions is reviewed briefly in the next two sections. An elementary interpretation of the conclusions in terms of dislocation models is presented in the final section.

HIGH STRAIN RATE PRESSURE-SHEAR EXPERIMENTS

The experimental configuration for these experiments is shown in Fig. 1. The projectile is guided by a 63.5 mm barrel with a keyway to prevent rotation. Its velocity V_0 is obtained from the times at which the flyer contacts five accurately spaced pins mounted at the muzzle of the barrel. Impact occurs in a vacuum chamber. The impact faces are lapped flat to approximately one wavelength of light over the diameter. These faces are aligned to be parallel by means of an optical technique developed by Kumar and Clifton. Parallelity at impact,

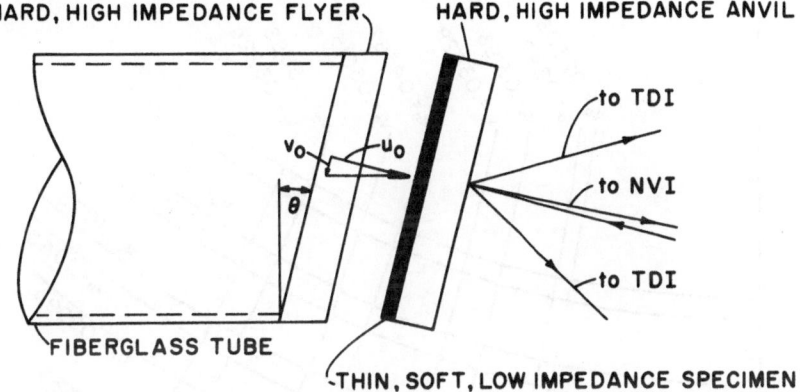

HARD, HIGH IMPEDANCE FLYER   HARD, HIGH IMPEDANCE ANVIL

FIBERGLASS TUBE

THIN, SOFT, LOW IMPEDANCE SPECIMEN

Fig. 1.   Schematic of high strain-rate experiment.

determined by recording the times at which four voltage-biased
contact pins on the target make contact with the flyer, is
generally better than 10^{-3} radians. The specimen is mounted either
on the anvil (conventional test) or on the flyer (reversed test).
Adhesive is used only at the periphery of the specimen to avoid
lubrication of an interface that is to transmit a shear traction.
The motion of the rear surface of the anvil is monitored with a
combined NVI and TDI. A 200 lines/mm diffraction grating copied
onto the rear surface of the anvil is used to provide the
symmetrically diffracted beams required for the TDI. Further
details on the experimental techniques are given in the thesis by
Li (1982).

    The nominal longitudinal strain rate $\dot{\varepsilon}$ and nominal shear
strain rate $\dot{\gamma}$ are obtained from the normal and in-plane components,
u_o and v_o, of the flyer plate velocity and the normal and in-plane
components, u_{fs} and v_{fs}, of the free surface velocity at the rear
of the anvil. Once a nominally homogeneous state of stress is
established in the specimen the nominal strain rates are

$$\dot{\varepsilon} = \frac{u_o - u_{fs}}{h} \tag{1}$$

$$\dot{\gamma} = \frac{v_o - v_{fs}}{h} \tag{2}$$

where h is the specimen thickness. Linear elastic wave propagation
in the anvil and flyer plates is assumed in relating the traction
and motion at the faces of the specimen to the velocity at the free

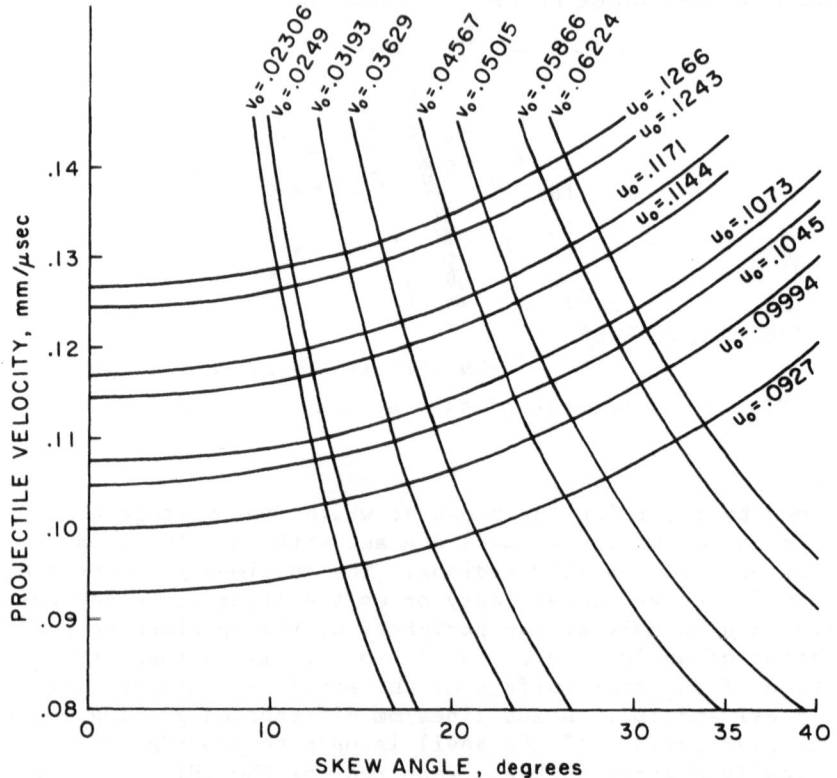

Fig. 2.   Constant u_0 and constant v_0 lines in the $V_0 - \theta$ plane.

surface of the anvil. The normal and shear tractions at the rear face of the specimen are given by

$$\sigma = \tfrac{1}{2}\rho c_1 u_{fs} \qquad (3)$$

$$\tau = \tfrac{1}{2}\rho c_2 v_{fs} \qquad (4)$$

where ρ, c_1, c_2, are, respectively, the mass density, the elastic longitudinal wave speed, and the elastic shear wave speed for the anvil. A dynamic stress-strain curve for shearing deformation is obtained by plotting the shear stress τ from (4) versus the shear strain γ obtained from the integration of (2). This stress-strain response is at essentially constant pressure since, because of the specimen's elastic stiffness to volume changes, the volume becomes essentially constant (i.e. $\dot{\varepsilon} \to 0$) after a few reverberations of longitudinal waves through the thickness of the specimen. Numerical simulation of the experiment shows that the value of the hydrostatic pressure is approximately equal to the value of the normal stress σ obtained from (3) as $u_{fs} \to u_o$ (Li, 1982).

From (1)-(4), and the preceding discussion, the shear strain rate is determined primarily by the component v_o of the projectile velocity and the hydrostatic pressure is determined primarily by the component u_o. Thus, the effects of shear strain rate and pressure on the flow stress in shear can be examined independently by systematic variation of v_o or u_o. Such systematic variation is obtained by choosing values of the angle θ and projectile velocity V_o which lie along contours of constant v_o or u_o, shown in Fig. 2. Angles θ used are (14°, 18.4°, 26.6°, 33.7°) corresponding to ratios u_o:v_o of (4:1, 3:1, 2:1, 3:2). No slip appears to occur at the interfaces even for the largest values of θ.

Representative dynamic stress-strain curves for annealed 1100-0 aluminum, as-received 6061-T6 aluminum, and annealed OFHC copper are shown in Figs. 3-5.

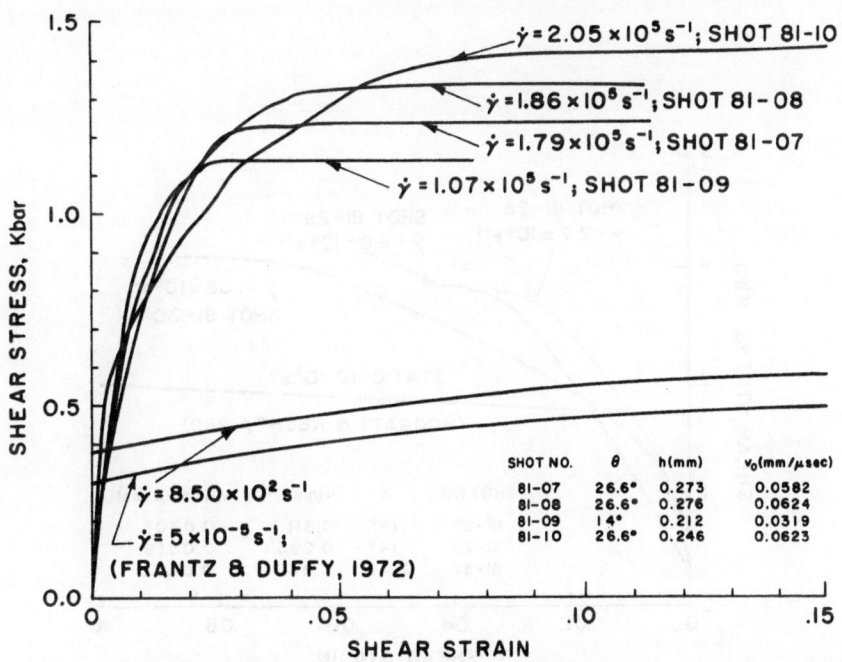

Fig. 3. Dynamic stress-strain curves for 1100-0 Aluminum at P = 28.1 Kbar.

The pressure-shear experiments reported in each figure were conducted at approximately the same value of u_o to exclude variations in hydrostatic pressure that could affect the interpretation of the results. However, for the case of 1100-0 aluminum in which 27 pressure-shear experiments were conducted, the effect of hydrostatic pressure was found to be statistically insignificant. Thus, the marked difference between the stress-strain curves at $\sim 10^5$ s^{-1} and those at strain rates of 5 x 10^{-5}s^{-1} and 8.5 x 10^2s^{-1} is believed to be due to the differences in strain rate, although there is approximately a 28.1 kbar difference in pressure between the experiments reported here and the dynamic torsion tests by Frantz and Duffy (1972). The series of experiments on 6061-T6 aluminum (8 tests) and OFHC copper (2 tests) were not sufficiently comprehensive to determine whether or not the flow stress in shear is affected significantly by hydrostatic pressure. However, it is expected that the increase in flow stress shown in Figs. 4 and 5 for increases in strain rate from $\sim 10^3$ s^{-1} (Hogatt and Recht, 1969; Senseny, Richman and Duffy, 1975) to $\sim 10^5$ s $^{-1}$ is primarily a strain rate effect and not the effect of the increased hydrostatic pressure.

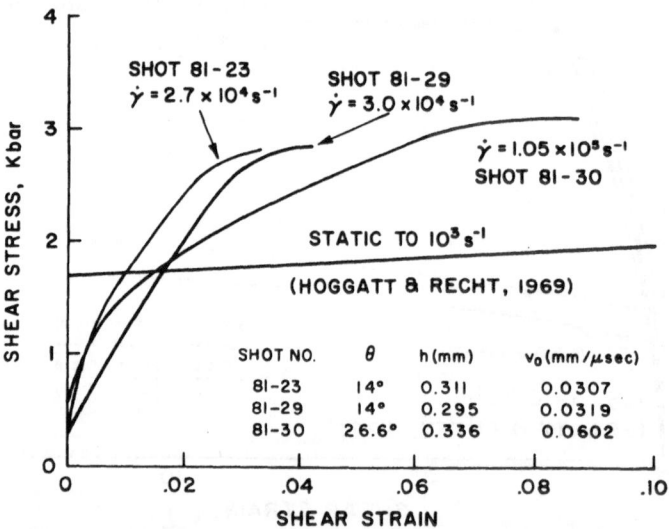

Fig. 4.   Dynamic stress-strain curves for 6061-T6 Aluminum at P = 28.2 Kbar.

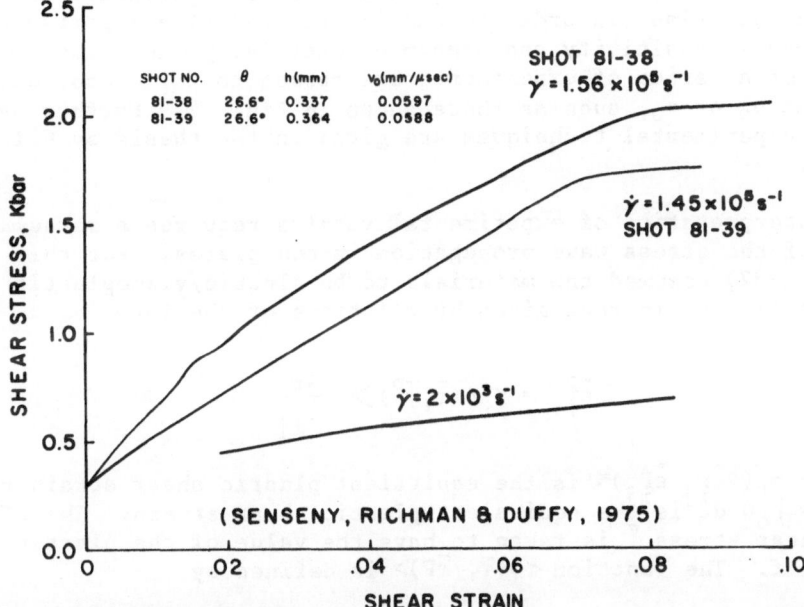

SHOT NO.	θ	h(mm)	v_0(mm/μsec)
81-38	26.6°	0.337	0.0597
81-39	26.6°	0.364	0.0588

Fig. 5.   Dynamic stress-strain curves for OFHC Copper at
          P = 60 Kbar.

As in other experiments in which several wave reverberations
through the specimen thickness are required before nominally homogeneous states of stress are obtained (Kolsky, 1949) the early
parts of the stress-strain curves in Figs. 3-5 are not indicative
of the material response for homogeneous deformations. The steeply
rising parts of the stress-strain curves in these figures should be
discounted for this reason. Finally, the stress-strain curves at
the lower strain rates in Figs. 3-5 have been shifted to the left
by approximately a 1.5% shear strain to provide more direct
comparison with the curves obtained from the pressure-shear experiment in which the longitudinal compressive waves cause an additional effective shear strain of approximately 1.5%.

SYMMETRIC IMPACT EXPERIMENTS

The experimental configuration for the symmetric impact
experiments is similar to that shown in Fig. 1 with the thin
specimen removed. Direct impact of the flyer and anvil plates
causes these plates to be deformed plastically. Instrumentation
for recording the projectile velocity, the parallelity of the
impact faces at impact, and the velocity of the rear surface of the

anvil is the same as for the high strain rate experiments described
previously. Also, in order to examine the relative importance of
strain-rate sensitivity and pressure sensitivity the values of V_o
and θ for a series of experiments are chosen to follow contours of
constant u_o or v_o, such as those shown in Fig. 2. Further details
on the experimental techniques are given in the thesis by Gilat
(1982).

Interpretation of experimental results requires a mathematical
model of the stress wave propagation in the plates. For this model
Gilat (1982) assumed the materials to be elastic/viscoplastic with
the plastic strain rate given by relations of the form

$$\dot{\varepsilon}^P_{ij} = \langle \phi(\bar{\tau}, \bar{\gamma}^P) \rangle \frac{\partial f}{\partial \sigma_{ij}} \tag{5}$$

where $\phi =_t (2\dot{\varepsilon}^P_{ij} \dot{\varepsilon}^P_{ij})^{\frac{1}{2}}$ is the equivalent plastic shear strain rate
and $\bar{\gamma}^P = \int_o \phi \, dt$ is the equivalent plastic shear strain. The effec-
tive shear stress $\bar{\tau}$ is taken to have the value of the plastic potential f. The function $\langle \phi(\bar{\tau}, \bar{\tau}P) \rangle$ is defined by

$$\langle \phi(\bar{\tau}, \bar{\gamma}^P) \rangle = \begin{cases} 0 & \text{for } \bar{\tau} < k \\ \phi(\bar{\tau}, \bar{\gamma}^P) & \text{for } \bar{\tau} > k \end{cases} \tag{6}$$

where k is a quasi-static yield stress in pure shear. Two flow
potentials were used:

$$f(\sigma_{ij}) = (\tfrac{1}{2}s_{ij} s_{ij})^{\frac{1}{2}} \text{ (isotropic hardening)} \tag{7a}$$

and

$$f(\sigma_{ij}, \alpha_{ij}) = [\tfrac{1}{2}(s_{ij} - \alpha_{ij})(s_{ij} - \alpha_{ij})]^{\frac{1}{2}} \text{ (kinematic hardening)} \tag{7b}$$

where s_{ij} denotes the components of the stress deviator and α_{ij}
denotes the coordinates of the center of the translated yield
surface. The summation convention of summation over repeated in-
dices is used in (7) and throughout. The yield surfaces defined by
$\bar{\tau} = k$ are

$$\tfrac{1}{2}(s_{ij} s_{ij}) = k^2(\bar{\gamma}P) \text{ (isotropic hardening)} \tag{8a}$$

$$\tfrac{1}{2}(s_{ij} - \alpha_{ij})(s_{ij} - \alpha_{ij}) = k^2_o \text{ (kinematic hardening)} \tag{8b}$$

where $k(\bar{\gamma}P)$ is the quasi-static flow stress in pure shear at a
plastic shear strain $\bar{\gamma}P$, and k_o is the initial yield stress in shear.
The yield surface (8b) translates in the direction of the

normal to the flow potential according to the relation

$$\dot{\alpha}_{ij} = \dot{\mu}\, \frac{\partial f}{\partial \sigma_{ij}} \qquad (9)$$

where μ is a scalar function of $\bar{\gamma}P$ -- chosen to provide agreement with experiments for quasi-static proportional loading paths.

Representative velocity-time profiles for the transverse component of the velocity at the rear surface of the anvil plate (thickness d) are shown in Figs. 6 and 7 for a series of experiments on
6061-T6 aluminum specimens (7 tests) and alpha-titanium specimens
(10 tests). In making the comparisons between theory and experiment
the plastic strain rate function $\phi(\bar{\tau}, \bar{\gamma}P)$ for 6061-T6 aluminum is
taken to be the one used by Read, Triplett and Cecil (1971), based
on data from quasi-static stress-strain curves, dynamic stress-
strain curves at strain rates of 10^3 s^{-1} and plate impact experiments at normal incidence (i.e. $\theta = 0$). For alpha-titanium the
function $\phi(\bar{\tau}, \bar{\gamma}P)$ is, for strain rates up to 10^3 s^{-1}, taken to be
the one used by Hsu and Clifton (1974). At high strain

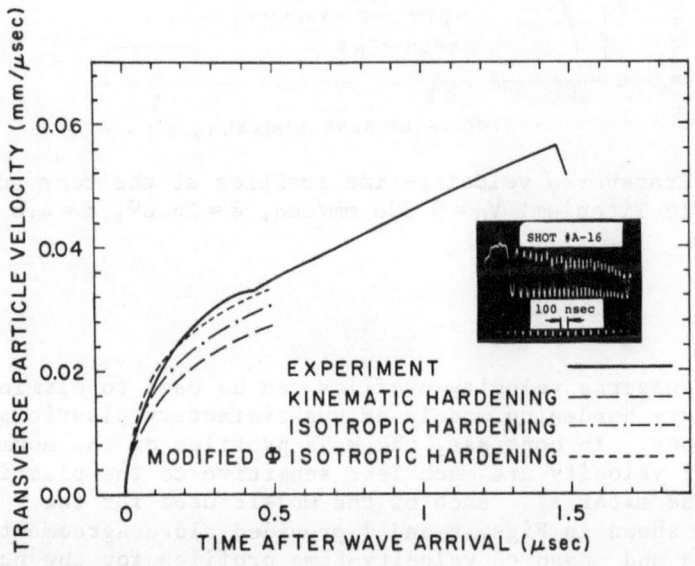

Fig. 6.   Transverse velocity-time profiles at the rear surface.
(6061-T6 Aluminum, $V_0 = 0.22$ mm/sec, $\theta = 26.6^{\circ}$, d = 3.2 mm)

rates the function is adjusted in order to improve agreement
between the theoretical and experimental velocity-time profiles for
the normal component of particle velocity.

Several characteristic features emerge when comparisons such
as those in Figs. 6 and 7 are made for a series of experiments.
First, the wave profiles of the transverse component of the free
surface velocity are sensitive to the plastic response of the
material. Thus, for example, comparisons of calculated and

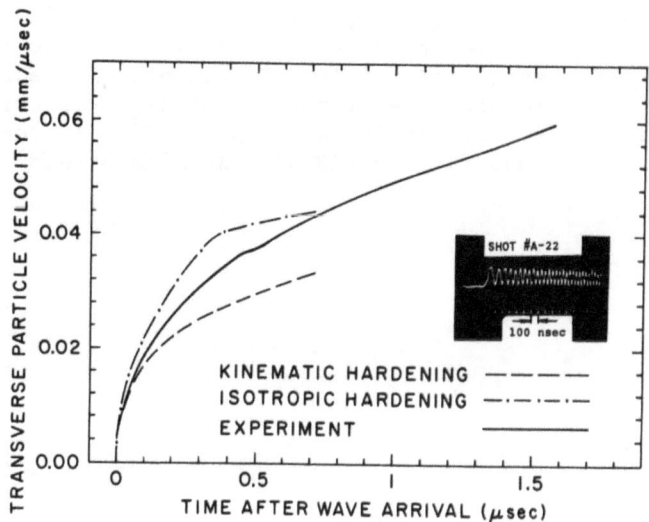

Fig. 7.   Transverse velocity-time profiles at the rear surface.
(α-Titanium, $V_O = 0.208$ mm/sec, θ = 26.6°, d = 4.6 mm)

measured transverse velocity profiles can be used to eliminate
unsatisfactory hardening models or unsatisfactory plastic strain
rate functions. In contrast, the wave profiles of the normal
component of velocity are much less sensitive to the plastic response of the material. Each of the models used for the
comparisons shown in Figs. 6 and 7 provided close agreement between
the computed and measured velocity-time profiles for the normal
component of the free surface velocity for all experiments
conducted on the respective materials.

A second general feature of the comparisons between theory and experiment is that for all experiments on 6061-T6 aluminum specimens the calculated transverse velocity profiles lie below the measured ones. The overall downward shift of the calculated profiles is believed to be due to the inadequacy of the strain rate function $\phi(\bar{\tau}, \bar{\gamma}P)$ (Read et al., 1971) for modeling plastic response at strain rates greater than 10^4 s^{-1}. At these high strain rates this function is such that for a given plastic strain rate, the flow stress decreases with increasing plastic strain. When the dependence on plastic strain in the function $\phi(\bar{\tau}, \bar{\gamma}P)$ is modified to obtain agreement with high strain rate pressure-shear experiments at strains of 10-15% (Li, 1982), the flow stress increases with increasing plastic strain and the agreement between theory and experiment is improved as shown in Fig. 6 for the modified $\phi(\bar{\tau}, \bar{\gamma}P)$.

For all experiments on alpha-titanium specimens the measured transverse velocity profiles are below the profiles computed for an isotropic hardening model and above those computed for a kinematic hardening model. This result suggests that the plastic strain rate function $\phi(\bar{\tau}, \bar{\gamma}P)$ used for the calculations is reasonably satisfactory and that satisfactory agreement can be obtained by a modified hardening model -- perhaps by a combination of isotropic and kinematic hardening. The marked difference between the calculated profiles for the two different hardening models is due largely to differences that occur when the unloading wave reflected from the rear surface interacts with the oncoming plastic wave. Because of the translation of the yield surface in the case of kinematic hardening the stress state after unloading is relatively far outside the yield surface. This causes a relatively large plastic strain rate and the corresponding softness of the materials' response leads to the reduced levels of transverse velocity shown in Fig. 7.

Finally, inclusion of an effect of hydrostatic pressure on plastic response does not appear to be critical in obtaining good agreement between theory and experiment. For the experiments on 6061-T6 aluminum the angle θ was varied from 14° to 33.8° with no apparent effect of hydrostatic pressure on the differences between computed and measured transverse velocity profiles. For the experiments on alpha-titanium the angle θ was varied from 9.5° to 33.8° with, at most, a small effect of increased hydrostatic pressure tending to raise the measured values of transverse velocity relative to the computed values.

INTERPRETATION OF RESULTS

The marked increase in flow stress obtained for the high strain rate pressure-shear experiments is viewed as indicative of a change in the rate controlling mechanism of dislocation motion. For 1100-0 aluminum the flow stress at strain rates 10^{-3}s^{-1} to 10^3s^{-1} satisfy an Arrhenius relation:

$$\dot{\gamma}^P = \omega \, \exp \left[\frac{-(\tau^*(\gamma^P) - \tau)v^*}{KT} \right] \qquad (10)$$

where, for a model of thermally activated motion of dislocation past obstacles, $\tau^*(\gamma P)$ is the stress barrier, τ is the applied shear stress, v^* is the activation volume, K is Boltzmann's constant and T is the absolute temperature. The pre-exponential factor can be viewed as a product $\omega = bNL\omega_0$ of the Burgers vector b, the mobile dislocation density N, a characteristic distance L between obstacles, and an attempt frequency ω_0. Equation (10) gives a linear de-pendence of the flow stress τ on the logarithm of the strain rate $\dot{\gamma}^P$.

Over the range of strain rates (0.5–$3.1 \times 10^5 s^{-1}$) spanned by the high strain rate pressure-shear experiments the flow stress at the maximum strain obtained in these experiments can be fit by the linear relationship

$$\tau \simeq \tau_o + \mu_o \dot{\gamma}^P \qquad (11)$$

with $\tau_o = 0.77$ Kbar and $\mu_o = 0.27 \times 10^{-5}$ Kbar s^{-1}. Equation (11) should be viewed as a strictly empirical relation that shows the trend of the experimental results over the indicated range of strain rates. Plastic strain is not regarded as an independent variable in obtaining the fit (11), even though it varies approximately in proportion to the strain rate since the duration of the loading pulse was comparable in all the experiments. Attribution of the change in flow stress to the change in strain rate and not to the change in strain is justified by the lack of stress increase with increasing strain at constant strain rate in the individual tests (See Figure 3) and by the fact that the observed changes in flow stress are much larger than expected for strain hardening in the strain regime of the experiments.

The empirical relation (11) does not have the form expected from an elementary model of viscous drag. According to such a model the dislocation velocity v_d is related to the local resolved shear stress τ_ℓ by

$$\tau_\ell b = B v_d \qquad (12)$$

where B is the drag coefficient and b is again the Burgers vector. For single slip the plastic strain rate is

$$\dot{\gamma}^P_\ell = bNv_d \qquad (13)$$

where N is the mobile dislocation density. The macroscopic plastic shear strain rate $\dot{\gamma}P$ and the macroscopic shear stress τ are related to the local plastic shear strain rate $\dot{\gamma}_\ell^P$ and the elastic stress field $\hat{\tau}(\underset{\sim}{x})$ due to the tractions τ by the volume average (Rice, 1970):

$$\tau\dot{\gamma}^P = \frac{1}{V} \int_V \hat{\tau}(\underset{\sim}{x})\dot{\gamma}_\ell^P(\underset{\sim}{x})dV \tag{14}$$

in which, for multiple slip, $\hat{\tau}(\underset{\sim}{x})\dot{\gamma}^P(\underset{\sim}{x})$ denotes the sum over all slip systems of such products. If, the local quantitites are written in the form

$$f_\ell(\underset{\sim}{x}) = f + \delta f(\underset{\sim}{x}) \tag{15}$$

where δf denotes the variation from the mean value f, then substitution of (12) and (13) into (14) gives

$$\tau\dot{\gamma}^P \approx \frac{b^2N\tau^2}{B} + \frac{b}{V} \int_V \{\tau\delta N(\underset{\sim}{x})\delta v(\underset{\sim}{x}) + N\delta\hat{\tau}(\underset{\sim}{x})\delta v(\underset{\sim}{x}) + v\delta N(\underset{\sim}{x})\delta\hat{\tau}(\underset{\sim}{x})\}dV \tag{16}$$

where τ, N, v denote mean values of the shear stress, dislocation density and dislocation velocity. The approximation in (16) is the neglect of the volume average of $\delta v(\underset{\sim}{x})\delta\hat{\tau}(\underset{\sim}{x})\delta N(\underset{\sim}{x})$ as being of higher order. Using (12) to eliminate $\delta v(\underset{\sim}{x})$ in (16) one obtains

$$\tau = \frac{B\dot{\gamma}^P}{b^2N(1 + \beta)} \tag{17}$$

where

$$\beta = \frac{1}{V} \int_V \left\{\frac{\delta N(\underset{\sim}{x})\delta\tau_\ell(\underset{\sim}{x})}{N\tau} + \frac{\delta\tau_\ell(\underset{\sim}{x})\delta\hat{\tau}(\underset{\sim}{x})}{\tau^2} + \frac{\delta N(\underset{\sim}{x})\delta\hat{\tau}(\underset{\sim}{x})}{N\tau}\right\}dV \tag{18}$$

and $\delta\tau_\ell(\underset{\sim}{x})$ denotes the deviation of the actual local stress field from the mean value. The variation $\delta\tau_\ell(\underset{\sim}{x})$ is expected to correlate positively with both the elastic stress variation $\delta\hat{\tau}(\underset{\sim}{x})$ and the variation $\delta N(\underset{\sim}{x})$ in the mobile dislocation density. Thus, the term β in (17) is expected to be positive. Equation (17) is the relationship between the macroscopic shear stress and shear strain rate for dislocation motion satisfying the linear drag model (12).

Although experimental data $(\tau, \dot{\gamma}P)$ that fit (11) are sometimes regarded as indicating that viscous drag is the operative rate controlling mechanism, such a conclusion is not justified in view of the differences in form of (11) and (17). Indeed, Equation (11) includes a "cut-off" stress τ_0 that does not appear in (17) and is too large to attribute to the Peierls barrier, which is neglected in (12) and, consequently, in (17). It appears preferable to

interpret the fit of the data by (11) as indicating that the
viscous drag model (12) does not describe the operative rate controlling mechanism. Instead, the response is indicative of a transition regime between the regime where thermally activated processes
are rate controlling and the regime where viscous drag processes
are rate controlling. In this transition regime the dislocation
velocity is expected to increase with increasing stress in such a
way that a tangent to the stress-velocity curve would give a positive
cut-off stress. For small variations (δv, $\delta \hat{\tau}$, δN) the curve of τ
vs. $\dot{\gamma}P$ would be similar in shape to the curve of τ vs. v_d. Thus,
over a finite interval of strain rates the curve would be fit by a
relation of the form (11).

An indication of the expected effect of hydrostatic pressure
on the flow stress can be obtained for the dislocation models (10)
and (12,13). For this, consider a thought experiment in which the
pressure is changed from its reference value while temperature,
plastic strain and plastic strain rate are unchanged. Let the
lengths of line elements be changed by the factor λ due to the change
in pressure. Parameters that include length dimensions would be
changed according to

$$b = \lambda b_o, \qquad \rho = \rho_o \lambda^{-3}, \qquad N = N_o \lambda^{-2} \qquad (19)$$

where a subscript 'o' denotes the value of the corresponding quantity
at the reference pressure (i.e. for $\lambda = 1$). When viscous drag is
the rate controlling mechanism it follows from (17) and (19) that
after the change in pressure the flow stress, at fixed strain rate,
would be

$$\tau = \frac{B\dot{\gamma}^P}{b_o^2 N_o (1 + \beta)} \qquad (20)$$

where B, β are the values of the corresponding quantities in (17) at
the new pressure. For conditions under which viscous drag mechanisms are dominant it appears that β depends weakly on pressure.
Contributions to β from the quotient $(\delta \hat{\tau}(\underset{\sim}{x})/\tau)$ depend on ratios of
elastic moduli which are relatively insensitive to pressure. The
quotient $(\delta \tau(\underset{\sim}{x})/\tau)$ includes $(\delta \hat{\tau}(\underset{\sim}{x})/\tau)$ plus $(\delta \tau_d(\underset{\sim}{x})/\tau)$ where $\delta \tau_d(\underset{\sim}{x})$
is the residual elastic field due to defects such as dislocations,
impurities, and grain boundaries. When viscous drag mechanisms are
dominant the quotients $(\delta \tau_d(\underset{\sim}{x})/\tau)$ should be much less than unity
and the resulting contribution to β should be insignificant. Thus,
neither of the quotients in (18) appears to contribute an important
pressure dependent term to β for the viscous drag case governed by
(17). Consequently, for this case the variation of τ with pressure
follows the variation of the drag coefficient B with pressure.

For dislocation motion governed by phonon viscosity the value of the drag coefficient can be estimated by (e.g. Hirth and Lothe, 1968)

$$B \simeq \frac{3KT}{10b^2(G/\rho)^{\frac{1}{2}}} = \frac{3KT}{10b_o^2} G^{-\frac{1}{2}} \lambda^{-7/2} \tag{21}$$

where all quantities have been defined previously, and (19) is used to rewrite the first expression in terms of λ, ρ_o and b_o. Differentiation of (21) with respect to the pressure p and evaluation at the reference pressure gives

$$\frac{dB/dp}{B} = -\frac{dG/dp}{2G} + \frac{7C_v}{6} \tag{22}$$

where $C_v = -3(d\lambda/dp)$ is the bulk compliance. Thus, the fractional change in flow stress can be written as

$$\frac{d\tau}{\tau} = \left\{ -\tfrac{1}{2} \frac{dG}{dp} + \frac{7(1 - 2\nu)}{4(1 + \nu)} \right\} \frac{dp}{G} \tag{23}$$

where ν is Poisson's ratio and C_v has been replaced by $3(1-2\nu)/(2G(1+\nu))$.

For aluminum, the values of ν and (dG/dp) are, respectively, 1/3 and approximately 17/8 (Jung, 1981) so that (23) reduces to

$$\frac{d\tau}{\tau} \simeq -\frac{5}{8} \frac{dp}{G} \tag{24}$$

where G = 274 kbar for aluminum. Equation (24) gives the unexpected result that, for dislocation motion controlled by phonon viscosity, the flow stress is predicted to decrease with increasing pressure. The sensitivity of the flow stress to changes in pressure is, however, predicted to be weak since an increase in pressure of 10 kbar would result in only a 2.3% reduction in the flow stress. Such variations in flow stress are below the resolution of the experiments described here. Independent measurements of the variation of the drag coefficient B with pressure do not appear to be available.

For the case of dislocation motion controlled by thermal activation past barriers the effect of hydrostatic pressure can be analyzed by considering the effect of pressure on the various

quantities in (10). Due to changes in pressure these quantities
become

$$v^* = v_o^* \lambda^3, \qquad \tau^* = G\tau_o^*/G_o, \qquad \omega = \omega_o \left(\frac{G\lambda}{G_o}\right)^{\frac{1}{2}} \qquad (25)$$

where subscripts 'o' again denote evaluation at the reference pressure
for which $\lambda = 1$. Substitution of (25) into (10), differentiation
with respect to the pressure p, and evaluation at the reference
pressure gives

$$\frac{d\dot{\gamma}^P/dp}{\dot{\gamma}^P} = \frac{dG/dp}{2G_o} - \frac{C_v}{6} - \frac{v_o^* \tau_o}{KT} \left[\frac{\tau_o^*(dG/dp)}{\tau_o G_o} - \frac{d\tau/dp}{\tau_o} - C_v\left(\frac{\tau_o^*}{\tau_o} - 1\right)\right] \qquad (26)$$

where C_v again denotes the bulk compliance. At fixed plastic strain
rate the left side of (26) is zero and the fractional change in
flow stress becomes

$$\frac{d\tau}{\tau_o} = \left\{\frac{\tau_o^*}{\tau_o}\frac{dG}{dp} - \frac{3(1-2\nu)}{2(1+\nu)}\left(\frac{\tau_o^*}{\tau_o} - 1\right)\right.$$

$$\left. + \frac{(\partial\tau/\partial\ell n\dot{\gamma}^P)}{\tau_o}\left[\frac{(1-2\nu)}{4(1+\nu)} - \frac{1}{2}\frac{dG}{dp}\right]\right\}\frac{dp}{G_o} \qquad (27)$$

where C_v has been replaced as in (23) and $(\partial\tau/\partial\ell n\dot{\gamma}^P)$ has been substituted for KT/v_o^*. Substitution of $\nu = 1/3$ and $dG/dp = 17/8$ in
(27) to model aluminum (cf. (24)) gives

$$\frac{d\tau}{\tau_o} \approx \left\{\frac{7}{4}\frac{\tau_o^*}{\tau_o} + \frac{3}{8} - \frac{(\partial\tau/\partial\ell n\dot{\gamma}^P)}{\tau_o}\right\}\frac{dp}{G_o} \qquad (28)$$

for aluminum. The strain rate sensitivity term in (28) is
generally negligible for aluminum and aluminum alloys (e.g. for
1100-0 aluminum this term is less than 0.02 based on data shown in
Fig. 3 for strain rates less than 10^3 s^{-1}). The ratio (τ_o^*/τ_o) of
the barrier stress, or the flow stress at 0° Kelvin, to the flow
stress at the reference temperature, pressure, and strain rate is
greater than unity. From the results of Lindholm (1977) this ratio
appears to be approximately 1.5 - 2.3 for plastic deformation of
1100-0 aluminum at strain rates of 10^3 s^{-1} (at room temperature and
at atmospheric pressure). Thus, from (28), the flow stress is
predicted to increase with increasing pressure when the thermally
activated motion of dislocations past obstacles is the rate-
controlling mechanism of plastic deformation. The sensitivity of

the flow stress to changes in pressure is predicted to be much greater than for the case of viscous drag, with an increase in pressure of 10 kbar causing an increase in flow stress of 10-20% for 1100-0 aluminum.

The predictions of both qualitative and quantitative differences in the effect of pressure on the flow stress when different dislocation mechanisms are rate controlling allows the possibility of reconciling the results of the pressure-shear experiments with other experiments on the pressure sensitivity of plastic response. Thus, the significant positive influence of pressure on the flow stress of steels reported by Richmond and Spitzig (1980) at quasi-static strain rates can be attributed to the thermally activated motion of dislocations for which (27) applies. Recent quasi-static experiments by Richmond (1982) on aluminum alloys show a positive effect of pressure on flow stress that is more pronounced than for steels. This result, as well as results of related investigations, (e.g. Yoshida and Oguchi, 1970); can also be interpreted as indicating that the plastic response is in a regime for which (27) is applicable. At very high strain rates, such as occur in the pressure-shear experiments, the pressure sensitivity predicted from (27) or (28) is expected to decrease as the quotient (τ^*_o/τ_o) decreases and the quotient $(\partial\tau/\partial\ell n\dot\gamma P)/\tau_o$ increases. Furthermore, at sufficiently high shear stresses and shear strain rates the rate controlling mechanism is expected to become viscous drag for which a small negative effect of pressure on the flow stress is predicted by (23). Thus, the lack of pressure sensitivity observed in the pressure-shear experiments on aluminum can be viewed as consistent with the relatively weak pressure sensitivity predicted for the plastic response in the regime of transition between thermal activation and viscous drag as the governing dislocation mechanism. The previous discussion of the comparison of (11) with (17) suggests that the plastic repsonse in the high strain-rate pressure-shear experiments is in such a transition regime. Finally, the effects of pressure discussed here are for moderate changes in pressure. Qualitatively different effects of pressure on flow stress are reported for the high pressures used in shock wave studies such as those conducted by Chhabildas and Asay (1981).

ACKNOWLEDGEMENTS

This research was supported by the Army Research Office and the National Science Foundation, including support from the NSF Brown University Materials Research Laboratory. The computations reported here were carried out on the Brown University, Division of Engineering, VAX-11/780 computer. The acquisition of this computer was made possible by grants from the U.S. National Science Foundation, the General Electric Foundation and the Digital Equipment Corporation.

REFERENCES

Abou-Sayed, A.S., Clifton, R.J., and Hermann, L., 1976, The
    Oblique-Plate Impact Experiment, Exp. Mechs., 16:127.
Barker, L.M., and Hollenbach, R.E., 1972, Laser Interferometer for
    Measuring High Velocities of any Reflecting Surface, J. Appl.
    Phys., 43:4664.
Chhabildas, L.C., and Asay, J.R., 1981, Determination of the Shear
    Strength of Shock Compressed 6061-T6 Aluminum, in: "Shock
    Waves and High-Strain-Rate Phenomena in Metals," Meyers, M.A.
    and Murr, L.E. eds., Plenum, New York.
Chhabildas, L.C., Sutherland, H.J., and Asay, J.R., 1979, Velocity
    Interferometer Technique to Determine Shear-Wave Particle
    Velocity in Shock-Loaded Solids, J. Appl. Phys., 50:5196.
Frantz, R.A. Jr., and Duffy, J., 1972, The Dynamic Stress-Strain
    Behavior in Torsion of 1100-0 Aluminum Subjected to a Sharp
    Increase in Strain Rate, J. Appl. Mech., 39:939.
Gilat, Amos, 1982, An Experimental Investigation of Pressure-Shear
    Waves in 6061-T6 Aluminum and Alpha-titanium, Ph.D. thesis,
    Brown University, Providence.
Hirth, J.P., and Lothe, J., 1968, "Theory of Dislocations," McGraw-
    Hill, New York, p. 195.
Hogatt, C.R., and Recht, R.F., 1969, Stress-Strain Data Obtained at
    High Rates Using an Expanding Ring, Exp. Mech., 9:441.
Hsu, J.C.C., and Clifton, R.J., 1974, Plastic Waves in a Rate
    Sensitive Material, Part I: Waves of Uniaxial Stress, J. Mechs.
    Phys. Solids, 22:253.
Jung, J., 1981, A note on the influence of hydrostatic pressure on
    dislocations, Phil. Mag. A, 43:1057.
Kim, K.S., Clifton, R.J., and Kumar, P., 1977, A Combined Normal
    and Transverse Displacement Interferometer with an Application
    to Impact of Y-cut Quartz, J. Appl. Phys., 48:4132.
Kim, K.S., and Clifton, R.J., 1980, Pressure-Shear Impact of 6061-
    T6 Aluminum, J. Appl. Mech., 47:11.
Kolsky, H., 1949, An Investigation of the Mechanical Properties of
    Materials at Very High Rates of Loading, Proc. Phys. Soc.
    London, B, 62:676.
Kumar, P., and Clifton, R.J., 1977, Optical Alignment of Impact
    Faces for Plate Impact Experiments, J. Appl. Phys., 48:1366.
Li, C.H. and Clifton, R.J., 1981, Dynamic Stress-Strain Curves at
    Plastic Shear Strain Rates of 10^5 s^{-1}, in: "Shock Waves in
    Condensed Matter - 1981", W.J. Nellis, L. Seaman, and R.A.
    Graham, eds., American Institute of Physics, New York.
Li, C.H., 1982, A Pressure-Shear Experiment for Studying the
    Dynamic Plastic Response of Metals at Shear Strain Rates of
    10^5 s^{-1}, Ph.D. thesis, Brown University, Providence.
Lindholm, U.S., 1977, Deformation Maps in the Region of High
    Dislocation Velocity in: "IUTAM Symposium on High Velocity
    Deformation of Solids," K. Kawata and J. Shioiri, eds.,
    Springer-Verlag, Berlin, p. 26.

Read, H.E. Triplett, J.R., and Cecil, R.A., 1971, A Rate Dependent
    Constitutive Equation for Plastic Flow in Metals, Technical
    Report No. 3SR-308, Systems, Science and Software.
Rice, J.R., 1970, On the Structure of Stress-Strain Relations for
    Time Dependent Plastic Deformation in Metals, J. Appl. Mechs.,
    37:727.
Richmond, O., and Spitzig, W.A., 1980, Pressure Dependence and
    Dilatancy of Plastic Flow, in: Proceedings of the 15th
    International Congress of Theoretical and applied Mechanics,
    F.P.J. Rimrott, B. Tabarrok, eds., North-Holland, Amsterdam,
    p.377.
Richmond, O., 1982, Personal Communication.
Senseny, P.E., Richman, M.H., and Duffy, J., 1975, The Influence of
    Annealing Temperature on the Strain Rate Sensitivity of Copper
    in Torsion, J. Appl. Mech., 42:245.
Yoshida, S. and Oguchi, A., 1970, Influence of High Hydrostatic
    Pressure on the Flow Stress of Aluminum Polycrystals, Trans.
    Japan Inst. Metals, 11:424.

STRAIN RATE HISTORY EFFECTS AND DISLOCATION SUBSTRUCTURE

AT HIGH STRAIN RATES

J. Duffy

Division of Engineering
Brown University
Providence, RI 02912

ABSTRACT

A brief review is presented of strain rate history and temperature history effects during the high strain rate deformation of
metals with special emphasis on the BCC metals. Possible explanations to account for history effects are reviewed. Experiments are
described in which single crystals of aluminum are deformed at
various strain rates. Observations by transmission electron
microscopy are used to relate microscopic to macroscopic behavior.
Results show that cell size influences the strain rate sensitivity
and may be a cause of history effects.

INTRODUCTION

Strain rate history and temperature history effects during
high strain rate deformation of metals have been studied for a
number of years for the purpose of understanding material behavior
and describing this behavior through appropriate constitutive
equations. An experiment frequently employed for this purpose
involves prestraining the specimen at a particular temperature and
strain rate, followed by continued straining after a change either
in the temperature or in the strain rate. In early tests of this
nature, the change in temperature or strain rate was effected in
an interrupted test, i.e., by unloading the specimen before reloading under the new conditions of deformations. Generally, the
unloading was made necessary by a change to a new testing machine
or new instrumentation. In more recent years the incremental or
jump test has been employed in which the prestrain is followed
either by a sharp increase in strain rate or by a drop in temperature with no intermediate unloading. While either method can

demonstrate the presence of history effects, the latter technique
imposes simpler temperature and strain rate histories as functions
of strain (or of time) and therefore may be more conducive to
modeling. The original motivation for these investigations was the
hope that material behavior could be described through an equation
of state. However, Dorn et al., showed that an equation of state
could not be used to describe the effects of a temperature change
during deformation of aluminum, copper or steel (1). Lindholm
showed that reloading at a different strain rate also produces
history effects (2), and since that time it has been shown that
history effect may be quite significant, depending on the metal,
its degree of purity, grain size, prior heat treatment, current
temperature, strain rate, etc. It is not necessary to review
history effects here, since such a review was presented not long
ago (3). Lately, however, in tests including deformation at high
strain rates it has been shown that two additional factors determining history effects are (a) the lattice structure of the particular metal involved, and (b) the development of a microstructural
substructure during deformation. The purpose of this paper is to
review briefly some recent results concerning these two subjects.

a) <u>Influence of Lattice Structure on History Effects</u>

        In our experiments the specimens are deformed in shear. For
this purpose, steel and other polycrystalline metal specimens are
machined as thin-walled tubes and mounted in a torsional Kolsky
(split-Hopkinson) bar. As may be seen from the schematic sketch
in Figure 1, the bar used is adapted to loading the specimen either
dynamically, i.e. at about 10^3 s^{-1}, or quasi-statically, i.e. at
about 10^{-4} s^{-1}. To impose a dynamic strain rate on the specimen

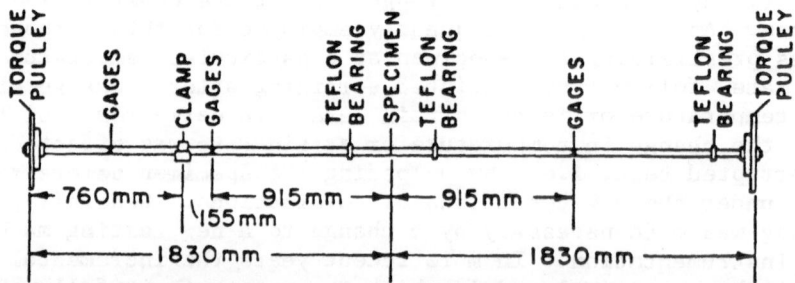

Figure 1   Schematic diagram of torsional Kolsky bar. Chiem and
            Duffy (21).

a torque is stored between the left end of the bar and the clamp.
With sudden release of the clamp, a torsional pulse propagates
toward the specimen. The subsequent stress- and strain-time histories are measured as described in Reference 4. In contrast, quasi-
static loading is imposed by twisting the bar slowly from the right
end, the stress being measured by gages mounted on the bar and the
strain by a pair of DCDT's which measure the rotation of the bar
to either side of the specimen. This latter measurement requires
a subsequent subtraction for the elastic deformation of the portion
of the bar between the DCDT's. Finally, by making use simultaneously
of the loading mechanisms at both ends of the bar, it becomes
possible to impose a sudden increment in strain rate during deformation at a quasi-static rate. This is done by loading first quasi-
statically from the right end of the bar then, with no change in
the quasi-static rate and hence no unloading of the specimen, re-
leasing the stored torque at the left end to increase the strain
rate abruptly. Tests also can be performed at temperatures above
or below the ambient by using an environmental chamber. In addition,
sudden drops in temperature can be imposed easily during quasi-
static deformation.

Results of incremental strain rate tests in polycrystalline
specimens of four metals, as obtained with this apparatus, are
shown in Figure 2, (5). These results agree, at least qualitatively,
with those of other investigators and may be considered

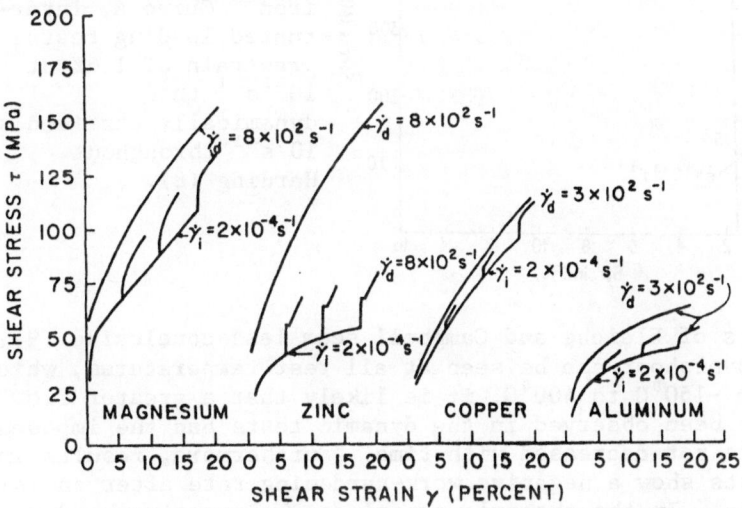

Figure 2   Shear stress – shear strain curves for four metals
           Curves show constant strain rate results as well as
           results of incremental strain rate tests. Senseny
           et al. (5).

typical of the deformation behavior in two common lattice structures: FCC and HCP, for temperatures less than about half the
melting temperature. For these metals it is seen that the flow
stress is greater the higher the strain rate, and that a sudden
increment in strain rate produces an increase in the flow stress
which, however, falls short of the stress found in a test conducted
entirely at the higher rate. Titanium shows a similar behavior,
although evidence points to very little in the way of history effects in this metal in spite of a pronounced sensitivity to strain
rate itself (6). The situation for BCC metals is not as clear. In
early experiments of the interrupted type, in which complete unloading occurs between succeeding strain rates, history effects
appear quite evident (7). Nicholas, obtained a small overshoot in
stress, that is, he found that the dynamic flow stress after a
quasi-static prestrain exceeds the stress found in a test conducted
entirely at the higher rate (6). This overshoot did not occur
immediately upon application of the dynamic strain rate, since
Nicholas did not see an upper and lower yield point. Harding in
experiments performed with iron, found a pronounced overshoot,
Figure 3, (8).

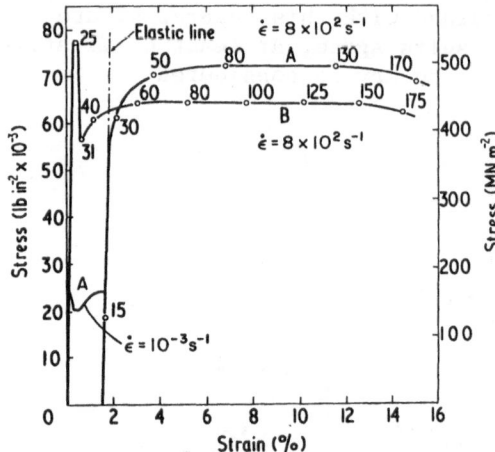

Figure 3
Stress-strain curves for
iron. Curve A, inter-
rupted loading tests:
prestrain of 1.6% at
$10^{-3}s^{-1}$ then
dynamically strained at
$10^{3}s^{-1}$ throughout.
Harding (8).

The results of Eleiche and Campbell seem less conclusive (9).
While an overshoot can be seen at all test temperatures, which
range from $-150^{0}C$ to $400^{0}C$, it is likely that a greater flow stress
would have been observed in the dynamic tests had the imposed
strain rate not decreased with time. Furthermore, results in many
of the tests show a negative work-hardening rate after an initial
deformation. In the author's experience, a negative hardening
rate during dynamic loading in shear implies the formation of a
shear band within the specimen. More recent experiments with steel
indicate that history effects are absent in incremental strain
rate tests conducted at room temperature but that an overshoot

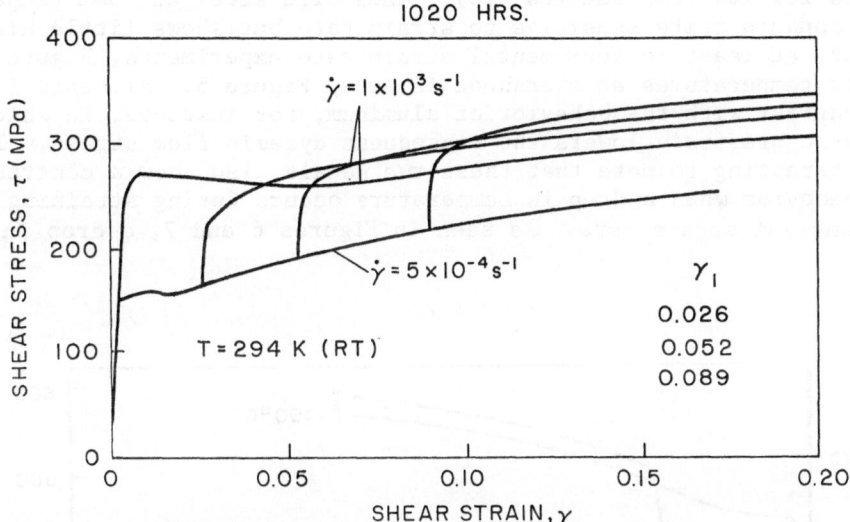

Figure 4   Stress-strain curves for 1020 hot-rolled steel,
deformed at room temperature. Klepaczko and
Duffy (10).

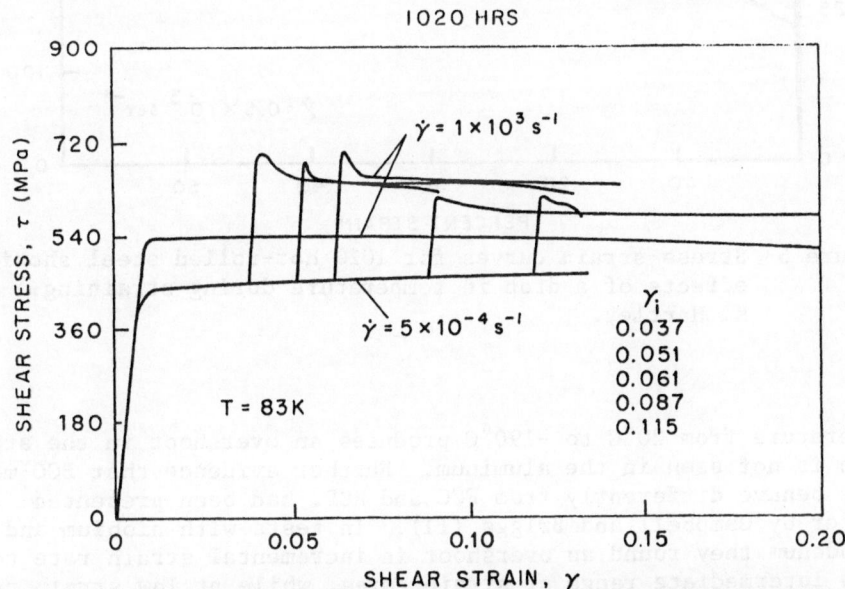

Figure 5   Stress-strain curves for 1020 hot-rolled steel,
deformed at 83K. Klepaczko and Duffy (10).

occurs for low temperatures (10). Thus mild steel at room temperature appears quite sensitive to strain rate but shows little history
effect, at least in incremental strain rate experiments, Figure 4.
At low temperatures an overshoot is seen, Figure 5. All this is
in contrast with the behavior of aluminum, for instance, in which
a static prestrain lowers the subsequent dynamic flow stress. It
is interesting to note that these two metals also show a contrasting behavior when a drop in temperature occurs during straining at
at constant strain rate. As seen in Figures 6 and 7, a drop in

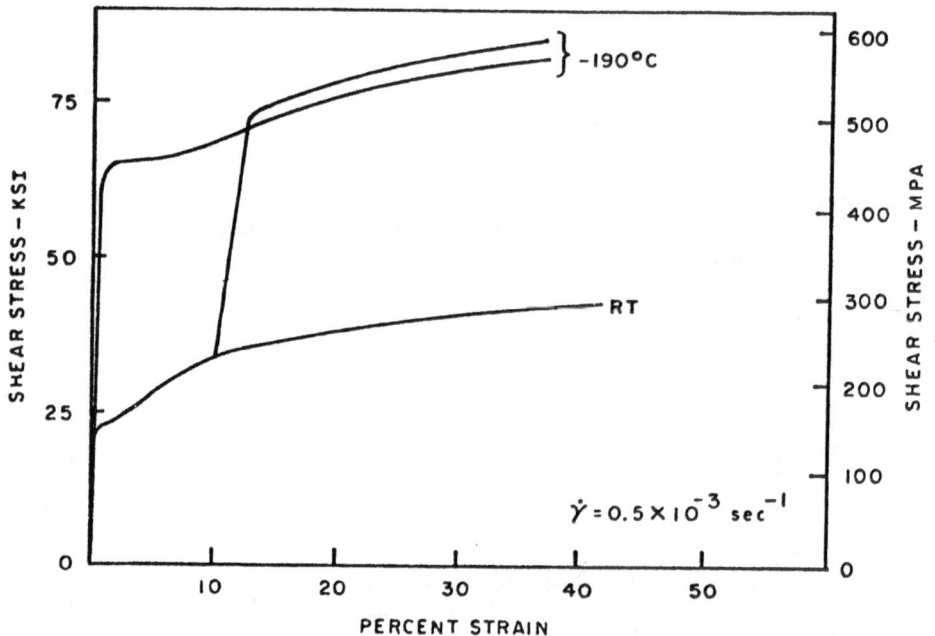

Figure 6   Stress-strain curves for 1020 hot-rolled steel showing
           effects of a drop in temperature during straining.
           K. Hartley.

temperature from 20°C to -190°C produces an overshoot in the steel
which is not seen in the aluminum. Further evidence that BCC metals
might behave differently from FCC and HCP, had been presented
earlier by Campbell and Briggs (11). In tests with niobium and
molybdenum they found an overshoot in incremental strain rate tests
in an intermediate range of strain rates, while at low strain rates
the observed behavior agrees qualitatively with that seen in FCC
metals.

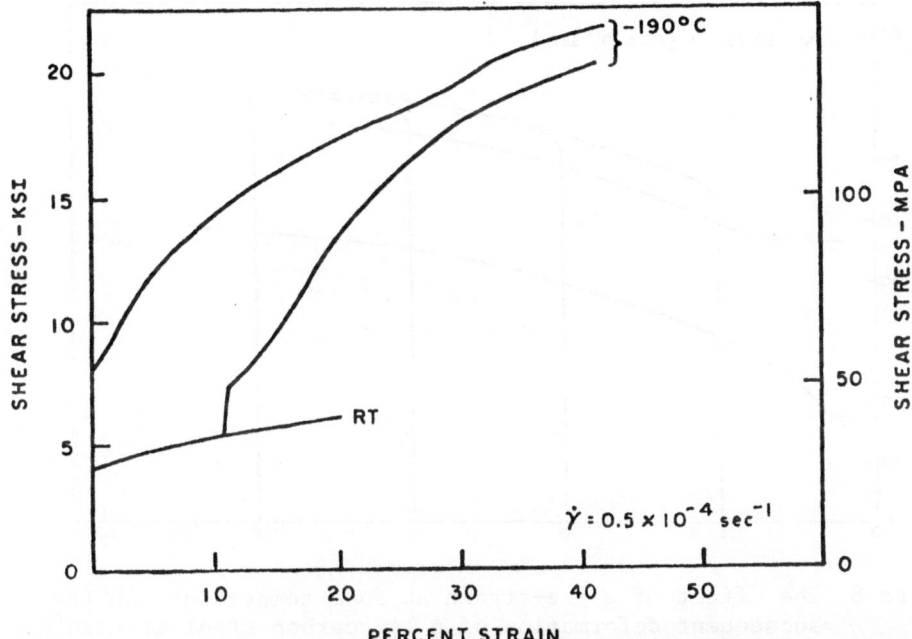

Figure 7   Stress-strain curves for 1100-0 aluminum showing effects
of a drop in temperature during straining. K. Hartley.

        The reasons for the different behavior of BCC metals are still
uncertain. A recent review of history effects in BCC metals presents
a number of possible explanations (12). Among these are adiabatic
heating and dynamic strain aging. The former occurs at higher
strain rates and results in a softening of the metal which generally
can be estimated, at least approximately, while the latter effect
is unlikely to be of consequence at the low temperatures in question.
More important are differences occurring on the microstructural
scale. The overshoot due to a sudden drop in temperature which is
described above, has been investigated by Lindley (13). His results,
Figure 8, show that the difference in hardening characteristics
occurs at small strains. Metallographic examination indicated that
the prestrain at room temperature inhibits twinning in subsequent
deformation at -120^{0}C whereas prestrain at -120^{0}C does not, or at
least does so only to a far lesser extent. In addition to twinning,
it may be expected that there will be differences in dislocation
multiplication rates between high and low strain rates or temperatures. Smith argues that history effects in steel during deformation at low temperatures may be accounted for by a greater number
of dislocation sources generated at the higher strain rate resulting
subsequently in the activation of a larger number of mobile dislocations upon reloading and hence in a lower flow stress (14). He

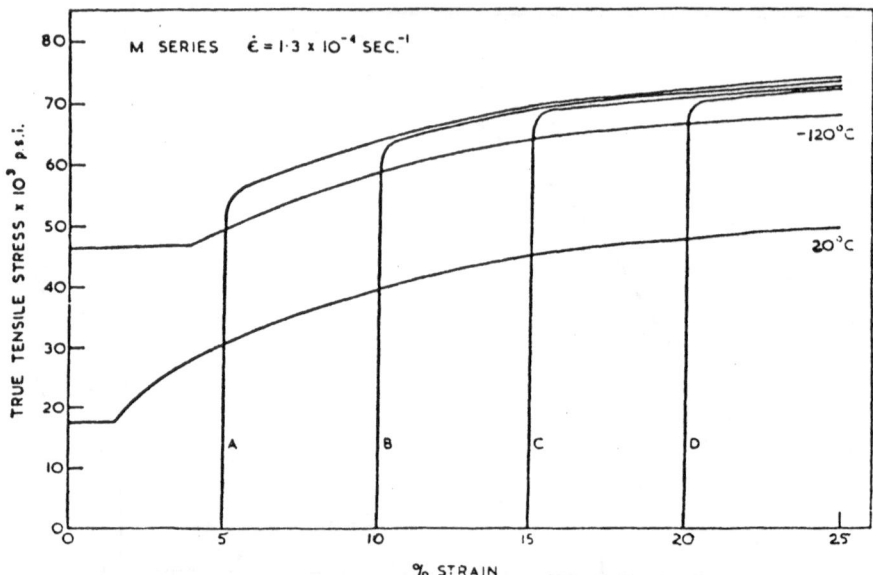

Figure 8   The effect of a pre-strain at room temperature on the
            subsequent deformation of a low carbon steel at -120°C.
            Lindley (13).

points out that this is consistent with the results of Campbell and
Duby who observed fine slip lines after high rate deformation of
steel and coarser lines after quasi-static deformation (7).

b)   Development of Substructure during Deformation

       While the above probably provide the principal causes of
history effects in BCC metals, it is likely that dislocation distribution within individual grains also plays a role. For instance,
it has been known for some time that testing temperature influences
the dislocation arrangement which develops as a result of plastic
flow (15-17). At low temperatures the dislocations tend to distri-
bute themselves uniformly throughout the crystal or the individual
grains. At room temperature, in contrast, the dislocations tend to
form a cellular pattern which becomes more pronounced as straining
continues and frequently leads to well-defined subgrains. Rate of
strain also affects the dislocation distribution. At dynamic
strain rates, cells do not form as readily, and are smaller in
size; furthermore the spacing of the dislocations is larger within
the cell walls (18-20). As regards history effects, it was noted
quite early by Keh and Weissmann that results such as those obtained
by Lindley or Harding could be explained on the basis of a difference in the dislocation arrangement (15). These investigators found
that a cell structure formed in iron as a result of a prestrain at

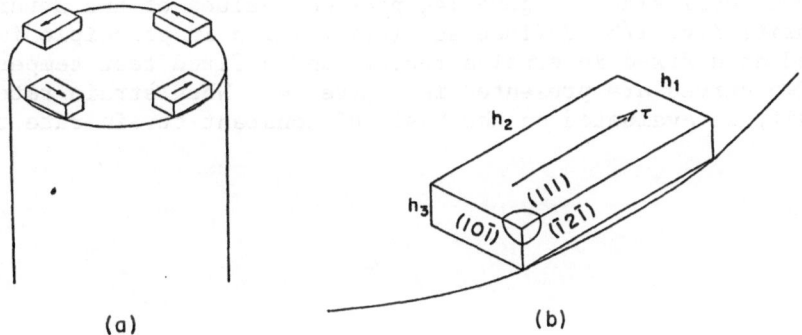

Figure 9   Schematic diagram showing arrangement of single
           crystals relative to Kolsky bar. Chiem and Duffy (21)

        Recently, a study was conducted into the role of dislocation
arrangement on history effects and on the strain rate sensitivity
of aluminum (21). The specimens consisted of high purity single
crystals deformed in shear in the torsional Kolsky bar described
previously. The apparatus was modified to accommodate single
room temperature. However, upon re-straining at a lower temperature
the distribution of dislocations became more uniform.

crystals by insuring that each of the four crystals used per test
had the same crystallographic orientation relative to the applied
stress, Figure 9. The imposed strain rates ranged from $10^{-4} s^{-1}$ to
$1600\ s^{-1}$, to strains of about 20%. The constant strain rate tests
were supplemented by a series of incremental strain rate experiments.
After testing, thin foils of the specimens were examined by transmission electron microscopy. The constant strain rate tests show
an increase in flow stress with strain rate, Figure 10. It is
evident that the strain rate sensitivity in these experiments is
greater than generally found in polycrystalline aluminum, and this is
undoubtedly due to the relatively high degree of purity of the
aluminum, 99.99%. For strain rates greater than about $500\ s^{-1}$ the
flow stress increases more rapidly indicating the possibility of a
change in the dominant deformation mechanism. Transmission electron
micrographs show more distinct dislocation cells after quasi-static
straining, Figure 11, than after dynamic straining, Figure 12. A
cell pattern similar to that shown in Figure 11 had already been
observed by a number of investigators. One need only compare the
results of Swann (17) in aluminum deformed to about 20% or those of
Keh and Weissmann (15) in iron. Furthermore, as expected, the flow
stress, τ, is shown to vary inversely with cell size, d, Figure 13.
In addition, it is found that strain rate sensitivity also is a

function of cell size. Figure 14, provides values of the strain
rate sensitivity, $1/\beta$, defined as $\partial\tau/\partial\ell n\dot{\gamma}$ which in principle is
evaluated at a fixed internal structure and a fixed test tempera-
ture. Two curves are presented in Figure 14. When strain rate
sensitivity is evaluated on the basis of constant strain rate tests,

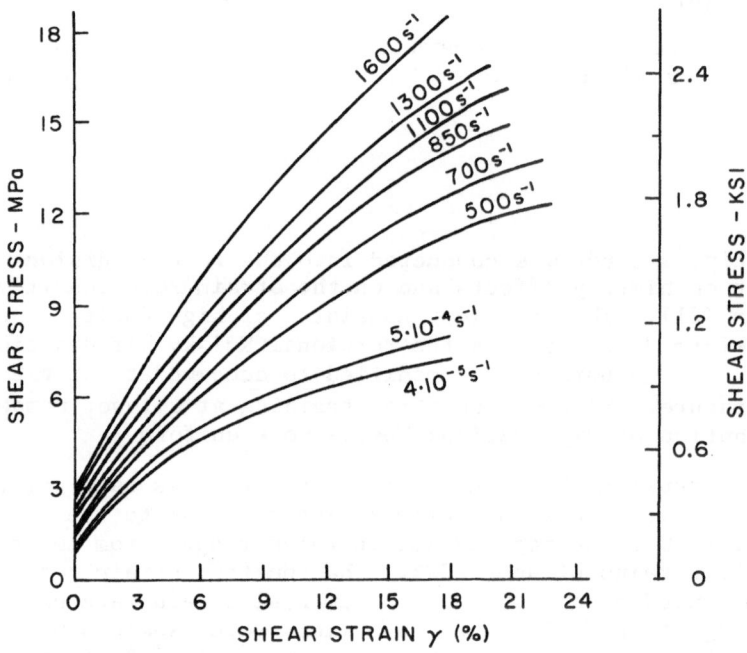

Figure 10   Shear stress-shear strain for aluminum single crystals
            tested at various rates of strain. Chiem and Duffy (21).

it is termed an apparent value, $1/\beta_a$. On the other hand, an increment in strain rate provides an increment in flow stress which is
used to find a "true" strain rate sensitivity, $1/\beta_t$. While the
measured dislocation density following quasi-static straining

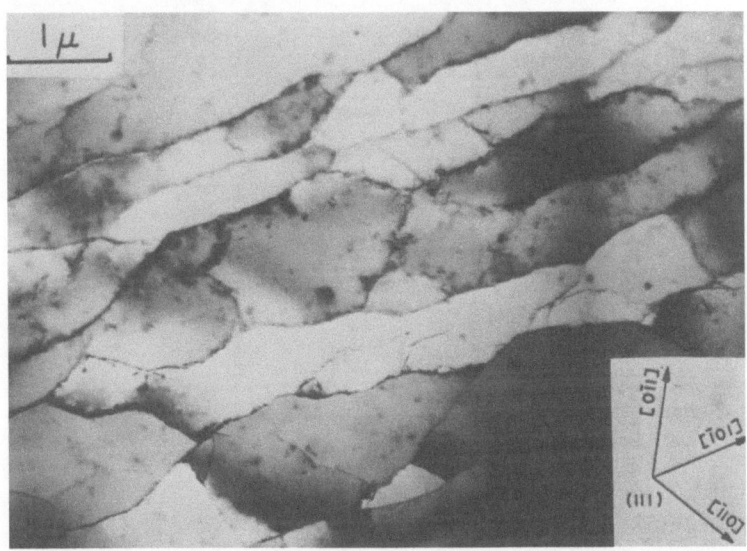

Figure 11   Transmission electron micrograph of dislocation sub-
             structure in aluminum following quasi-static defor-
             mation in shear to 18%. Chiem and Duffy (21).

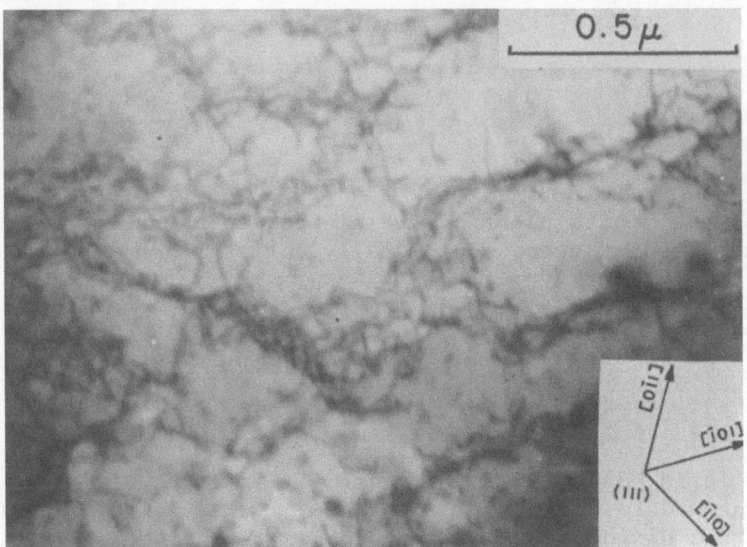

Figure 12   Transmission electron micrograph of dislocation sub-
             structure in aluminum following deformation at 500 s^{-1}
             in shear to 22.7%. Chiem and Duffy (21).

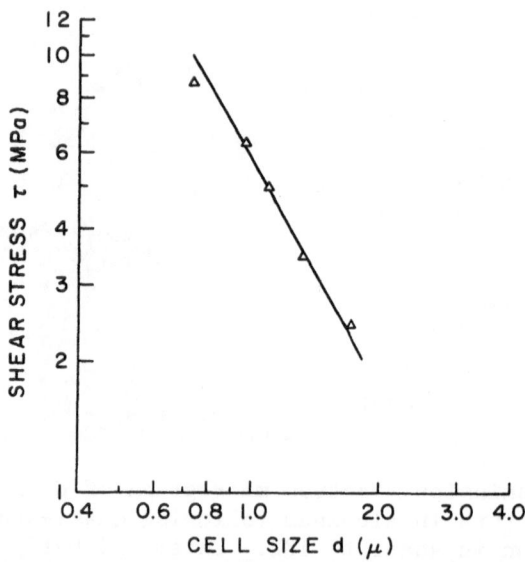

Figure 13   Shear stress vs. dislocation cell size in aluminum
             single crystals loaded quasistatically.   Chiem and
             Duffy (21).

varies linearly with plastic strain, the dependence of stress on
cell size can be fitted about equally well through

$$\tau_s = 7.45 \ d^{-1} - 1.84$$

as predicted by Holt (23), or through a Hall-Petch dependence

$$\tau_s = 14.66 \ d^{-\frac{1}{2}} - 8.92$$

where τ_s is in MPa and d in microns.   The results of incremental
strain rate tests could not be related as clearly to cell formation
although the dislocation distribution appears to be distributed uniformly irrespective of the amount of prestrain that precedes the
increment in strain rate.   However, incremental strain rate tests
do allow for the calculation of an activation volume based on the
true strain rate sensitivity and a comparison of this value to one
based on observations of dislocation density, Figure 15.   Details
of the calculations are given in Reference 21.

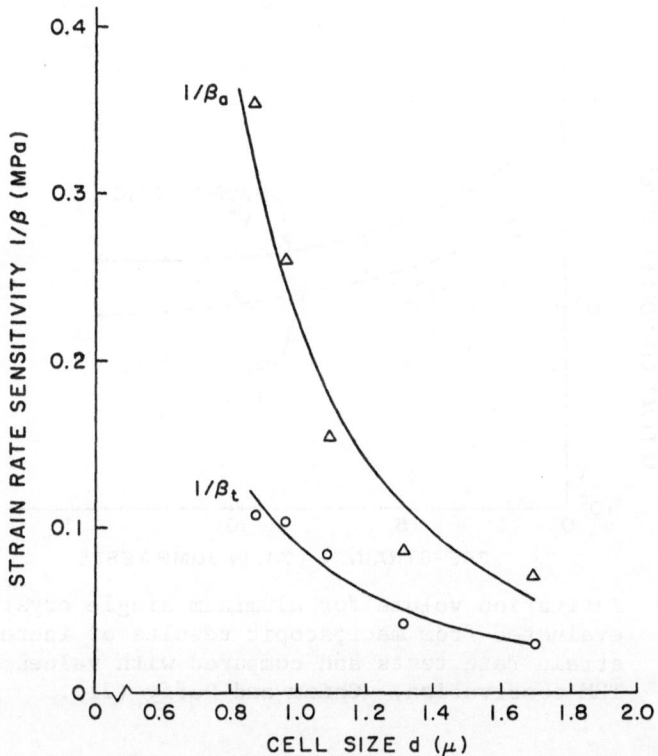

Figure 14   Strain rate sensitivities, apparent and true, as
            functions of dislocation cell size in aluminum single
            crystals.   Chiem and Duffy (21).

    Dislocation cells may also influence history effects in their
persistence under changed loading conditions. Lipkin et al. deformed OFHC copper specimens at an initial dynamic strain rate,
850 s^{-1}, which they reduced very rapidly after about 15% strain to
a lower, constant dynamic rate, 200 s^{-1} (22). Their results show
that the work hardening rate does not change immediately upon
dropping the strain rate. There is a delay in the response which
the authors hypothesize is due to the dislocation arrangement.
According to their explanation, the initial deformation produces a
dislocation arrangement which is thermodynamically stable under
the initial deformation conditions but unstable after the change
in strain rate. However, the change from one dislocation arrangement to another is effected only with continued straining. An
instability of this nature had been predicted by Holt and is discussed by Alden (23,24).

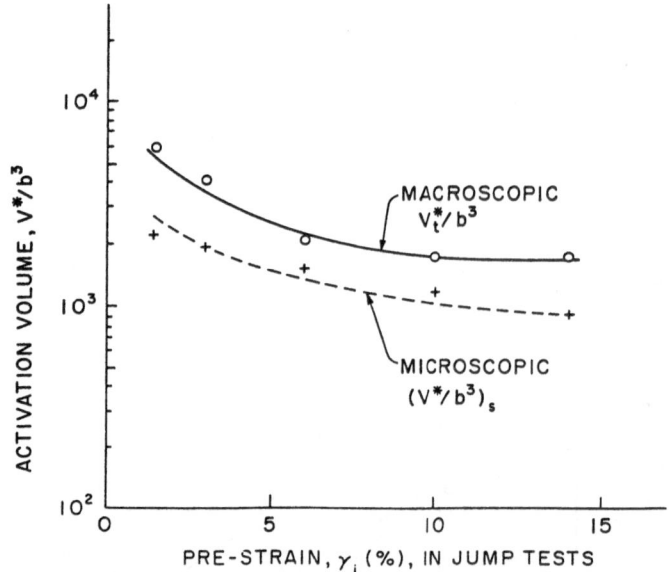

Figure 15   Activation volume for aluminum single crystals as
            evaluated from macroscopic results of incremental
            strain rate tests and compared with values based on
            TEM observation.   Chiem and Duffy (21).

    Continued research into history effects seems warranted in
view of the fact that in many applications metals frequently are
deformed prior to use and often under different temperature conditions or at different strain rates.   They are, furthermore, of
value from a basic viewpoint since the relation of macroscopic
behavior to the evolving microstructure is not well established.
While obviously this relation depends on dislocation density,
dislocation arrangement probably plays a role as well, and one
which is far from well understood today (25–26).

CONCLUSIONS

    In a brief review, it is shown that the strain rate history
and temperature history effects during deformation of BCC metals
are not the same as in FCC or HCP metals.   The reasons for this
contrasting behavior have not been determined.   Dynamic strain
aging may account for some differences, but cannot be expected to
play a significant role in the low temperature range.   The occurrence of twinning during low temperature deformation in BCC metals
is perhaps a more likely cause and should be investigated more
thoroughly.   Another strong possibility probably lies in

differences in the number of dislocation sources created during deformation at higher strain rates and hence in differences in mobile dislocation densities during subsequent deformation. Finally, experiments are described combining macroscopic and microscopic observations. These show that strain rate determines dislocation arrangement including cell size. It is shown also that strain rate sensitivity depends on dislocation cell size and suggested that dislocation arrangement may account in part for certain history effects. Additional research is needed if a full understanding of history effects and indeed of strain rate and temperature effects themselves is to be achieved. Such research should include observations of dislocation density and dislocation arrangement. A complicating factor is the softening of the metal due to work-hardening which generally accompanies deformation at high strain rates.

ACKNOWLEDGEMENTS

The author gratefully acknowledges the support of the National Science Foundation under Grant MEA 79-23742, and the many helpful discussions with Professors R. J. Clifton and R. J. Asaro of Brown University. Figures 6 and 7 represent results of experiments performed by Ms. Kathleen Hartley and are to appear in a future report. The author is grateful for permission to use them.

REFERENCES

1. J. E. Dorn, A. Goldberg and T. E. Tietz, "The Effect of Thermal-Mechanical History on the Strain Hardening of Metals", AIME Trans., Vol. 180, pp. 205-224, 1949.
2. U. S. Lindholm "Some Experiments with the Split Hopkinson Pressure Bar", J. Mech. Phys. Solids, Vol. 12, pp. 317-335, 1964.
3. J. Duffy, "The J. D. Campbell Memorial Lecture: Testing Techniques and Material Behaviour at High Rates of Strain",Proc. Conf. on Mechanical Properties at High Rates of Strain, 1979, Edited by J. Harding, The Institute of Physics, Bristol and London, 1980, pp. 1-15.
4. R. A. Frantz, Jr. and J. Duffy, "The Dynamic Stress-Strain Behavior in Torsion of 1100-0 Aluminum Subjected to a Sharp Increase in Strain Rate", J. Appl. Mech., Vol. 39 pp. 939-945, 1972.
5. P. E. Senseny, J. Duffy, and R. H. Hawley, "Experiments on Strain Rate History and Temperature Effects During the Plastic Deformation of Close-Packed Metals", J. Appl. Mech., Vol. 45, pp. 60-66, 1978.
6. T. Nicholas, "Strain-Rate and Strain-Rate-History Effects in Several Metals in Torsion", Experimental Mechanics, Vol. 11, No. 8, pp. 370-374, 1971.

7. J. D. Campbell and J. Duby, "Delayed Yield and Other Dynamic
    Loading Phenomena in a Medium-Carbon Steel", Proc. Conf. on
    Properties of Materials at High Rates of Strain, The Institu-
    tion of Mechanical Engineers, London, pp. 214-220, 1957.

8. J. Harding, Discussion of paper by Klepaczko and Duffy, in Pro.
    Conf. on Mechanical Properties at High Rates of Strain, The
    Institute of Physics, London, p. 191, 1974.

9. A. M. Eleiche and J. D. Campbell, "The Influence of Strain-
    Rate History and Temperature on the Shear Strength of Copper,
    Titanium and Mild Steel", University of Oxford, Report
    AFML-TR-76-90, 1976.

10. J. Klepaczko and J. Duffy, "Strain Rate History Effects in
    Body-Centered-Cubic Metals", Proc. Conf. on Mechanical Testing
    for Deformation Model Development, Ed. by R. W. Rohde and J.
    C. Swearengen, ASTM STP 765, pp. 251-268, 1982.

11. J. D. Campbell and T. L. Briggs, "Strain-Rate History Effects
    in Polycrystalline Molybdenum and Niobium", J. Less Common
    Metals, Vol. 40, pp. 235-250, 1975.

12. J. Klepaczko and J. Duffy, "History Effect in Polycrystalline
    BCC Metals and Steel Subjected to Rapid Changes in Strain
    Rate and Temperature", Brown University Technical Report, June
    1982. To be published in Archives of Mechanics, Warsaw,
    Poland, No. 4, 1982.

13. T. C. Lindley, "The Effect of a Pre-Strain on the Low
    Temperature Mechanical Properties of a Low Carbon Steel", Acta
    Met., Vol. 13, pp. 681-689, 1965.

14. R. C. Smith, "Studies of Effect of Dynamic Preloads on
    Mechanical Properties of Steel", Experimental Mechanics, Vol.
    1, No. 11, pp. 153-159, 1961.

15. A. S. Keh and S. Weissmann, "Dislocation Substructure in Body-
    Centered-Cubic Metals", Chapter 5 in Electron Microscopy and
    Strength of Crystals, Ed. by G. Thomas and J. Washburn,
    Interscience, New York, 1963.

16. W. C. Leslie, J. T. Michalak, and F. W. Aul, "The Annealing
    of Cold-Worked Iron", Proc. Conf. on Iron and Its Dilute Solid
    Solutions, Ed. by C. W. Spencer and F. E. Werner, Interscience,
    1963.

17. P. R. Swann, "Dislocation Arrangements in Face-Centered Cubic
    Metals and Alloys", Chapter 3 in Electron Microscopy and
    Strength of Crystals, Ed. by G. Thomas and J. Washburn, Inter-
    science, New York, 1963.

18. H. J. McQueen and J. E. Hockett, "Microstructures of Aluminum
    Compressed at Various Rates and Temperatures", Met. Trans. A,
    Vol. 1, pp. 2997-3004, 1970.

19. A. Korbel and K. Swiatkowski, "The Role of Strain Rate in
    Formation of Dislocation Structure and Its Influence on the
    Mechanical Properties of Aluminum", Met. Sci. J., Vol. 6, pp.
    60-63, 1972.

20. M. R. Staker and D. L. Holt, "The Dislocation Cell Size and
    Dislocation Density in Copper Deformed at Temperatures between
    25 and 700 C", Acta Met., Vol. 20, pp. 569-579, 1972.

21. C. Y. Chiem and J. Duffy, "Strain Rate History Effects and Observations of Dislocation Substructure in Aluminum Single Crystals Following Dynamic Deformation", Brown University Technical Report, NSF MEA 79-23742/3 and MRL E-137, October 1981.

22. J. Lipkin, J. D. Campbell and J. C. Swearengen, "The Effects of Strain-Rate Variations on the Flow Stress of OFHC Copper", J. Mech. Phys. Solids, Vol. 26, pp. 251-268, 1978.

23. D. L. Holt, "Dislocation Cell Formation in Metals", J. Appl. Phys., Vol. 41, No. 8, pp. 3197-3201, 1970.

24. T. H. Alden, "Microstructural Interpretation of Work Softening in Aluminum", Met. Trans. A, Vol. 7, pp. 1057-1063, 1976.

25. A. P. L. Turner and T. Hasegawa, "Deformation Microstructures and Mechanical Equations of State", Proc. Conf. on Mechanical Testing for Deformation Model Development, Ed. by R. W. Rohde and J. C. Swearengen, ASTM STP 765, pp. 322-341, 1982.

26. R. W. Rohde, W. B. Jones and J. C. Swearengen, "Deformation Modeling of Aluminum: Stress Relaxation, Transient Behavior and Search for Microstructural Correlations", Acta Met., Vol. 29, pp. 41-52, 1981.

MATERIALS RESPONSE TO LARGE PLASTIC DEFORMATION

M. G. Stout and S. S. Hecker

Materials Science and Technology Division
Los Alamos National Laboratory
Los Alamos, N.M. 87545

INTRODUCTION

Many important practical applications of metals require a knowledge of the plastic response at large deformations and at high strain rates. For example, metal forming processes and impact or penetration problems combine the effects of large strain, high rate and temperature. Accurate modeling of such processes requires a good constitutive description of material behavior. However, controlled laboratory experiments at large strains are difficult because most involve large geometry changes accompanied by either deformation gradients (such as barreling in compression) or plastic instability (such as necking in tension). High rate deformation adds the complication of an uncontrolled temperature rise. In the strain rate regime of 1 to 10^3 sec^{-1}, deformation may be neither completely isothermal nor adiabatic, but a combination.

Because of these experimental complications few good laboratory experiments have been conducted at high rates to large plastic strains. Lindholm describes some high-rate torsion results in this volume. In this paper, we concentrate on large strain behavior without the complication of high rates. Gil Sevillano, van Houtte, and Aernoudt[1] have recently published a comprehensive review of large strain deformation. We are adding some of our recent experimental results to a review we presented at a recent workshop.[2]

REVIEW OF EXPERIMENTAL TECHNIQUES

    Uniaxial tension is restricted to small strains by plastic
instability (necking). Few metals can stretch more than 50%
before necking. Early attempts to remachine specimens to remove
the neck have not become popular. With the advent of closed
loop testing machines it has become more popular to use a diametral extensometer at the neck to control the tensile test with
diametral strain. Area and triaxial stress corrections are then
used to reconstruct a uniaxial tensile curve.[3,4] This method
involves a number of assumptions and has yet to be proven. Uniaxial compression is limited by barreling and end effects from
the platen. To achieve a true uniaxial stress state excellent
lubrication and a length/diameter ratio of ~1.6 are required.
Large changes in geometry require remachining, which has been
practiced much more extensively in compression than in tension.
It obviously requires starting with a very large specimen. We
will present some data of interrupted compression tests to
strain levels of 4.

    Biaxial tension experiments on sheet material are able to
produce effective strain levels approximately twice those attainable in uniaxial tension. The added stability under biaxial
tension has been discussed by a number of authors.[5-8] Hydraulic
bulging offers the best experimental technique for stress-strain
measurements of sheet specimens. Unfortunately, there is considerable controversy over the potential errors introduced by
small bulge diameters, biaxial extensometers, and varying strain
rates during bulging.[9,10] A much more accurate and versatile
technique for biaxial tension is the axial loading/internal
pressurization of thin-walled tubes. However, geometric instabilities limit the strain levels to values even lower than those
attainable in uniaxial tension.[11]

    Torsional deformation offers the best hope for large-strain
experiments because it is accompanied by very small geometry
changes. However, torsion also suffers from several important
restrictions. Experimentally, specimens elongate during torsion
and care must be exercised not to restrain their length. Torsion of solid rods also produces an inherently non-uniform
stress state, varying from zero at the axis to a maximum at the
surface. There has been considerable discussion over the years
about how to properly convert a measured torque-angle curve to
an effective stress-effective strain curve.[12,13] Much of the

torsional data in the literature is inconsistent because of the different methods of analysis employed. Most recently, Canova et al.[14] have developed a technique using several specimens of slightly different diameters to establish an accurate stress-strain curve. The torque-angle conversion to stress-strain is simpler for thin-walled tubes, but torsional buckling limits the strain levels attainable. Apparently some large strain tests have been conducted successfully on very short tubes[15] without adverse end effects.

Most of the large-strain information available in the literature was obtained by indirect tests. These tests are conducted by imparting large prestrains in a deformation mode relatively insensitive to plastic instability (such as wire drawing or sheet rolling) and then testing the prestrained samples in tension. Tensile tests define a stress-strain curve as shown for our rolling + tension experiments on 1100 aluminum in Fig. 1. Because sheet rolling approximates a state of plane strain it is necessary to adjust the rolling prestrain to an effective strain. We used the von Mises effective strain criterion which gives the correction of $\varepsilon_{eff} = 1.155 \times \varepsilon_t$, where ε_t is the thickness strain during rolling. Similar curves can be constructed for wire and strip drawing. For the case of wire drawing the reduction in area equals the effective strain. Although very large strains have been achieved by these techniques (up to 7 in rolling + tension and 10 in wire drawing + tension), these techniques have some obvious drawbacks. The tests are not direct; deformation is incurred under one stress state and the flow stress measured under another. Tests are also interrupted and, in many cases, require remachining of specimens. An effective strain criterion must be assumed for proper comparison and, in some cases, the deformation zone geometry may change during the very large prestrains. The large deformation also affects the development of crystallographic texture. Nevertheless, these tests have been used extensively to provide a measure of hardening at large strains.

We will focus on low strain rate data obtained by methods discussed above. From these data we hope to show trends which are also important for large strain, high strain rate predictions. Specifically, we will address the influence of deformation mode, crystal structure, material purity, and alloying on stress-strain behavior.

RESULTS

Aluminum and Aluminum Alloys

Our large strain deformation results for commercially pure
aluminum are shown in Fig. 2.[16,17]   Tests were performed by

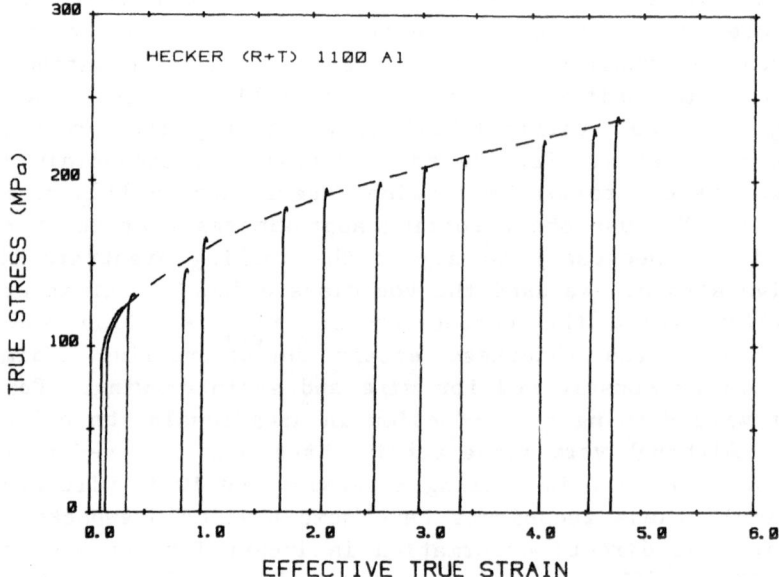

Fig. 1.   Construction of a flow stress curve from rolling prestrain followed by uniaxial tension (R+T). The thickness reduction is converted to a von Mises effective strain.

tension, rolling + tension (R+T), and incremental compression.
Hardening continues to very large strains with no evidence of
saturation. The flow curve for R+T is described accurately by
the parabolic hardening expression $\sigma = 155\ \varepsilon_p^{0.27}$ (MPa). The
hardening rate in compression appears somewhat lower at large
strains, but these data showed considerable scatter. Hence,
there appears to be very little dependence on deformation mode.

However, no torsion data for this lot of material are available for comparison.

LeFevre and coworkers[18,19] measured the flow behavior of a variety of aluminums of different purities using the technique of wire drawing + tension (WD+T). Some of these results are illustrated in Fig. 3 along with results of Luthy et al.[20] for very high purity aluminium tested in torsion at -20°C. The results clearly show the strong influence of purity on hardening rates and flow stress levels. High purity leads to saturation, regardless of deformation mode. We believe that the major role of impurities at large strains is to impede the dynamic recovery process. Without impurities dislocation annihilation occurs readily and balances dislocation multiplication to produce a steady state saturation flow stress.

Alloying can produce a variety of behavior. The results of Lloyd and Kenny[23] on Al-6% Ni exhibit saturation at relatively small strains (Fig. 4). The addition of nickel results in a

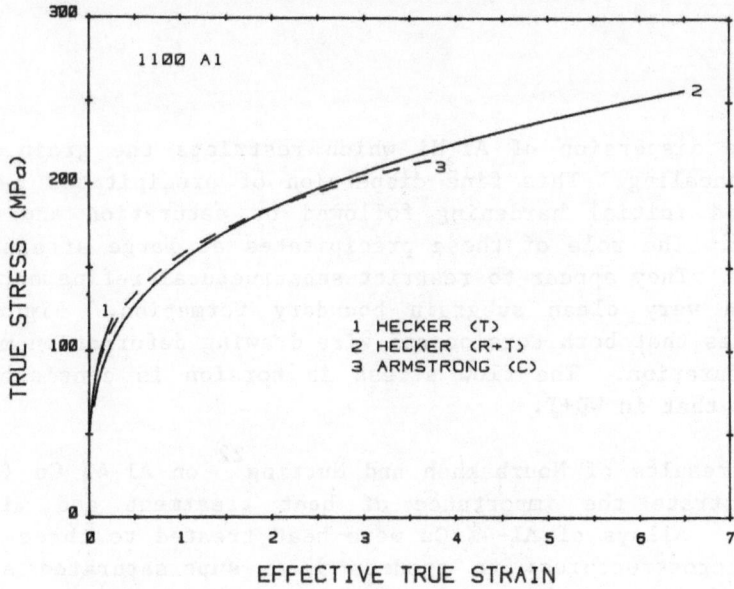

Fig. 2.   Comparison of stress-strain curves as determined by tension, rolling + tension, and compression of 1100 aluminum. (References 16 and 17.)

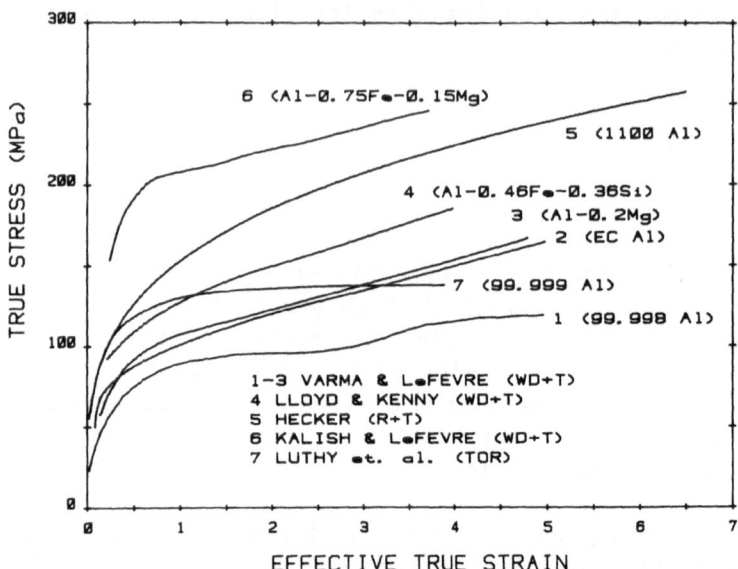

Fig. 3. Stress-strain curves for aluminum of different purities. (R+T) denotes rolling + tension and (WD+T), wire drawing
+ tension. (References 16-20.)

very fine dispersion of Al_3Ni which restricts the grain size
during annealing. This fine dispersion of precipitates causes
very rapid initial hardening followed by saturation and work
softening. The role of these precipitates at large strains is
not clear. They appear to restrict substructural refinement and
result in very clean subgrain boundary formation. Figure 4
illustrates that both torsion and wire drawing deformation modes
cause saturation. The flow stress in torsion is consistently
less than that in WD+T.

The results of Nourbakhsh and Nutting[22] on Al-4% Cu (Fig.
4) demonstrate the importance of heat treatment and microstructure. Alloys of Al-4% Cu were heat treated to three different microstructures: to produce 1) a supersaturated solid
solution of Cu in Al (Curve 1), 2) GP zones (Curve 2), and 3) θ'
precipitates (Curve 3). The hardening of the supersaturated
solid solution continues at all strain levels, similar to Al-5%
Mg. The alloy with GP zones showed initial hardening to a much

higher flow stress because of the GP zones contribution. However, at larger strains the GP zones were disrupted and the extra hardening increment lost. Hence, the flow stress actually leveled off and approached that of the supersaturated alloy at strains of 0.3. However, at this stress level the dislocations cut through the θ' precipitates and the flow stress decreases. Work softening stops at a strain of ~1, where most θ' precipitates are cut to result in a fine dispersion, at which point hardening resumes at a rate similar to the supersaturated alloy.

Lloyd[24] has also determined the effect of different initial grain sizes on the same Al-6% Ni alloy, Fig. 5. The influence of the initial grain size remains even to large strains. This indicates that initial grain size contributes to plastic behavior not only at small strains (Hall-Petch) but that it must also influence dislocation substructure development.

Copper

The large-strain flow behavior of copper for a variety of purities and a number of deformation modes is shown in Fig. 6. Again a tendency towards saturation for high-purity copper is evident. Only the low-purity ETP copper and phosphorus-deoxidized copper exhibit distinct, continued hardening. The two curves plotted for the data of Cairns et al.[27] represent their and our interpretation of their data. The highest purity (99.999% Cu) copper of Truckner and Mikkola[28] saturates at a very low stress level at a low strain.

The effect of deformation mode on the stress-strain behavior of copper is shown in Fig. 7. Although these data are from different material heats they indicate a lower flow stress for torsional deformation. For both axisymmetric and plane strain deformation there is a tendency towards saturation.

Nickel

Two large-strain flow curves for 200 nickel[21,30] are plotted in Fig. 8. Both curves are for R+T experiments. They show a plateau in flow stress between strains of 3.0 and 4.5. After a strain of 4.5 hardening continues with no tendency towards saturation. There is an absence of high purity nickel data, thus we do not know if nickel displays the same impurity effects as those found in aluminum and copper, where commercially pure

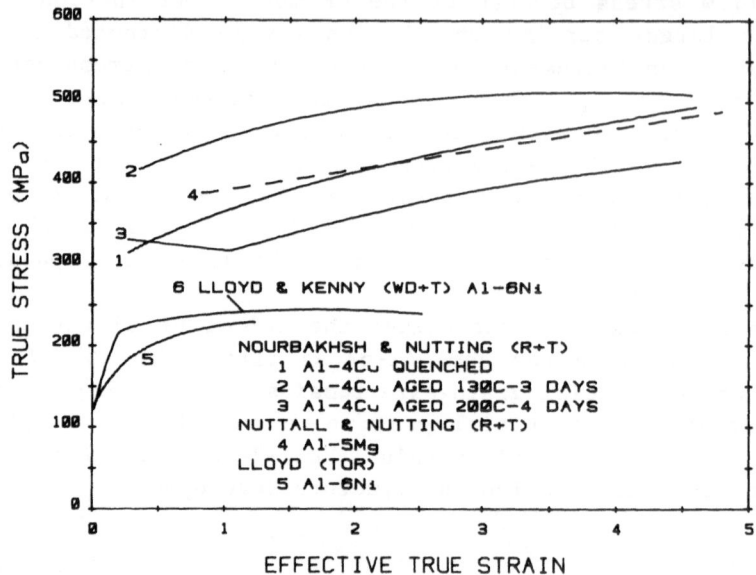

Fig. 4. Stress-strain curves for various aluminum alloys. (References 21-24.)

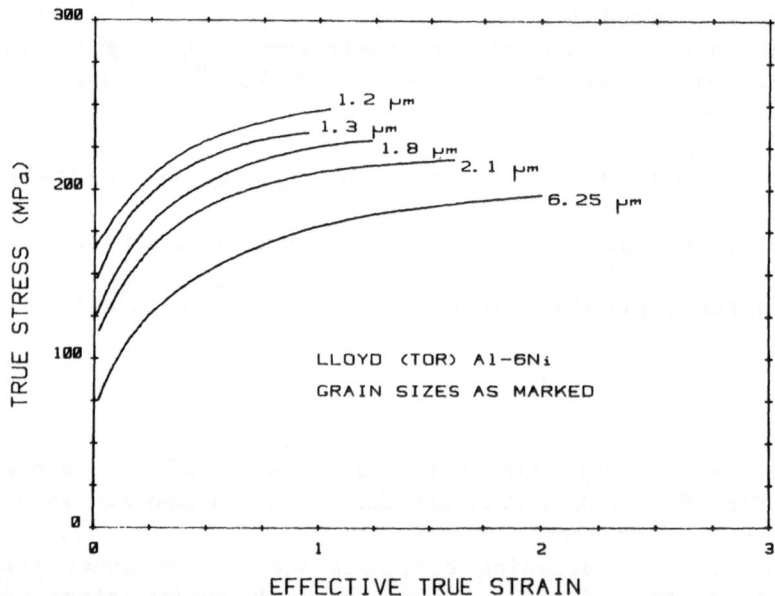

Fig. 5. Stress-strain curves of Al-6% Ni at different grain sizes as marked. (Reference 24.)

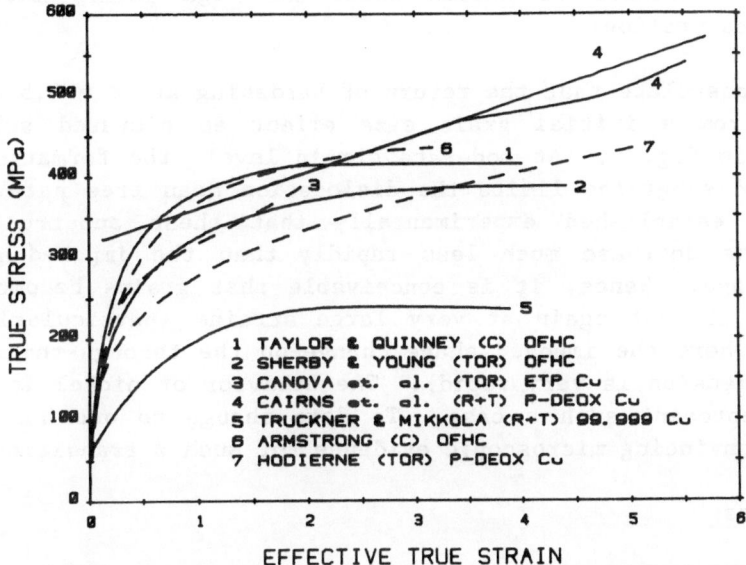

Fig. 6. Stress-strain curves for copper of different purities.
(References 14, 15, 25-29.)

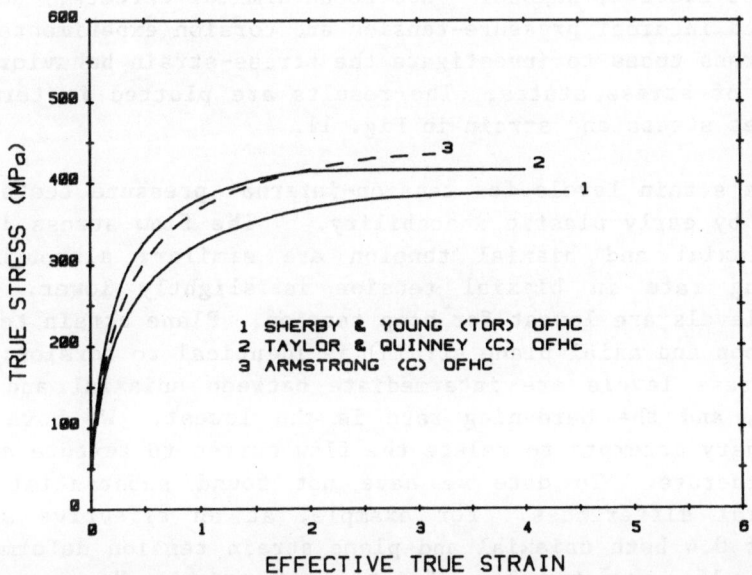

Fig. 7. Stress-strain curves for copper comparing compression
versus torsion stress-strain curves. (References 25, 26, 29.)

material continues to harden while the high purity material
shows a saturation.

We postulate that the return of hardening at ε = 4.5 might
result from a initial grain size effect as pictured schema-
tically in Fig. 9. At moderate strain levels the formation of
cells and subgrains limits the dislocation mean free path. It
is well established experimentally that these substructural
dimensions decrease much less rapidly than the imposed grain
size change. Hence, it is conceivable that grains become the
limiting element again at very large strains (particularly in
rolling where the imposed shape change on the through-thickness
grain dimension is very rapid). The behavior of nickel in Fig.
8 may represent such a case. To date though no one has presented convincing microscopic evidence for such a transition.

70-30 Brass

Experimentally the forming limit diagram of 70-30 brass has
less biaxial and plane strain ductility than is predicted by a
Marciniak analysis.[5] Ghosh[31] explained this in terms of a lower
work hardening exponent for plane strain deformation. The plane
strain work hardening exponent was determined from plane strain
punch stretching and plane strain tensile data shown in Fig.
10. Most recently Wagoner[32] has found similar effects. We have
conducted internal pressure-tension and torsion experiments with
70-30 brass tubes to investigate the stress-strain behavior in a
variety of stress states. The results are plotted in terms of
von Mises stress and strain in Fig. 11.

The strain levels for tension-internal pressure tests were
limited by early plastic instability.[11] The flow stress levels
for uniaxial and biaxial tension are similar, although the
hardening rate in biaxial tension is slightly lower. Flow
stress levels are lowest for hoop tension. Plane strain tension
(both hoop and axial plane strain) is identical to torsion; the
flow stress levels are intermediate between uniaxial and hoop
tension, and the hardening rate is the lowest. We have made
preliminary attempts to relate the flow curves to texture and
microstructure. To date we have not found substantial substructural differences. For example, at an effective strain
level of 0.4 both uniaxial and plane strain tension deformation
exhibit ~15 percent volume fraction of twins. However, pole
figure measurements indicate that the flow curves are qualitatively consistent with the initial textures.

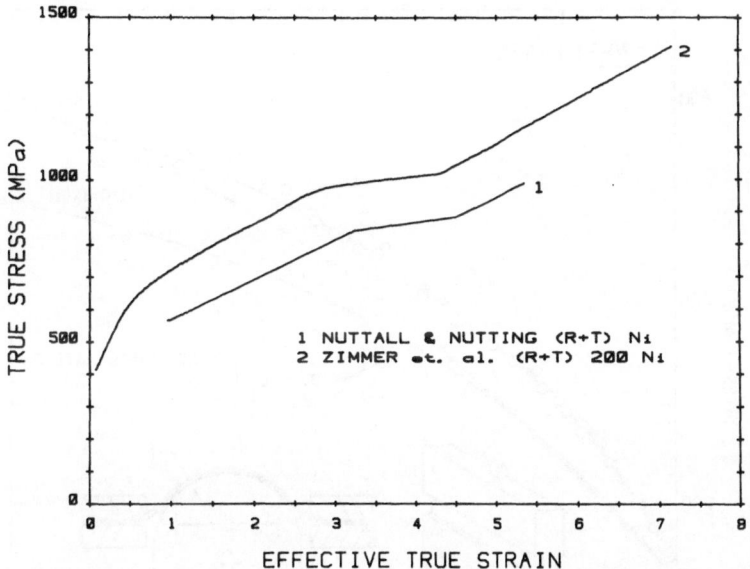

Fig. 8. Stress-strain curves for nickel. (References 21, 30.)

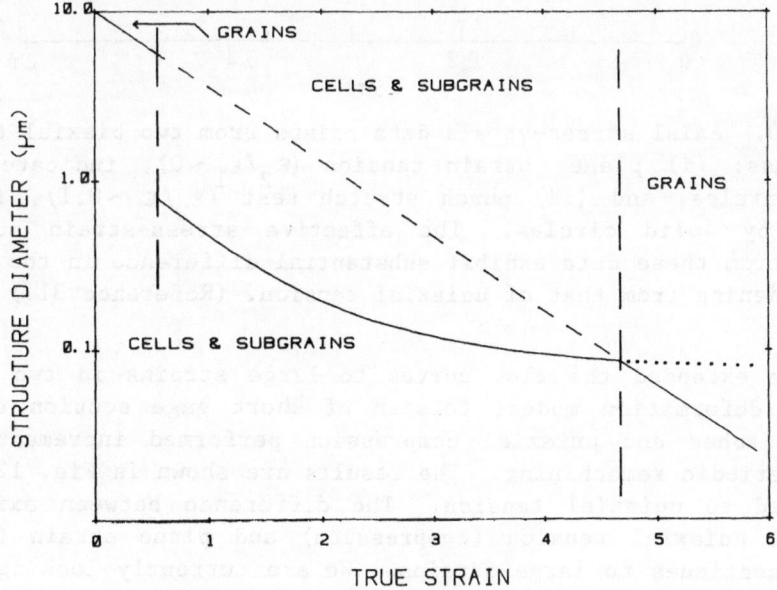

Fig. 9. The potential role of grain and substructure sizes controlling the flow stress. (Reference 1.) The dashed diagonal line represents the imposed decrease in transverse grain size by the external shape change due to rolling. The solid curve for cells and subgrains is schematic, but typical for most materials.

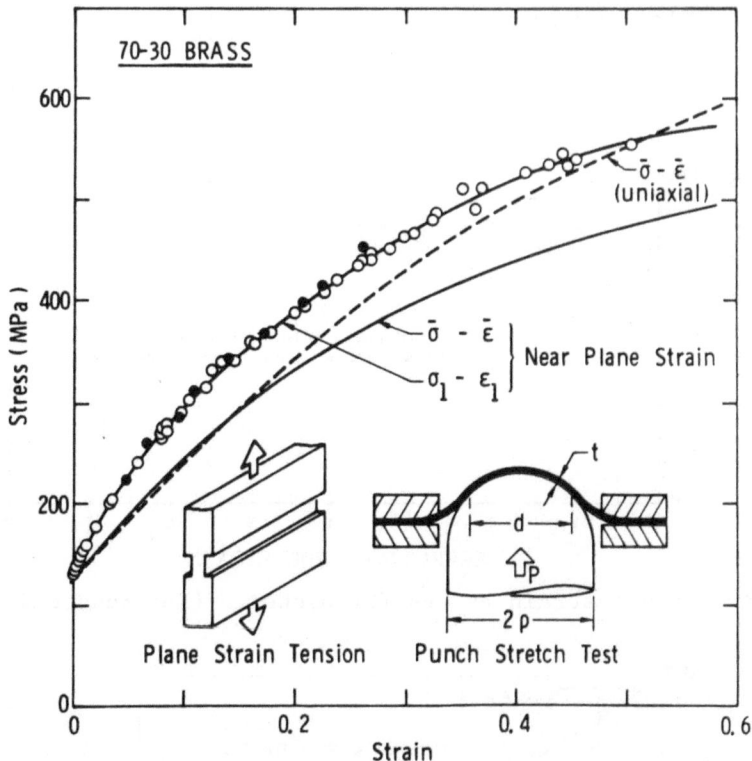

Fig. 10.   Axial stress-strain data points from two biaxial tests on brass: (i) plane strain tension ($\varepsilon_2/\varepsilon_1 \sim 0$), indicated by open circles, and (ii) punch stretch test ($\varepsilon_2/\varepsilon_1 \sim 0.1$), indicated by solid circles.   The effective stress-strain curves drawn from these data exhibit substantial difference in the rate of hardening from that of uniaxial tension. (Reference 31.)

We extended the flow curves to large strains in two different deformation modes; torsion of short gage section thin-walled tubes and uniaxial compression performed incrementally with periodic remachining.   The results are shown in Fig. 12 and compared to uniaxial tension.   The difference between axisymmetric, uniaxial tension (compression) and plane strain (torsion) continues to large strains.   We are currently looking for a microstructural or textural explanation for this difference.

BCC Iron and Iron Alloys

Most of the available literature on bcc metals and alloys was summarized by GVA.[1]   We will present only the highlights here.   Perhaps the most dramatic of the bcc results are shown in

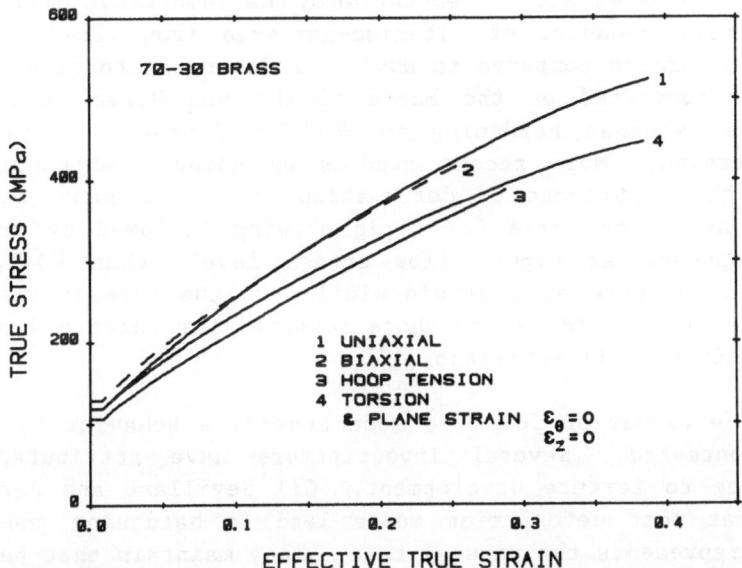

Fig. 11.   Stress-strain curves for 70-30 Brass obtained from tension-internal pressure, and torsion loadings of thin-walled tubes.   Note: ε_z = 0 is a plane strain state with no length change. ε_θ = 0 is a plane strain state with no change in the hoop dimension.

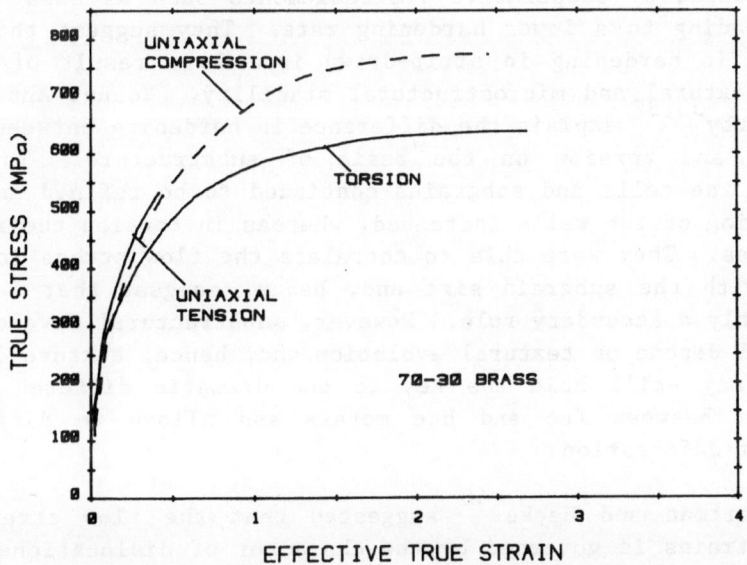

Fig. 12.   A comparison of torsion (on thin-walled) tubes and uniaxial compression (on solid bars) stress-strain data for 70-30 Brass to an effective strain of 2.5.

Fig. 13. Young et al.[33] demonstrated the remarkable difference
in hardening behavior of titanium-gettered iron (Fe-0.17% Ti)
tested in torsion compared to WD+T. Hardening in torsion (solid
rod data converted on the basis of the von Mises criterion)
saturates, whereas hardening in WD+T is linear ($\sigma = K\varepsilon$) at
large strains. More recent studies by Razavi and Langford[34]
confirm the importance of deformation mode. As shown in Fig.
14, the hardening curve for strip drawing followed by tension
(SD+T) appears at higher flow stress levels than WD+T, but
starts to saturate at a strain similar to the torsion results.
These result are similar to those reported for eutectoid steels
by Aernoudt and Gil Sevillano.[35,36]

    The explanation for this mode sensitive behavior is still
being contested. Several investigators have attributed the
difference to texture development. Gil Sevillano and Aernoudt
claim that most deformation modes lead to hardening and that
torsion represents the unusual case. They maintain that because
of the texture developed in torsion the slip distance remains
unchanged at moderate strains and at large strains dynamic
recovery actually increases the slip distance. Razavi and
Langford[34] relate the continued hardening during wire drawing to
redundant strain (curling of grains) necessary to maintain grain
continuity. In strip drawing and torsion, deformation may be
accommodated by cooperative rearrangements such as shear banding, leading to a lower hardening rate. They suggest that the
decline in hardening in strip-drawn iron is a result of achi-
eving textural and microstructural stability. Young, Anderson,
and Sherby[33,37] explain the difference in hardening between wire
drawing and torsion on the basis of substructure. In wire
drawing the cells and subgrains continued to be refined and the
perfection of the walls increased, whereas in torsion their size
saturates. They were able to correlate the flow stress in both
cases with the subgrain size and, hence, suggest that texture
plays only a secondary role. However, substructural development
may well depend on textural evolution and, hence, texture development may still hold the key to the dramatic differences in
response between fcc and bcc metals and alloys to different
modes of deformation.

    Weertman and Hecker[40] suggested that the flow stress at
large strains is governed by the character of dislocations. In
torsion most of the slip occurs on systems with a common slip

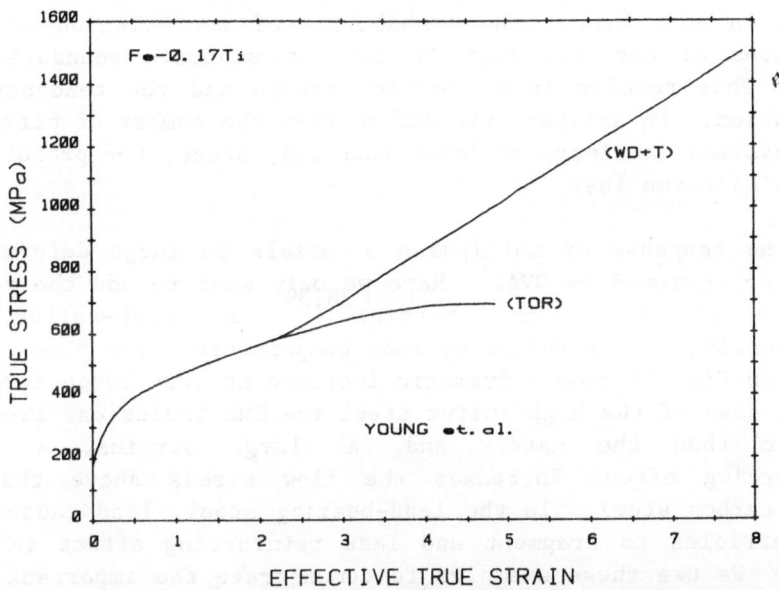

Fig. 13. Comparison of stress-strain curves for Fe-0.17% Ti deformed by torsion and wiredrawing + tension. (Reference 33.)

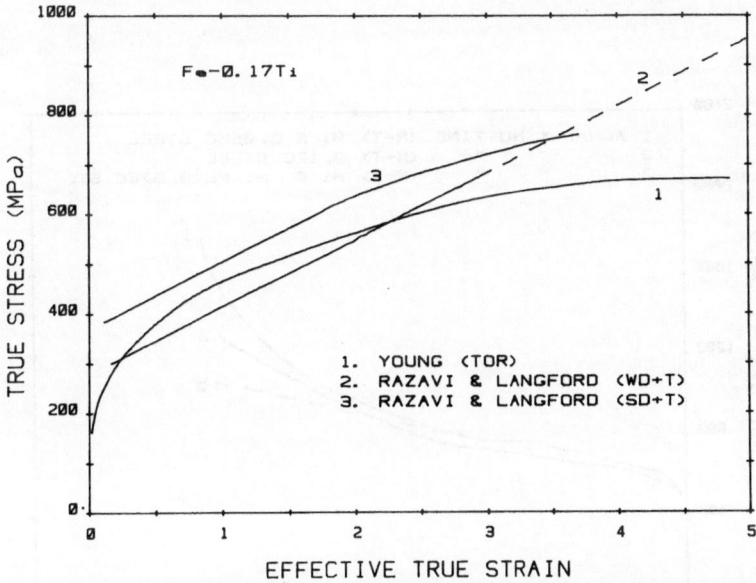

Fig. 14. Comparison of stress-strain curves for Fe-0.17% Ti deformed by torsion, wiredrawing, and strip drawing. The dashed line is a linear extrapolation of Razavi and Langford's data based on other literature values. (References 33, 34.)

direction and, hence, the probability of annihilation of dislocations of opposite sign in cell or subgrain boundaries is
large. This results in a low flow stress and the tendency for
saturation. In axisymmetric deformation the number of different
slip systems is always at least four and, hence, the probability
of annihilation less.

The response of multiphase materials to large deformation
was also reviewed by GVA.[1] Here we only want to add the recent
results of Aghan and Nutting[38,39] on high-sulfur and
lead-bearing steels rolled at room temperature. The flow curves
shown in Fig. 15 show a dramatic increase at very large strains.
In the case of the high-sulfur steel the MnS inclusions are more
plastic than the matrix and, at large strains, a fiber
reinforcing effect increases the flow stress above that of
plain carbon steel. In the lead-bearing steel, lead causes the
MnS particles to fragment and less reinforcing effect is realized. We use these examples to demonstrate the important role
that second-phase particles can have on the flow stress at large
strains without having much influence at small strain.

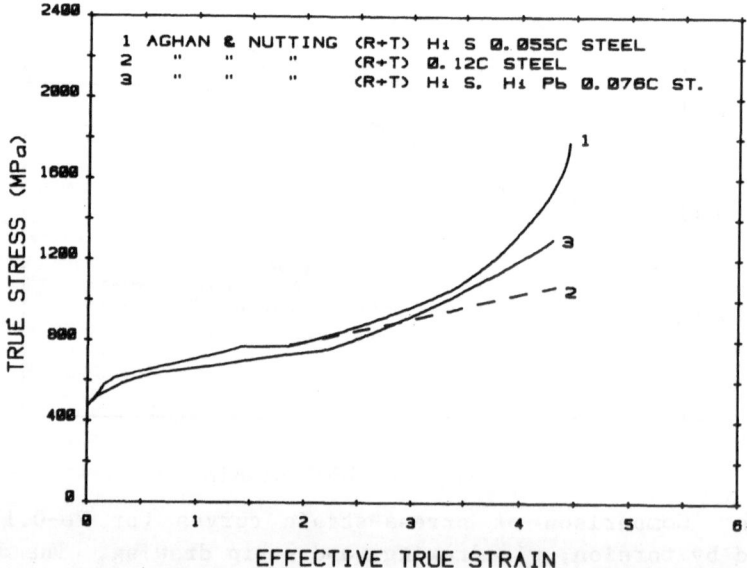

Fig. 15. Stress-strain curves for low-carbon steel (#2), high-
sulfur steel (#1), and high-sulfur, leaded steel (#3). (References 38, 39.)

DISCUSSION

The results presented here and those reviewed by GVA[1] and Hecker, Stout, and Eash[2] indicate that saturation (steady state behavior) of the flow stress in fcc materials is not universal. Steady state behavior should be expected if the evolution of substructure is controlled only by undisturbed dislocation interaction. However, as pointed out by Mecking and Grinberg,[41] many potential disturbing influences may appear at large strains. These include i) grain size effects, ii) deformation bands, iii) surface effects, iv) strain-induced transformations (twinning or martensite) v) changing deformation mode, vi) deformational instabilities such as shear bands, vii) texture development, and viii) second-phase particles. As a result of these disturbing influences we find that continued hardening is generally observed. High purity and torsional deformation modes favor steady state and saturation.

The effects of material purity is demonstrated convincingly above. As explained, solute atoms retard the recovery process, shifting the balance between dislocation generation and anni-hilation to larger strains. Alloying can have similar effects or introduce much greater complexities by the interaction of dislocations with complex microstructures such as those demon-strated for Al-4% Cu alloys (Fig. 4). The influence of grain size remains much stronger at large strains than expected (Fig. 5). Also, as indicated in Fig. 9 grain boundaries may again become important structural elements at very large strains and, perhaps, explain the peculiar hardening behavior of nickel (Fig. 8).

The evolution of texture most likely plays a significant role in controlling the flow behavior. Our experiments on brass indicate that the difference between axisymmetric and torsional deformation at ε_{eff} = 0.4 can be explained qualitatively on the basis of texture development. A quantitative comparison awaits the type of rigorous analysis of texture prediction being con-ducted by Jonas et al.[42] Results for bcc materials also indi-cate potential textural effects. If texture plays a major role in hardening, then the likelihood of finding a single, intrinsic hardening curve for polycrystalline materials is small because deformation geometry will influence the results.

Twinning, deformation bands, and shear bands all have been studied microscopically. The role that these mechanisms play in hardening has not been established quantitatively. Second phase particles can play an important role in hardening. In Al-Fe or Al-Ni alloys they stabilize the size of the substructure. In steels containing high sulfur or lead additions they can produce a fiber reinforcing effect at very large deformations.

SUMMARY

      Strain hardening at large plastic strains cannot be inferred from small-strain tensile tests. Most metals and alloys at room temperature do not reach steady state saturation at strain levels of 3 to 5. Typically, some disturbing influence offsets the balance between dislocation generation and annihilation. The most prominent of these appears to be texture formation. However, grain size, second-phase particles, and deformation on shear bands are also important. The effect on hardening of most of these features depends on geometry (or deformation mode) and, hence, no single intrinsic hardening curve can be expected at large strains. It should be noted that high material purity and a torsional deformation mode favor saturation.

ACKNOWLEDGMENTS

      This work was sponsored by the Division of Materials Sciences, Office of Basic Energy Sciences, U. S. Department of Energy.

REFERENCES

1. J. Gil Sevillano, P. van Houtte, and E. Aernoudt, Large
       Strain Work Hardening and Textures, Prog. Mater. Sci.,
       25, pp. 69-412 (1980).

2. S. S. Hecker, M. G. Stout, and D. T. Eash, Experiments in
       Plastic Deformation at Finite Strains, Proc. of Work-
       shop on "Plasticity of Metals at Finite Strains:
       Theory, Experiment and Computation," held at Stanford
       University, CA, June 29, 30 and July 1, 1981.

3. C. S. Hartley, and D. A. Jenkins, Tensile Testing at
       Constant True Strain Rates, in:"Proceedings of 6th
       International Conference on Experimental Stress Anal-
       ysis", Munich, W. Germany, pp. 379-383 Sept. 18-22
       (1978).

4. C. S. Hartley, D. A. Jenkins, and J-J. Lee, Strain Depend-
    ence of Strain-Rate Sensitivity in:"Proceedings of 5th
    International Conference on Strength of Metals and
    Alloys", P. Haasen, V. Gerold, and G. Kostorz, eds.
    Pergamon Press, pp. 523-528 (1980).

5. Z. Marciniak, and K. Kuczynski, Limit Strains in the
    Processes of Stretch-Forming Sheet Metal, Int. J.
    Mech. Sci., 9, pp. 609-620 (1967).

6. S. P. Keeler, and W. A. Backofen, Plastic Instability and
    Fracture in Sheets Stretched Over Rigid Punches,
    Trans. ASM, 56, pp. 25-48 (1963).

7. M. Azrin, and W. A. Backofen, The Deformation and Failure
    of Biaxially Stretched Sheet, Met. Trans., 1, pp.
    2857-2856 (1970).

8. A. K. Ghosh, and W. A. Backofen, Strain Hardening and
    Instability in Biaxially Stretched Sheets, Met.
    Trans., 4, pp. 1113-1123 (1973).

9. A. J. Ranta-Eskola, Use of the Hydraulic Bulge Test in
    Biaxial Tensile Testing, Mech. Sci., 21, pp. 457-465
    (1979).

10. R. Bell, J. L. Duncan, and I. H. Wilson, A Sheet-Bulging
    Machine with Closed Loop Control, J. Strain Anal., 2,
    pp. 246-253 (1967).

11. M. G. Stout, and S. S. Hecker, Comparison of Plastic
    Instability in Sheet and Tubular Specimens of 70-30
    Brass, presented at Fall TMS/AIME Meeting, Louisville,
    Ky Oct. 13, 1981. Abstract in J. Metals, 33, p. 21
    (1981).

12. A. Nadai in: "Theory of Flow and Fracture", 2nd ed.,
    McGraw Hill Book Company Inc., 1, p. 349 (1950).

13. D. S. Fields, Jr., and W. A. Backofen, Determination of
    Strain-Hardening Characteristics by Torsion Testing,
    Proceedings ASTM, 57, pp. 1259-1272 (1957).

14. G. R. Canova, S. Shrivastava, J. J. Jonas, and C. G'Sell, The Use of Torsion Testing to Assess Material Formability, Prepared for presentation at the ASTM Symposium "Formability-2000," Chicago, Ill. (1980).

15. F. A. Hodierne, A. Torsion Test for Use in Metalworking Studies, J. Inst. Metals, 91, pp. 267-273, 1963.

6. S. S. Hecker, D. L. Rohr, and R. M. Aikin, Unpublished work Los Alamos National Laboratory (1978).

17. P. E. Armstrong, J. E. Hockett, and O. D. Sherby, Large Strain Multidirectional Deformation of 1100 Aluminum at 300 K, J. Mech. Phys. Sol., 30, pp. 37-58 (1982).

18. S. K. Varma, and B. G. LeFevre, Large Wire Drawing Plastic Deformation in Aluminum and Its Dilute Alloys, Met. Trans. A, 11A, pp. 935-942 (1980).

19. D. Kalish and B. G. LeFevre, Subgrain Strengthening of Aluminum Conductor Wires, Met. Trans. A, 6A, pp. 1319-1324 (1975).

20. H. Luthy, A. K. Miller, and O. D. Sherby, The Stress and Temperature Dependence of Steady-State Flow at Intermediate Temperatures for Pure Polycrstalline Aluminum, Acta Met., 28, pp. 169-17 (1980).

21. J. Nuttall, and J. Nutting, Structure and Properties of Heavily Cold-Worked fcc Metals and Alloys, Metal Sci., 12, pp. 430-437, 1978.

22. B. Nourbaksh and J. Nutting, The High Strain Deformation of of an Aluminum -4% Copper Alloy in the Supersaturated and Aged Conditions, Acta Met., 28, pp. 357-365, 1980.

23. D. J. Lloyd, and D. Kenny, The Structure and Properties of Heavily Cold Worked Aluminum Alloys, Prepublication paper from Aluminum Company of Canada, Ltd. Research Center, Kingston, Ontario, Canada.

24. D. J. Lloyd, Deformation of Fine-Grained Aluminum Alloys, Metal Sci., 14, pp. 193-198 (1980).

25. G. I. Taylor, and H. Quinney, Proceedings of the Royal
     Society of London, 143, p. 307 (1934).

26. O. D. Sherby, and C. M. Young, Some Factors Influencing the
     Strain Rate-Temperature Dependence of the Flow Stress
     in Polycrystalline Solids, in: "Rate Processes in
     Plastic Deformation of Materials", J. C. M. Li, and A.
     K. Mukherjee eds., ASM, pp. 497-541 (1975).

27. J. H. Cairns, J. Clough, M. A. P. Dewey, and J. Nutting,
     The Structure and Mechanical Properties of Heavily De-
     formed Copper, J. Inst. Metals, 99, pp. 93-97 (1971).

28. W. G. Truckner, and D. E. Mikkola, Strengthening of Copper
     by Dislocation Substructures, Met. Trans. A, 8A, pp.
     45-49, (1977).

29. P. E. Armstrong, Los Alamos National Laboratory, (1979),
     unpublished work.

30. W. H. Zimmer, S. S. Hecker, L. E. Murr, and D. L. Rohr,
     Large-Strain Plastic Deformation of Commercially-pure
     Nickel, accepted for publication in Metal Sci. (1980).

31. A. K. Ghosh, Plastic Flow Properties in Relation to Local-
     ized Necking in Sheets, in:"Mechanics of Sheet Metal
     Forming-Material Behavior and Deformation Analysis",
     D. P. Koistinen, and N-M. Wang, eds., pp. 287-312,
     Plenum Press (1978).

32. R. H. Wagoner, Plastic Behavior of 70-30 Brass Sheet,
     Met. Trans. A, 13A, pp. 1491-1500 (1982).

33. C. M. Young, L. J. Anderson, and O. D. Sherby, On the
     Steady State Flow Stress of Iron at Low Temperatures
     and Large Strains, Met. Trans., 5, pp. 519-520 (1974).

34. A. Razavi, and G. Langford, Strain Hardening of Iron:
     Axisymmetric vs. Plane Strain Elongation, in:"Pro-
     ceedings of 5th International Conference on Strength
     of Metals and Alloys", P. Haasen, V. Gerold, and G.
     Kostorz, eds., pp. 831-836, Pergamon Press (1980).

35. E. Aernoudt, and J. Gil Sevillano, Influence of the Mode of
     Deformation on the Hardening of Ferritic and Pearlitic
     Carbon Steels at Large Strains, J. Iron and Steel
     Inst., 211, pp. 718-725 (1973).

36. J. Gil Sevillano, and E. Aernoudt, On the Influence of the
     Mode of Deformation on the Hardening of Iron at Low
     Temperature and Large Strains, Met. Trans. A, 6A, pp.
     2163-2164 (1975).

37. C. M. Young, L. J. Anderson, O. D. Sherby, reply to "On the
     Influence of the Mode of Deformation on the Hardening
     of Iron at Low Temperature and Large Strain," Met.
     Trans. A, 6A, pp. 2164-2165 (1975).

38. R. L. Aghan, and J. Nutting, Structure and Properties of
     Free-Cutting Steels After Deformation to High Strains,
     Metals Tech., 8, pp. 41-45 (1981).

39. R. L. Aghan, and J. Nutting, Structure and Properties of
     Low-Carbon Steel After Deformation to High Strains,
     Metal Sci., 14, pp. 233-237 (1980).

40. J. Weertman and S. S. Hecker, Theory for Saturation Stress
     Difference in Torsion versus Other Types of Deforma-
     tion at Low Temperatures, submitted to J. Mech. Mater.
     (1982).

41. H. Mecking and A. Grinberg, Discussion on the Development
     of a Stage of Steady-State Flow at Large Strains, in:
     "Proceedings of 5th International Conference on Stre-
     ngth of Metals and Alloys", P. Haasen, V. Gerold, and
     G. Kostorz, eds., pp. 289-294, Pergamon Press, (1980).

42. J. J. Jonas, G. R. Canova, S. C. Shrivastava and N.
     Christodoulou, Sources of the Discrepancy Between the
     Flow Curves Determined in Torsion and in Axisymmetric
     Tension and Compression Testing, Proc. of Workshop on
     "Plasticity of Metals at Finite Strains: Theory,
     Experiment and Computation," held at Stanford
     University, CA, June 29, 30 and July 1, 1981.

STRAIN-RATE EFFECTS IN METALS AT LARGE SHEAR STRAINS

Ulric S. Lindholm and Gordon R. Johnson

Southwest Research Institute
San Antonio, TX 78284

Honeywell Inc.
Hopkins, MN 55343

INTRODUCTION

Many problems in ballistics involve very large shearing deformation occurring at high deformation rates. Such problems include, for instance, explosive formation of shaped charge jets or self-forming fragments and the defeat of thick targets by long rod penetration. The development of a materials testing capability, generation of experimental data and evaluation of constitutive models for use in such problems has been underway for several years. This paper will review the major results of this effort and suggest some areas for future work.

The physical problems of interest may involve inelastic shear strains in ductile alloys well in excess of unity and strain rates up to the order of 10^4 sec^{-1}. To achieve the strain requirement, it was concluded that the most appropriate specimen geometry and loading mode was a thin-walled, short gage length tubular specimen subject to pure torsional loading. At low deformation rates, this specimen is capable of maintaining a homogeneous state of pure shear deformation (assuming a small thickness to diameter ratio) to very large strains, limited only by the ductility of the metal being tested. It is free of the dilational effects which lead to geometrical instability (necking or barreling) with development of large strain in tensile or compression tests. The high strain-rate requirement is perhaps most closely met by the torsional Hopkinson bar technique (1) which utilizes propagating elastic stress pulses to load a short tubular specimen. However,

with this technique, it is difficult to generate the large angles
of twist required for deformation of the most ductile metals and
the range in strain-rate capability is limited to about one order
of magnitude, 10^3–10^4 sec^{-1}. For this reason, our initial efforts
have employed a hydraulic loading device which is capable of strain
rates from "static" to approximately 500 sec^{-1} in a continuous
fashion. This wide range in strain rate allows detailed study of
the transition from isothermal deformation to nearly fully adiabatic for most metals. This approach has proven of benefit in
establishing both the isothermal constitutive behavior and the conditions for adiabatic shear instability, a critical problem in
practice. Extrapolation of the strain-rate dependence to higher
rates encountered in ballistic problems must, of course, be done
with caution until experimental verification can be achieved, perhaps by supplemental Hopkinson bar testing.

At present, we have accumulated stress-strain-strain rate data
on a wide range of both ductile and nonductile alloys (2,3). A
simple thermoviscoplastic constitutive model has been fit to this
data and been used to qualitatively and quantitatively predict the
conditions for the onset of adiabatic shear instability. The
quantitative prediction (4) involves a finite element analysis of
the specimen including the coupled heat generation and conduction
problem associated with deformation. The localized shear bands
which develop rapidly after instability occurs will be illustrated.
Observations include both deformation and transformation bands.
In addition, the possible contribution of internal damage as well
as thermal softening as a de-stabalizing mechanism is considered.

Finally, some observations and comments on the differences in
strain hardening and total ductility between tensile and shear
modes of deformation will be made. Suggestions for further experimental work and analysis to clarify some of these issues are offered.

EXPERIMENTAL PROCEDURES AND RESULTS

The dynamic torsional test facility has been previously
described in Reference 5. It consists of a compact torsional
hydraulic actuator integrally mounted with a load transducer and
specimen as shown in Fig. 1. For rotational speeds yielding specimen strain rates from static to about 10 sec^{-1}, actuator pressure
is actively regulated by a servovalve and closed-loop control
system. For higher rotational speeds a solenoid triggered, quick-
opening valve is used to connect the actuator to a high pressure
(20 MPa) accumulator. In this range the machine is operating in
an open loop fashion with the angular velocity being only approximately constant dependent upon the inertia of the system and the
changing shear resistance of the specimen. Transient measurements
are made of the rotary displacement of the actuator shaft (base of
specimen) and the torsional load transmitted by the specimen.

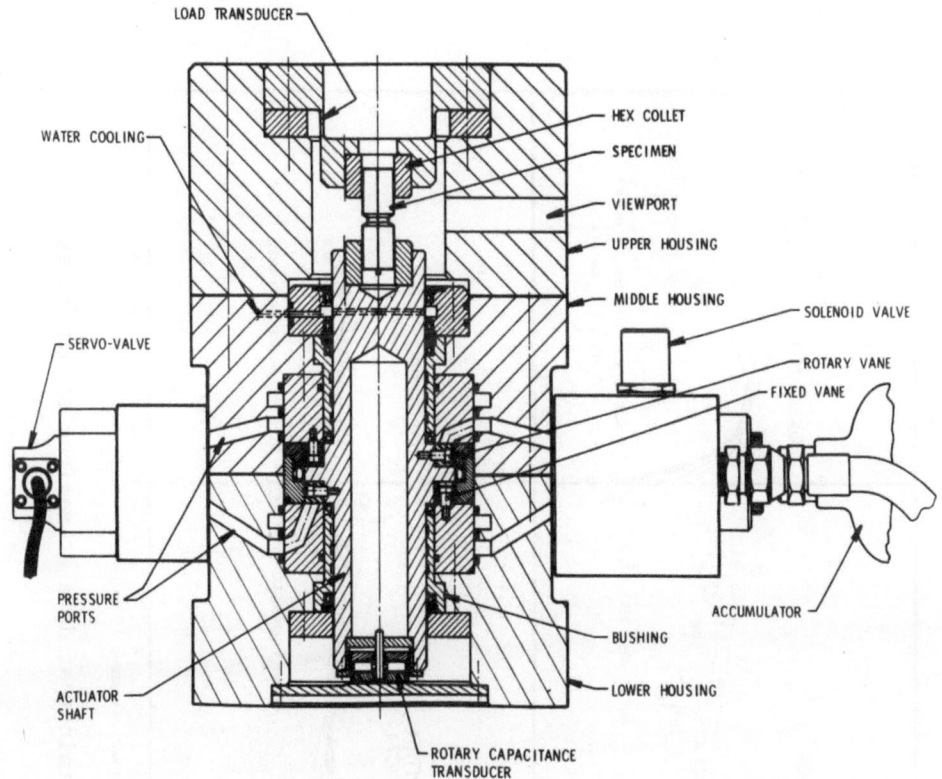

LOAD TRANSDUCER

WATER COOLING

HEX COLLET
SPECIMEN
VIEWPORT
UPPER HOUSING
MIDDLE HOUSING

SOLENOID VALVE

SERVO-VALVE

ROTARY VANE
FIXED VANE

PRESSURE PORTS

ACCUMULATOR

BUSHING

ACTUATOR SHAFT

LOWER HOUSING

ROTARY CAPACITANCE TRANSDUCER

Fig. 1.   Schematic of dynamic torsion test facility.

The unique design of the torsional actuator allows a positive rotational displacement of about 3.3 radians. When using a standard specimen of nominal radius R = 7 mm and gage length L = 3 mm, this rotation allows a total strain of $\gamma = R\theta/L \simeq 7$. Larger strains can be achieved by reducing the gage length.

In the axial direction the specimen is constrained only by friction in the grip. From estimates of this constraint and from post test remeasurement of axial gage length, the boundary conditions are essentially zero relative axial displacement rather than zero axial load. Thus, an axial load or stress component will develop with increasing torsional deformation; however, to date this axial load component has not been measured.

Examples of torsional stress-strain curves obtained at shear strain rates from 10^{-3} sec^{-1} to 500 sec^{-1} are shown in Figs. 2(a) and 2(b). Fig. 2(a) shows six ductile alloys ($\gamma > 3$) and Fig. 2(b) another six alloys which are less ductile ($\gamma < 1$). The alloys selected represent a wide range in both physical and mechanical

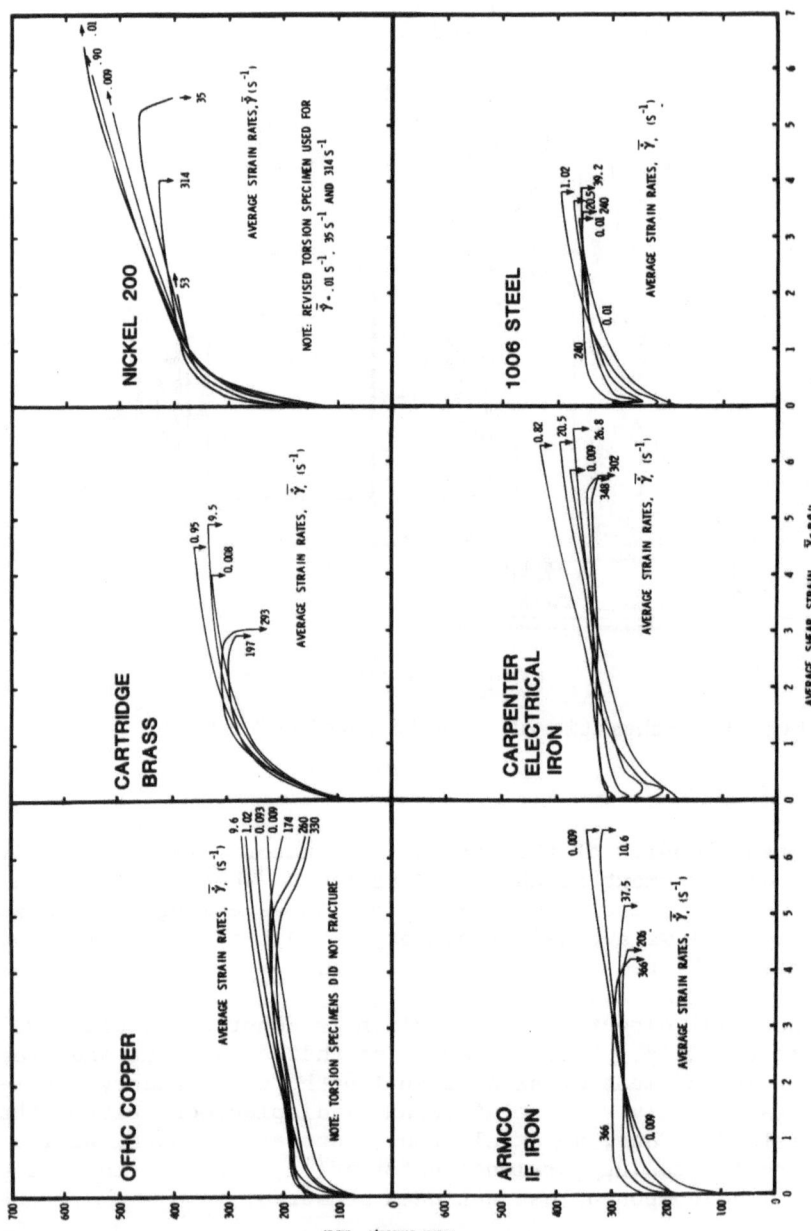

Fig. 2(a)   Torsional stress-strain test data at various strain rates.

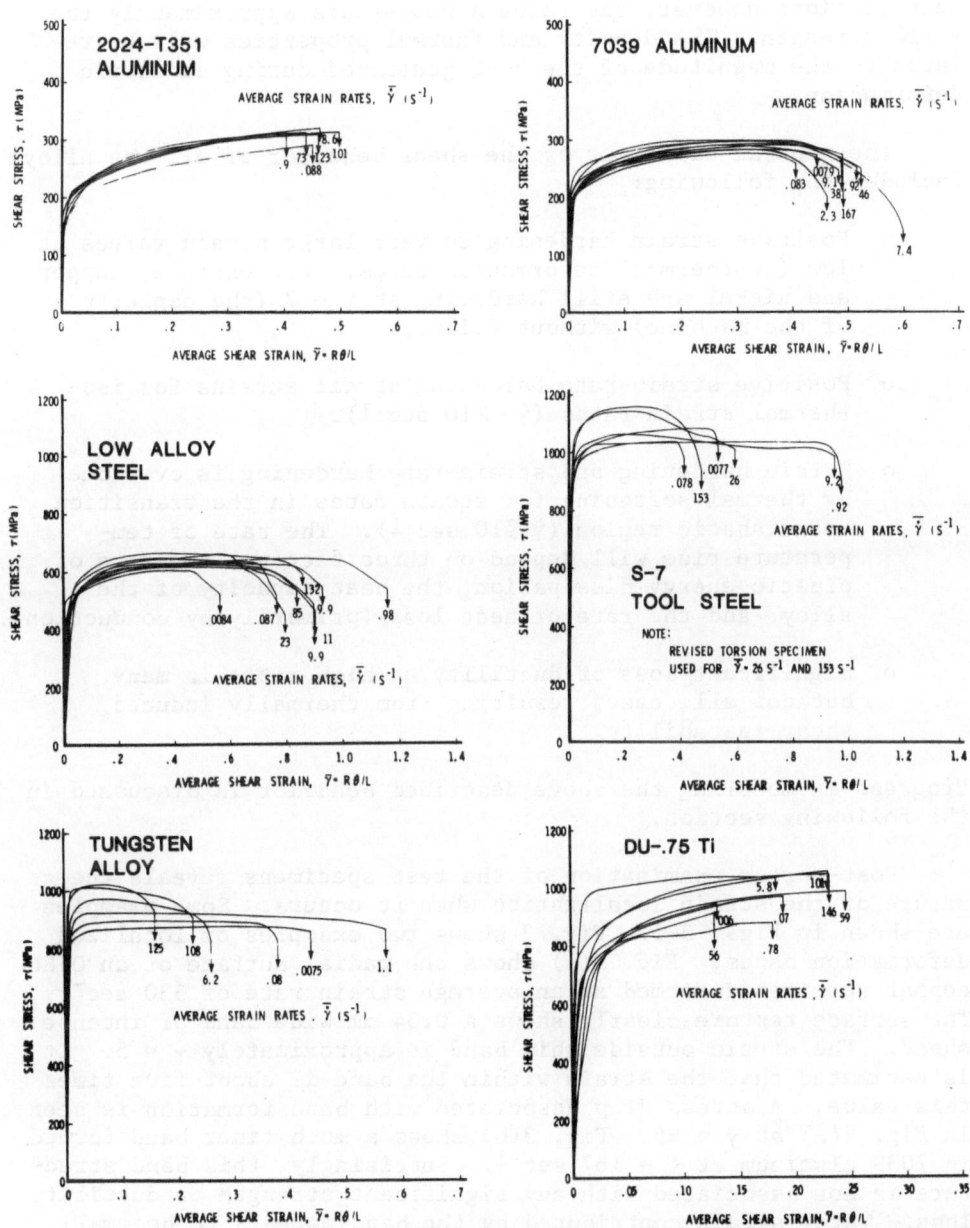

Fig. 2(b)   Torsional stress-strain test data at various
            strain rates.

strength properties. These properties are summarized in Table I.
The constitutive constants are those for a model presented in the
next section; however, the value A represents approximately the
yield strength. The density and thermal properties will be related to the magnitude of the heat generated during adiabatic
deformation.

The general character of the shear behavior of all the alloys
includes the following:

o Positive strain hardening to very large strain values at
  low (isothermal) deformation rates. For example, copper
  and nickel are still hardening at $\gamma = 7$ (the capacity
  of the machine) without failure.

o Positive strain-rate hardening at all strains for isothermal strain rates ($\dot{\gamma} \lesssim 10$ sec^{-1}).

o Strain hardening and strain-rate hardening is overcome
  by thermal softening for strain rates in the transition
  or adiabatic region ($\dot{\gamma} \gtrsim 10$ sec^{-1}). The rate of temperature rise will depend on three factors; the rate of
  plastic energy dissipation, the heat capacity of the
  alloy, and the rate of heat loss (primarily by conduction).

o Significant loss of ductility at high rates in many,
  but not all, cases resulting from thermally induced
  shear instability.

Progress in modeling the above described behavior is discussed in
the following section.

Post-mortem examination of the test specimens reveals the
nature of the strain localization when it occurs. Some examples
are shown in Figs. 3-5. Fig. 3 shows two examples of localized
deformation bands. Fig. 3(a) shows the radial surface of an OFHC
copper specimen deformed at an average strain rate of 330 sec^{-1}.
The surface texture clearly shows a 0.34 mm wide band of intense
shear. The strain outside this band is approximately $\gamma = 5$. It
is estimated that the strain within the band is about five times
this value. A stress drop associated with band formation is seen
in Fig. 2(a) at $\gamma = \approx 5$. Fig. 3(b) shows a much finer band formed
in 7039 aluminum at $\dot{\gamma} = 167$ sec^{-1}. Suprisingly, this band structure is not associated with any significant strength or ductility
loss. Total strain contributed by the band appears to be small.

In steels, the temperature and strain conditions within the
adiabatic shear band can be sufficient to induce a martensitic
transformation leaving a distinct white-etching zone. Such

TABLE I.   PROPERTIES OF MATERIALS TESTED

Material	Thermal Properties			Constitutive Constants				Temp. @ γ_c (C)	Critical Strain	
	Density $\rho(kg/_m3)$	Sp.Heat. $C_p(J/kgK)$	Melt.Temp. $T_M(K)$	A (MPa)	B (MPa)	n	C		Th.	Exp.
OFHC Copper (CDA 101)	8950	383	1355	69	106	0.32	0.027	252	5.3	5.8
Cartridge Brass (CDA 260)	8520	385	1189	62	186	0.34	0.007	241	3.7	3.0
Aluminum 2024-T 351	2770	875	775	152	202	0.34	0.015	89	0.66	0.50
Aluminum 7039	2770	875	877	193	157	0.41	0.010	100	0.77	0.55
Nickel 200	8900	446	1726	138	234	0.32	0.008	148	0.03	0.18
Armco IF Iron	7890	452	1811	76	196	0.25	0.028	296	4.3	4.1
Carpenter Electrical Iron	7890	452	1811	193	109	0.43	0.028	327	4.4	5.8
1006 Steel	7890	452	1811	200	129	0.36	0.022	269	3.3	3.5
RHA Steel	7840	477	1793	455	237	0.37	0.006	192	1.2	1.1
AMS 6418 Steel	7750	477	1763	896	200	0.18	0.010	65	0.16	0.20
S-7 Tool Steel	7750	477	1763	883	248	0.18	0.012	67	0.16	0.50
Tungsten Alloy (7% Ni, 3% Fe)	17000	134	1723	862	94	0.12	0.016	38	0.03	0.18
Depleted Uranium (0.75 Ti)	18600	117	1473	621	561	0.25	0.007	65	0.23	0.25

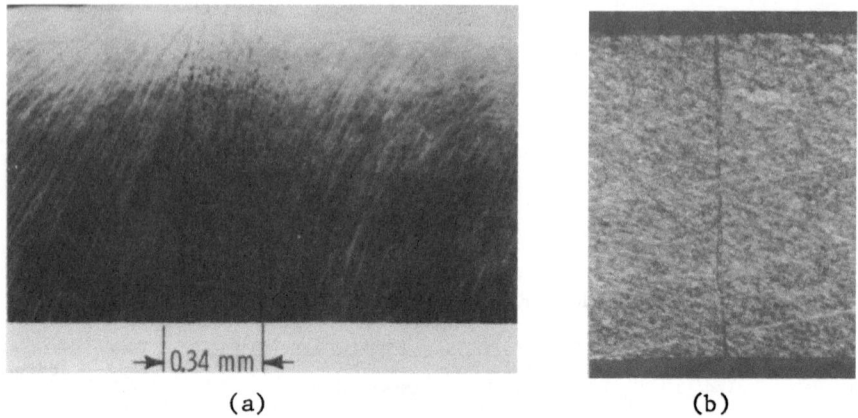

(a)   (b)

Fig. 3.   Deformation Bands in OFHC Copper (a) and 7039 Aluminum (b).

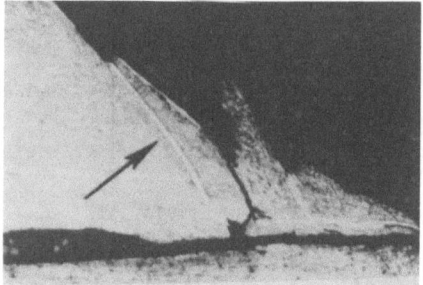

Fig. 4.   Transformation Bands in AMS 6418 Steel.

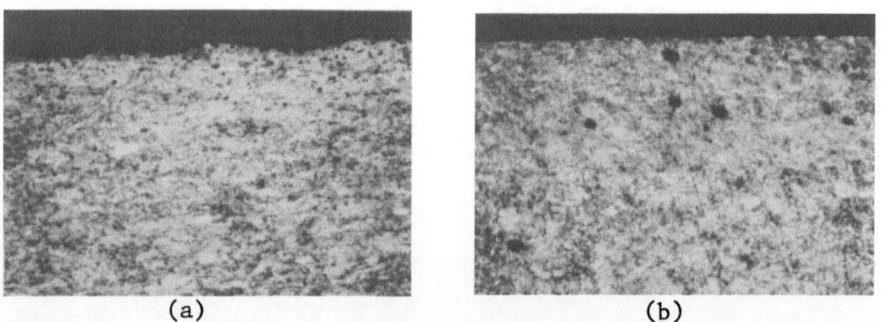

(a)   (b)

Fig. 5.   Cavitation in Cartridge Brass (a) and Tool Steel (b).

transformation bands were observed in AMS 6418 steel as shown in Figure 4. These bands occurred at a strain rate of 100 sec^{-1}.

An additional observation near the fracture plane in several alloys was the occurrence of significant cavitation or void growth. Two examples are shown in Figs. 5(a) for cartridge brass and 5(b) for tool steel. This cavitation damage does not appear to be as distinctly deformation rate dependent as is thermal softening. However, localized damage of this type is another potential strain-softening mechanism which may contribute to the onset of insta-bility or, more likely, to the rate of strain localization after instability occurs.

ANALYTICAL MODELING

In order to model the material behavior described in the pre-ceding section, the following steps were followed:

o   Develop a simple isothermal constitutive model for each
    material including strain hardening, strain-rate
    hardening and thermal softening.

o   Use this constitutive model in an approximate analysis
    of the critical strain for instability under fully
    adiabatic conditions.

o   Use a finite element code to solve the transient heat
    conduction problem in the test specimen.

o   Use the finite element code to solve the coupled problem
    of deformation and heat conduction to determine the
    transient strain and temperature distribution within the
    specimen.

Results of these analyses will be summarized below.

For the purposes of modeling simple uniaxial monotonic behav-ior, a relationship of the form

$$\tau = [A + B \, \gamma^n] \, [1 + C \, \ln (\dot{\gamma}/\dot{\gamma}_o)] \, K(T) \qquad (1)$$

was used. The first term represents power law strain hardening, the second term a logarithmic strain-rate hardening and the third term a general thermal softening function. The constants A, B, C and n are given in Table I and were derived from the isothermal test data. The constants A, B and n define the shear stress strain curve at room temperature (T = T_o), and $\dot{\gamma} = \dot{\gamma}_o = 1 \ \text{sec}^{-1}$. The rate constant C is derived from the data at an average strain

value $\gamma = 0.50$. The constitutive constants derived are given in
Table I for each material tested. The testing to date has not
included elevated temperature isothermal tests. Therefore, the
thermal softening term $K(T)$ was obtained from handbook values for
small plastic strains or was simply taken to be a linear function,
decreasing from the value given by Eq. 1 at room temperature, T_o,
to zero at melting temperature, T_M. In this approximation,

$$K(T) \;=\; \frac{T_M - T}{T_M - T_o} \tag{2}$$

Eq. 1 with $K(T) = 1$ $(T = T_o)$ was found to give a good fit to the
isothermal data. It was then used for subsequent predictions of
adiabatic behavior using Eq. 2 or alternate thermal softening
rates.

The temperature rise in the specimen caused by the conversion
of plastic work into heat is given by

$$\Delta T = \frac{\alpha}{\rho C_p} \int_0^{\gamma} \tau \, d\gamma \tag{3}$$

where α is a proportionality constant and ρ and C_p are the density
and specific heat, respectively (see Table I for values). In subsequent calculations, $\alpha = 0.9$. The temperature rise as deformation
proceeds, leads to a gradual loss in strain hardening capacity.
The usual assumption is that instability will occur when $d\tau/d\gamma = 0$.
For fully adiabatic conditions, it is possible to estimate the
critical strain, γ_c, for instability. For this case,

$$\left(\frac{d\tau}{d\gamma}\right)_a = \left(\frac{d\tau}{\gamma}\right)_T + \left(\frac{\partial\tau}{\partial T}\right) \frac{dT}{d\gamma} + \left(\frac{\partial\tau}{\partial\dot\gamma}\right) \frac{d\dot\gamma}{d\gamma} = 0 \tag{4}$$

Evaluating the differential terms in Eq. 4 under the initial conditions $T = T_o$ and $\dot\gamma = \dot\gamma_o$, assuming a constant strain rate,
$(d\dot\gamma = 0)$, and utilizing Eqs. 1-3 yields:

$$\left(\frac{d\tau}{d\gamma}\right) = n B \gamma^{n-1}$$

$$\left(\frac{\partial\tau}{\partial T}\right) = -\frac{A + B}{T_m - T_o}$$

and

$$\left(\frac{dT}{d\gamma}\right) = \frac{\alpha}{\rho c} (A + B \gamma^n)$$

Substituting in Eq. 4 yields the following relation for the critical strain

$$\gamma_c = \frac{n \rho C (T_M - T_o)}{\alpha (A+B)} - \frac{A}{B} \gamma_c^{1-n} \qquad (5)$$

Using the constants contained in Table I and solving Eq. 5 numerically results in the theoretical values of critical strain listed in the next-to-last column of Table I. The last column gives the corresponding experimental value. The agreement is generally good and the comparative values are plotted in Figure 6. The calculated temperature rise at the critical strain is also given in Table I.

The approximate relation of Eq. 5 is based on the assumption of fully adiabatic conditions. The strain rate at which each specimen attains this condition depends primarily on the rate of conduction of heat from the short deforming gage section to the unstrained material at each boundary which acts as a heat sink. A finite element model for heat conduction from an initial, uniformly heated, gage section to surrounding material at a lower reference temperature gives the temperature vs. time histories shown in Figure 7 for each material tested. The time scale, t, can be related to test conditions by dividing the maximum strain γ_{max} by the nominal strain rate, $\dot{\gamma}$. Specific ratios are indicated on Figure 7 for reference. It can be seen that the transition from isothermal to fully adiabatic conditions for a given material occurs over roughly a decade in the time or strain rate scale. For a low conductivity material such as steel, the transition occurs between 10 and 100 sec^{-1}. For copper, having very high thermal conductivity, the transition occurs at a higher strain-rate range from 100 to 1000 sec^{-1}. While these strain rates are generally representative, any specific situation will depend upon component geometry and the heat loss mechanisms available, conduction usually being dominant.

For a complete analysis of the high strain-rate test, it is necessary to simultaneously solve the incremental deformation and heat conduction problem with local temperature generation in proportion to the incremental plastic work during each time step. This combined thermal-mechanical problem was solved (4) using the EPIC-2 computer code. The finite element mesh and results of this analysis are shown in Figure 8. The constitutive model of Eq. 1 was used in this computation with the material properties for copper. The data is plotted for a maximum average strain of $\gamma = 7$. The localization of strain and strain-rate at the center of the specimen due to the temperature gradient is apparent. This corresponds directly to the experimental observation in Fig. 3(a). The severity of the localization is underestimated, primarily due to the coarseness of the finite element grid utilized. Following the complete development of a shear band would require extensive

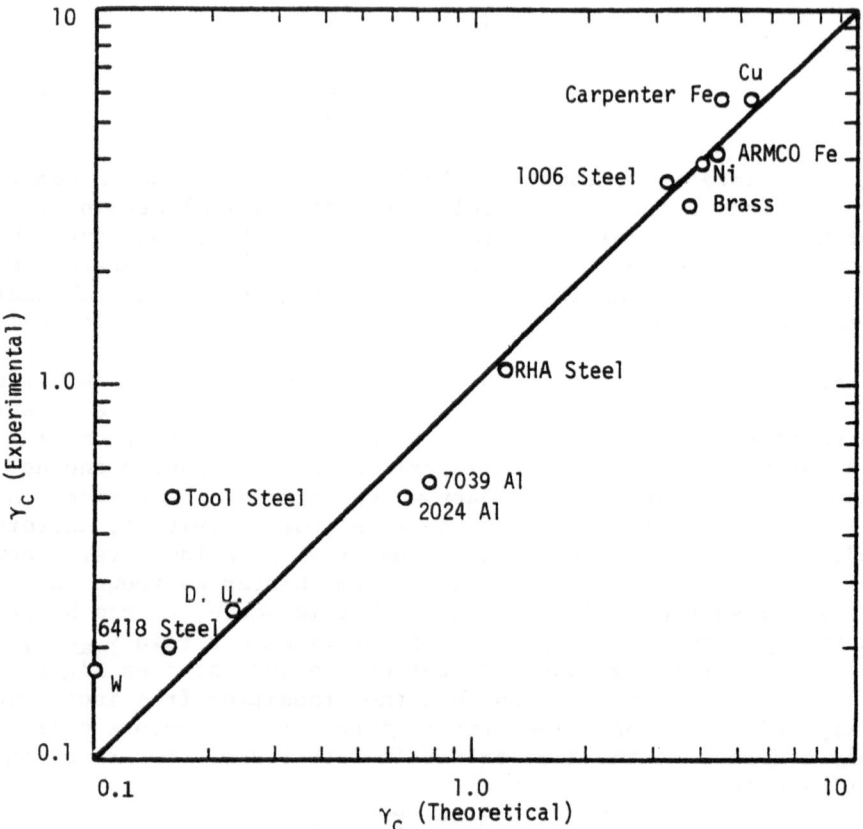

Fig. 6. Correlation of experimental vs. theoretical value for
         critical strain.

local rezoning or other numerical procedures. The results do
demonstrate, however, the essential features required to model
shear instability and strain localization.

     The curves in Fig. 8 labeled linear softening are based on a
K(T) in the form of Equation 2. The curves labeled bilinear softening are based on experimental data from the literature and represent
a faster rate of thermal softening. The result shown in Figure 8
and other similar computations demonstrate, that the rate of
thermal softening has a large effect on the rate of strain localization or, in effect, the strength of the instability.

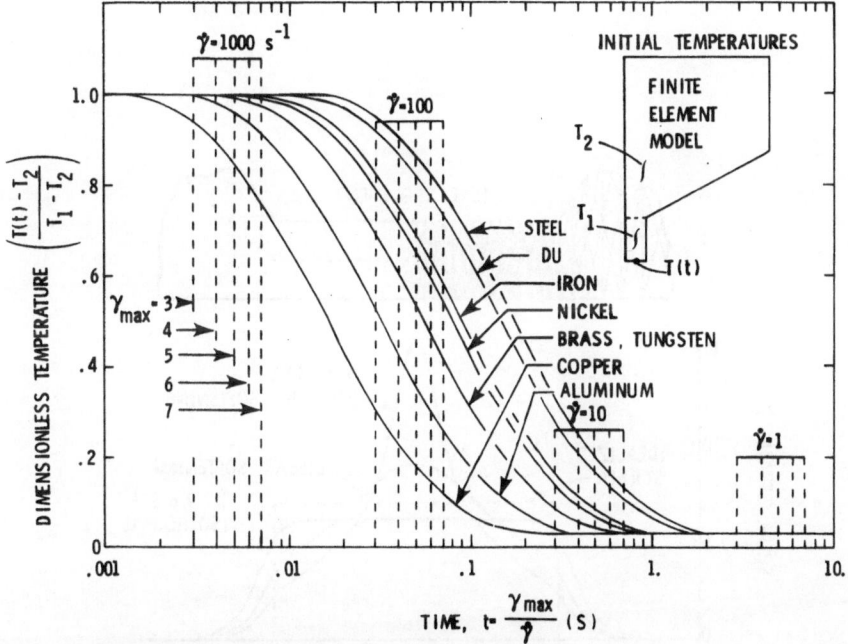

Fig. 7.   Transient temperature distributions in the torsion specimen
          due to an initially heated test section.

However, it should be noted, additionally, that a K(T) determined
from slow isothermal tests may not be representative of the true
softening effect experienced under the transient heating conditions
obtained during dynamic deformation. Equilibrium metallurgical
states may not be achieved with corresponding uncertainty regarding
thermal effect on instantaneous strain hardening rate.

    Another factor relating to the growth of the instability is
the effect of strain-rate hardening on the shear resistance of the
material in the shear band. Strain rates within the band are
greatly amplified, perhaps an order of magnitude above the nominal
rate before instability. Note that the strain rate outside the
band goes essentially to zero after instability. Recent data (6)
indicates a significant rate hardening change at strain rates
somewhere in excess of 10^3 sec^{-1}. The change is generally felt to
be from a logarithmic dependence as used in Eq. 1 to a linear
viscous dependence. The linear dependence, if operative, would
rapidly increase the viscous shear resistance and should tend to
stabilize the deformation. Better definition of the temperature
and rate effects in this region is obviously needed.

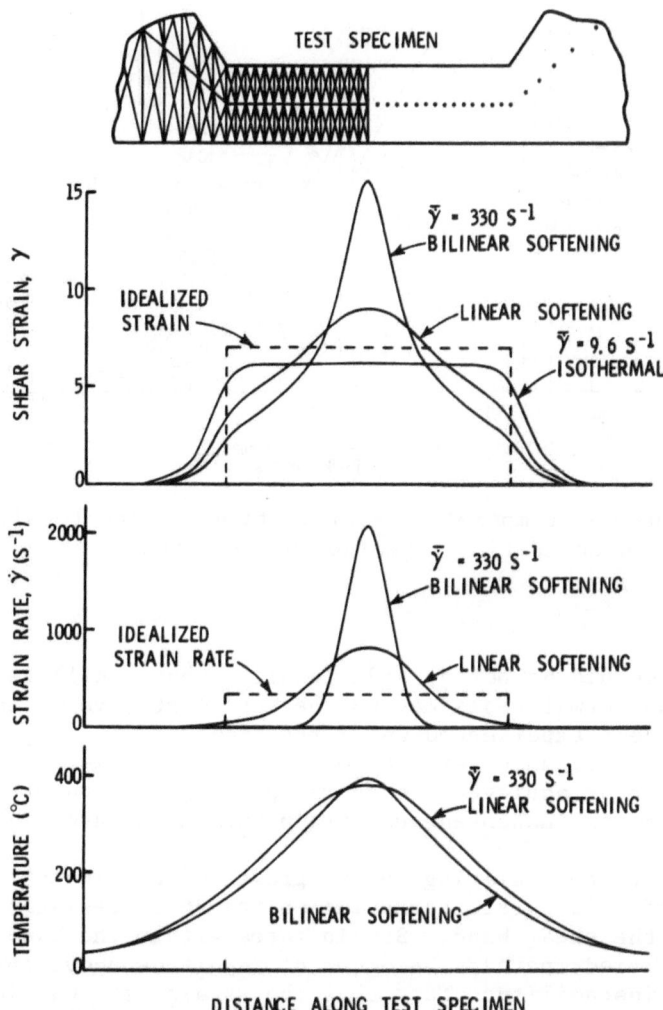

Fig. 8   Strain, strain rate and temperature distributions in
copper test specimen at $\bar{\gamma} = 7$ (4).

STRAIN HARDENING AND DUCTILITY UNDER NORMAL AND SHEAR LOADING

The condition for plastic instability depends strongly on the strain-hardening capacity of the material. It has been observed frequently (see Reference 7 for a review) that strain hardening is significantly greater during a large deformation tensile test than for a pure shear (torsion) test. The most plausible explanation appears to be based on a differing development of texture associated with the different deformation modes, although other factors have been discussed. Comparative tensile data for the materials shown in Fig. 2(a) are given in Reference 2 and support previous results with respect to higher hardening in tension. The differences were significantly greater for the copper, brass and nickel than for the irons and low carbon steel. Typical data for 1006 steel are shown in Fig. 9. For comparison purposes the torsion test data is converted to equivalent tensile stress and strain in accordance with a von Mises flow rule. At the same time, the tensile test data, based on current values of neck diameter and cross sectional area, is corrected in stress by the Bridgeman approximation which accounts for the developing triaxiality during necking. As previously noted, the stress difference for steel is not large and possibly within the limitations of the approximations made. A more significant difference is exhibited by the strain to failure. The equivalent strain at failure from the torsion test is about twice that obtained from the tension test. This raises the question of the effect of normal or mean stress on the criteria for ductile fracture.

Shaw (8) and colleagues have run tests in simple block shear with superimposed constant normal stress to obtain typical results as contained in Fig. 10. In this example, for low carbon steel, the shear stress-strain curve is relatively independent of the normal stress for shear strains up to approximately 1.5. This is in agreement with conventional plasticity theory when the inelastic shear (deviatoric) response is assumed uncoupled from the mean stress. For larger strains, strain softening is seen to occur, with increasing normal compressive stress retarding the rate of softening and delaying fracture to larger equivalent shear strains. This data supports the concept that mean pressure does not strongly influence strain hardening due to dislocation mobility but does play a major role in the development of damage, i.e., the development and growth of internal microcracks or voids. The development of voids near the fracture plane is clearly evident in Figs. 5(a) and 5(b).

The results of Rice and Tracy (9) for ductile void growth have been used by Hancock and Mackenzie (10) to estimate the combined effect of pressure and shear terms on failure strains. If

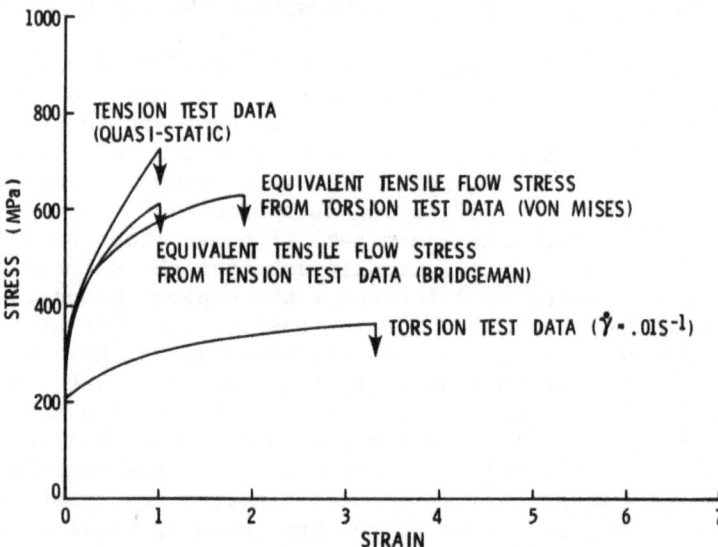

Fig. 9   Comparison of tension and torsion test data for
         1006 steel.

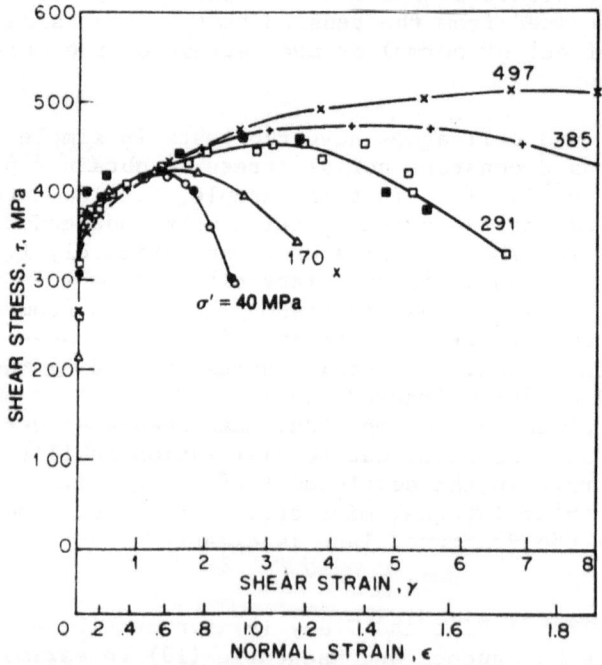

Fig. 10   Shear stress-strain results in simple shear for
          resulfurized low carbon steel with normal stress
          on shear plane (8).

the effective failure strain $\bar{\varepsilon}_f$ is inversely proportional to void
growth rate, then

$$\bar{\varepsilon}_f = K \exp (P/Y) \qquad\qquad (6)$$

where P is the mean pressure or stress (compression taken as
positive) and Y is the effective shear (deviatoric) stress. In
Fig. 11, data are plotted for the steels from Figs. 9 and 10 and
several other references (10,11,12). The correlations indicate
the general behavior predicted by Eq. 6 is valid.

The combined results presented here suggest that softening
resulting from thermal feedback and from damage accumulation may
both contribute to the plastic shear instability problem. Olson
(13) has demonstrated numerically the development of instabilities
using an isothermal, strain-softening model which could represent
the damage mode alone. These results suggest that the large deformation torsion experiments, if combined with static axial loading
(either tensile or compressive), could be effective in developing
combined plasticity and damage constitutive models, including the
effects of strain rate and temperature.

SUMMARY

It is hoped that this paper has indicated the general usefulness of the torsion test in defining large deformation behavior
without dilation effects over a wide strain-rate range including
the transition from isothermal to adiabatic conditions. Using
this test capability, the conditions required for adiabatic shear
instability have been verified by correlation with a finite element
analysis of the test conditions. The transient mechanical and
thermal response was well represented when the analysis included
a constitutive model incorporating strain hardening, strain-rate
hardening and thermal softening and the transient heat conduction
problem was solved simultaneously with the mechanical response.
Temperature gradients developed which led to localization of the
strain and strain rate. The strain localization was illustrated
by the formation of deformation or transformation bands in several
of the alloys tested. The deformation bands lead to a ductile
tearing mode of failure while the transformation bands in steel
have high hardness and generally result in cracking along the
band-matrix interface. Damage, in the form of distributed voids
in the region surrounding the fracture plane, was also observed.

More data are needed to define thermal softening rates at
large shear strains. An induction heating capability is being
added to the present facility in order to generate this type of

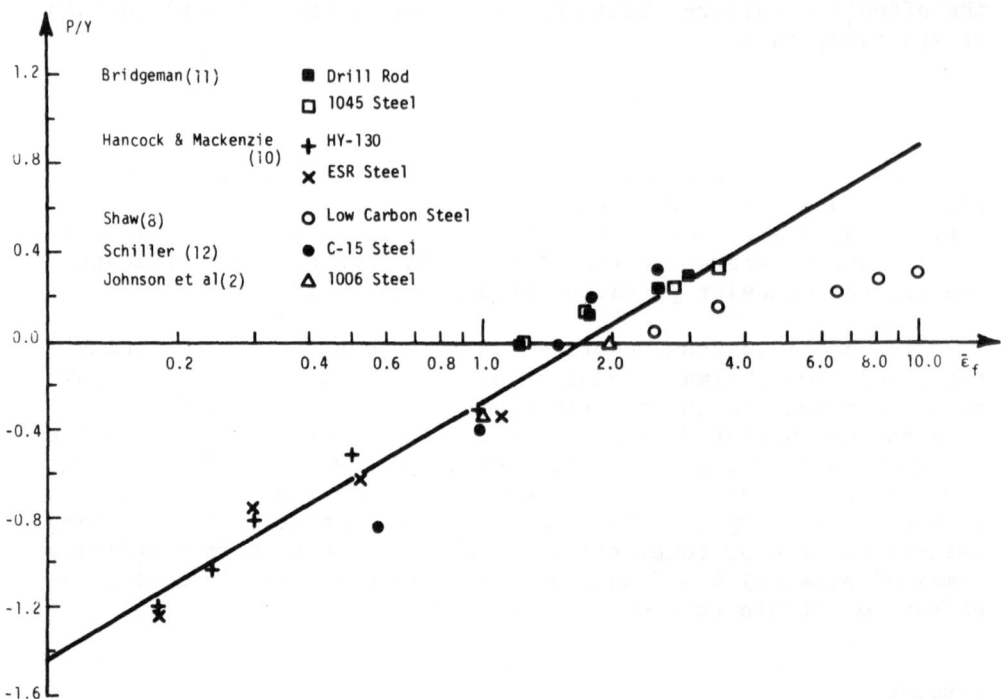

Fig. 11.   Failure strains as a function of P/Y.

data in the future. The effect of heating rate on mechanical response also needs better definition.

For large deformation problems under combined stress conditions there are a number of issues that need additional attention: kinematics and the definition of proper and consistent strain measures (7); the role of initial and deformation induced anisotropy on the flow rule; and finally, the proper coupling of pressure and shear in the formation of damage models and the role of damage in instability and failure mechanisms.

ACKNOWLEDGEMENT

The work reported herein was supported by the Honeywell Independent Development Program, Southwest Research Institute Internal Research Program and the U. S. Army, Ballistics Research Laboratory.

REFERENCES

1. U. S. Lindholm, "High Strain-Rate Tests," <u>Techniques in Metals</u>
    <u>Research</u>, Vol. V, Part 1, R. Bunshad (ed.), Interscience,
    New York (1971).

2. G. R. Johnson, J. M. Hoegfeldt, U. S. Lindholm and A. Nagy,
    "Response of Various Metals to Large Torsional Strains
    over a Large Range of Strain Rates--Part 1: Ductile
    Metals," ASME <u>Journal of Engineering Materials and</u>
    <u>Technology</u>, (In press 1983).

3. G. R. Johnson, J. M. Hoegfeldt, U. S. Lindholm and A. Nagy,
    "Response of Various Metals to Large Torsional Strains
    over a Large Range of Strain Rates--Part 2: Less Ductile
    Metals," ASME <u>Journal of Engineering Materials and</u>
    <u>Technology</u>, (In press 1983).

4. G. R. Johnson, "Dynamic Analysis of a Torsion Test Specimen
    Including Heat Conduction and Plastic Flow," ASME <u>Journal</u>
    <u>of Engineering Materials and Technology</u>, 103:201-211, No.3,
    July (1981).

5. U. S. Lindholm, A. Nagy, G. R. Johnson and J.M. Hoegfeldt,
    "Large Strain, High Strain Rate Testing of Copper,"
    <u>Journal of Engineering Materials and Technology</u>, Trans-
    actions of ASME 102:376-381 (1980).

6. U. S. Lindholm, "Deformation Maps in the Region of High
    Dislocation Velocity," <u>High Velocity Deformation of Solids</u>,
    K. K. Kawata and J. Shiori (eds.), Springer-Verlag
    New York, Inc., pp. 26-35 (1979).

7. S. C. Shrivastava, J. J. Jonas and G. Canova, "Equivalent
    Strain in Large Deformation Torsion Testing: Theoretical
    and Practical Considerations," <u>J. Mechs. Phys. Solids</u>,
    30:75 (1982).

8. M. C. Shaw, "A New Mechanism of Plastic Flow," <u>Int. J. Mech.</u>
    <u>Sci.</u>, 22:673 (1980).

9. J. R. Rice and D. M. Tracy, "On the Ductile Enlargement of
    Voids in Triaxial Stress Fields," <u>J. Mech. Phys. Solids</u>,
    17:201 (1969).

10. J. W. Hancock and A. C. Mackenzie, "On the Mechanisms of
    Ductile Failure in High-Strength Steels Subjected to
    Multiaxial Stress-States," <u>J. Mech. Phys. Solids</u>,
    24:147 (1976).

11. P. W. Bridgeman, "Studies in Large Plastic Flow and Fracture,"
    McGraw-Hill, New York (1952).

12. H. Schiller, Beitraz zum Einfluss der Art des Spannungszustands
    auf das Umformvermoegen Metallisher Wertstoffe, <u>Neue</u>
    <u>Huette</u>, 17:7 (1972).

13. G. B. Olson, J. F. Mescall and M. Azrin, Adiabatic Deformation
    and Strain Localization, in "Shock Waves and High-Strain-
    Rate Phenomena in Metals," ed. by M. A. Meyers and L. F.
    Murr, Plenum Press, New York, p. 221 (1981).

MICROSTRUCTURE AND MECHANICAL PROPERTIES OF PRECIPITATION

HARDENED ALUMINUM UNDER HIGH RATE DEFORMATION

Dennis E. Grady, James R. Asay, Richard W. Rohde and
Jack L. Wise
Sandia National Laboratories
Albuquerque, NM 87185

ABSTRACT

In recent years the shock deformation properties of a precip-
itation-hardened aluminum alloy (6061-T6) have been investigated
extensively with time-resolved wave profile gauges. Studies have
revealed complicating features of the shock deformation process
not easily explained by existing theories of plastic deformation.
Specifically, unusual behavior in the dynamic hardness or strength,
post-shock hardness, steady wave viscosity, and post-shock microstructure has been observed over the shock pressure range of 1-15
GPa. The deformation structure suggests that microscopic heterogeneity in the deformation and temperature state during dynamic
compression may be responsible for the observed effects. We have
undertaken a study to correlate the shock compression and quasistatic
deformation of 6061-T6 aluminum. Recovered specimens which have
been shock loaded are examined metallurgically, and results are
compared with both static and dynamic mechanical property measurements. Illuminating correlations are emerging from this study,
although an unambiguous determination of the importance of heterogeneous effects is yet to be achieved. A modeling effort of
dynamic deformation in aluminum based on heterogeneous deformation
and adiabatic thermal trapping is being pursued to guide the
experimental effort.

INTRODUCTION

The problem of describing the response of metals to shock and
high-rate deformation through models amenable to numerical calculations has occupied the efforts of numerous scientists over the
past several decades and is still of considerable interest. Models
of increasing sophistication have been developed as the abilities to

measure high-rate response phenomena improved. Recent advances in
methods for measuring time-resolved stress waves in solids, however,
have revealed features in the shock deformation of metals which are
not easily incorporated into present theories of shock deformation
and high-rate flow. Measurements of plastic wave profiles in
metals such as aluminum, copper and beryllium indicate strength
properties at the Hugoniot state and viscous effects within the
shock front which are unique in behavior and not readily explained.

Recently, attempts to rationalize metallographic studies of
shocked samples, which indicate strong heterogeneities in the microscale deformation, with the very high rate of flow determined from
the measured wave profiles have suggested the possibility of adiabatic shearing or thermal trapping within deformation features on
a microstructural level. Tentative models based on the ideas of
local transient thermal heterogeneities persisting during the shock
event suggest a possible explanation for the anomalous effects.

To explore these and related questions, a program focused on
the effect of microscale heterogeneous deformation on the shock
process has been undertaken. It is weighted toward metallographic
study of samples recovered from controlled shock-wave experiments
and correlation with measured time-resolved wave profiles. Although
we recognized the difficulties with recovering shocked samples with
unambiguous loading histories, this approach still appears to be
necessary if an understanding of the physics of the microscopic
deformation process during transient loading is to emerge. To
provide a basis against which to compare shock-wave results, samples
subjected to cold-working deformation are being studied. Here,
also, efforts have focused on metallography of the deformation
process.

This paper reports work in progress. Although a fairly large
effort is represented by the work to be discussed, many of the
results are preliminary. Most of the ideas have not yet been fully
explored and a number of points still remain unexplained. The work
described will focus strictly on experiments and analysis for
6061-T6 aluminum. Although efforts are being directed toward other
metals and alloys perhaps better suited for the study, we felt that
further work on this material was warranted because of the large
body of shock wave data already available.

The present report opens with a description of the experimental
methods currently used to explore the shock deformation process.
First, impact experiments are described which provide time-resolved
measurements of shock risetime and strength of the material shortly
after passage of the shock wave. Velocity interferometry provides
the key to this study. Secondly, a technique developed to recover
specimens shock loaded to 15 GPa for post-shock metallography
studies is described. Lastly, quasistatic cold-working methods and

metallography techniques used to investigate both the shock-loaded and cold-worked deformation are discussed.

In the next section, material properties inferred from wave profile measurements on aluminum are considered. These include the viscosity of flow associated with the material within the shock-wave risetime and the strength of the material in the Hugoniot state. The latter is measured from the structure of one-dimensional unloading or reloading waves which pass through the material within a microsecond after the initial shock wave but before complicating lateral release waves or normal thermal annealing effects can occur. The following section is focused on trends associated with hardness and deformation metallography in specimens which have been quasi-statically deformed by cold working over strains from a few percent to in excess of 70%. Emphasis is on heterogeneities in the deformation microstructure. Similarities and differences in the shock and quasistatically-deformed specimens include post-deformation hardness and features in the deformation microstructure.

The consequences of temperature as well as deformation hetero-geneities within the shock wave are explored next. This analytic study is based on the ideas of heterogeneous plastic flow within the shock wave leading to local hot regions in the microstructure. Although persisting for only a very short period of time, possible temperature effects on the Hugoniot stress and strength state as well as on the final metallurgy must be considered. In the final section, we close with our present interpretations of the shock deformation process, difficulties with these interpretations and the direction of future work.

EXPERIMENTAL PROCEDURES

In this section, we report the methods used to perform shock-wave loading and recovery experiments along with the metallographic techniques used to evaluate the shock-induced microstructure. The strength of aluminum in the shocked state is measured with controlled one-dimensional impact experiments where preselected impact mate-rials produce the input wave shapes necessary for the tests. Shock recovery experiments are performed on separate specimens; however, careful velocity and specimen dimension measurements assure know-ledge of the loading history through correlation with other wave-profile experiments. Recovered specimens are subjected to micro-hardness measurements along with optical and TEM analysis of the deformed microstructure. In addition, cold-working experiments on the same aluminum are being performed to provide a baseline defor-mation for comparison with the shock-induced microstructure.

Reshock and Unloading Experiments.

Plane shock loading of 6061-T6 aluminum specimens is achieved

with a powder gun[1] used to accelerate a projectile containing a
flat impact plate to a preselected velocity. A time-resolved
history of the shock wave is determined by measuring the particle
velocity at the interface between the surface of the specimen and
an optical window using a velocity interferometer.[2] Different
optical windows were used in these experiments. For stress levels
to 8 GPa, fused silica[3] windows were employed; whereas, for higher
shock stresses, either polymethylmethacrylate (PMMA) or single
crystal LiF[4] were used to measure the wave profiles. Through the
appropriate choice of impactor and backing material, it was possible
to produce initial shock loading at the specimen front surface,
followed by either unloading to a lower stress level or reloading
to a higher level. Below about 2 GPa, this was accomplished with
a single-crystal quartz impactor backed with either low density
foam or tungsten. At large impact stress levels, initial loading
was provided with an aluminum impactor again backed with foam or
tungsten.

Shock Recovery Techniques

The experimental soft recovery configuration used for the
present studies is illustrated in Figure 1. The sample is backed,
in turn, by a thin PMMA buffer plate and a low-density polyurethane
foam cushion. The target, buffer, and cushion are rigidly mounted
in a recovery capsule which has been fabricated from hardened 4340
steel. Impactors for these tests are fabricated from either 6061-
T6 aluminum or OFHC copper. After impact, the target

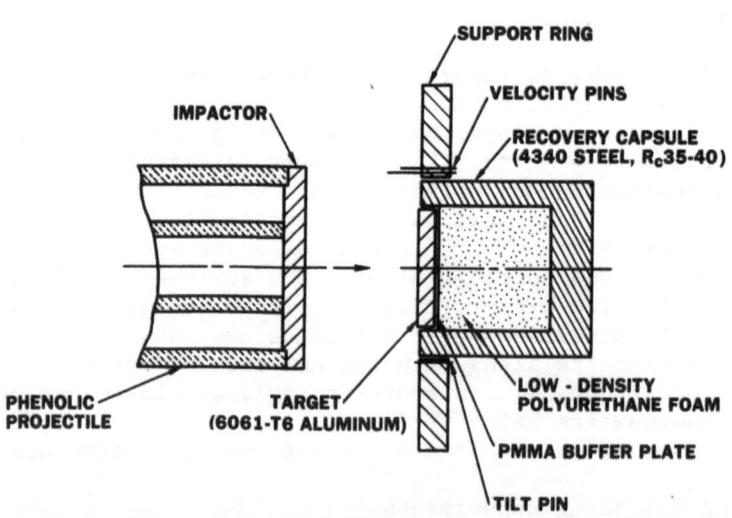

Fig. 1.   Impact and Sample Recovery System

and projectile components are decelerated by a catcher assembly
consisting of alternating layers of aluminum honeycomb and steel
plate.

Projectile velocities range from 0.366 to 1.217 km/s, producing
impact stresses in the aluminum samples from 0.9 to 14.7 GPa.
Impactor thicknesses in all experiments have been selected such that
release wave interactions occur within the impactor, thereby preventing spallation of the aluminum sample. During the impact
process, the central portion of the impactor plate is sheared off
along the inner edge of the recovery capsule and drives the target
plate into the capsule as the foam cushion is irreversibly crushed.
In all tests the targets and impactors have been recovered essentialy
intact.

Calculations have been performed using the one-dimensional
Lagrangian wavecode WONDY-IV,[5] which indicate an initial recompression of the sample following complete compaction of the foam,
and a subsequent recompression due to wave reflection from the
foam/steel interface. The magnitude of these recompressions is on
the order of 50% of the initial shock stress. As discussed by
Stevens and Jones[6] and, more recently, by Davison, Webb and
Graham,[7] it is not always appropriate to neglect the effects of
the additional plastic work on shock microstructure which is done
on the material during the radial release process. Although a
detailed two-dimensional analysis has not been performed for the
present recovery scheme, the absence of substantial concavity (or
dimpling) of the aluminum target plates suggests that tensile
stresses due to converging radial release waves were not of suffi-
cent strength to induce appreciable macroscopic plastic flow. In
addition, the absence of any noticeable variation in microstructure
either parallel or perpendicular to the target axis suggests that
changes in microstructure are not related either to plastic work
performed during radial release or to the duration of conditions
which hold during the uniaxial strain state. Rather, the present
results indicate that microstructural changes are predominantly
governed by processes occurring at, or near, the shock front.

Metallographic Techniques and Cold-Working Experiments

The aluminum alloy, 6061, was used for all experiments. This
alloy achieves most of its strength from precipitation of fine
Mg_2Si zones which hinder dislocation motion. As such, in normal
uniaxial stress experiments, the material exhibits limited work
hardening having a 0.2% offset yield strength of about 240 MPa.
The material was heat treated to the T6 condition by first annealing
at 550^0C for 8 hours then water quenching, which placed the Mg and
Si in solid solution, then aging at 180^0C for 8 hours to achieve
precipitate formation.

The recovered shock-loaded and quasistatically deformed materials were examined by optical microscopy, transmission electron microscopy (TEM) and microhardness. The surfaces examined were in all cases planes lying parallel to the axis of major strain. Microhardness measurements were conducted to ASTM specification E384 using a Knoop indenter, 200 g load and a 15 s hold time.

Cold-working experiments were performed to provide a basis for comparison of the hardness and microstructures developed in shock deformation. Samples experiencing quasistatic deformation from 10 to 70% strain were prepared by cold rolling methods. Single reductions of 25% strain could be achieved, so the larger strains were reached by sequential rolling passes. It is well known that small reductions by cold rolling impart nonuniform strains, with considerable strain occurring in the surface and little in the bulk. Therefore, plastic strains from 0.2 to 5.0% were reached by compressing specimens, made to ASTM E9, at a strain rate of 2×10^{-3} sec^{-1}.

DYNAMIC STRENGTH AND WAVE-PROFILE PROPERTIES

The strength of aluminum in the shocked state can be tested with a second wave passing through the material shortly behind the principal shock wave. It has been demonstrated that the shear stress state on the Hugoniot does not reside on the yield surface, and the offset cannot be evaluated from the initial shock conditions. Both reshock and unloading wave experiments are necessary to uniquely determine the stress state and strength at the shock state. The behavior of these properties as a function of the shock pressure has been found to be unique to the shock process.[8]

The shock state is achieved through the passage of a finite-risetime plastic wave. All of the shock dissipation processes and probably most of the shock-induced microstructure evolve within this brief time period. Traditionally, the strain rate and shear stress state within the plastic shock wave have been characterized with material viscosity. Recently, behavior of the shock viscosity in aluminum has also been found unique to the shock-wave process.[9]

Strength and Shear-Stress States on the Hugoniot

The objective of the time-resolved shock wave experiments was to determine the strength of aluminum in the shocked state. The analysis of strength under shock compression is based on the elastic-plastic model. Shock compression under plane loading in solids produces both longitudinal and lateral stresses. For an isotropic material, the longitudinal, σ_x, and lateral, σ_y, stresses are related to the resolved shear stress, τ, by $\sigma_x - \sigma_y = 2\tau$.

For the stress-volume response predicted by the simple elastic-plastic model, the dynamic response is elastic until the resolved shear stress produced by shock compression is equal to the constant critical resolved stress, τ_c, equal to one-half the quasistatic uniaxial yield stress, Y, which occurs at the Hugoniot elastic limit (HEL). Further compression is plastic with the Hugoniot stress offset from a state of hydrostatic pressure by 4/3 τ_c (or 2/3 Y). Upon unloading, the response is again elastic until reverse yielding ($\tau = - \tau_c$) occurs where further unloading is plastic.

In principle, the structure of an unloading stress-wave profile will determine the critical shear strength, τ_c, or the yield strength, Y, in the shocked state. Unfortunately, materials do not exhibit idealized response. Instead, effects such as strain-dependent hardening and Bauschinger effects during unloading produce a more complicated response which prevents determination of strength in the shocked state. Furthermore, the shear stress in the shocked state may not be equal to the critical shear strength because of time-dependent softening or strengthening effects.[8]

In a more general description of shock loading, it is expected that a shear stress τ_0 is present in the shocked state, and because of the time-dependent effects referred to earlier, may differ from the critical strength, τ_c, the material can support after shock compression. Thus, during either recompression or unloading from the shocked state, the shear stresses may increase in essentially an elastic manner until the critical strength is reached after which further compression is assumed to be plastic.

Elastic-plastic structure in both recompression and unloading waves suggests a method by which both the shear stress, τ_0, and the critical strength, τ_c, in the shocked state can be evaluated. To determine both quantities, it is necessary to make the following assumptions: (1) unloading and reloading from the shocked state follow rate-independent stress-strain paths, (2) a yield surface exists for the material after shocking which can be detected as a transition from elastic to plastic behavior during unloading and reloading, (3) the yield surface is symmetric and has an axis of symmetry which remains parallel to the hydrostat during subsequent deformation after shock loading, and (4) the shear strength may depend on some measure of plastic strain. Under these assumptions the yield surface and stress differences in the shocked state can be determined either graphically or analytically by extrapolating the plastic states achieved during unloading and reloading to the shocked state.[8] A series of experiments has been conducted on 6061-T6 aluminum to evaluate strength effects after shock compression.[8] Discussion here will relate to the metallurgical effects observed after shock loading, with the localized thermal effects occurring during shock compression being considered later.

Figure 2 illustrates a pair of typical unloading and reloading
wave profiles obtained for 6061-T6 aluminum. The initial impact
stress is nominally 21 GPa. As discussed above, strength and the
shear stress state at this stress level are determined from the
recompression and unloading wave structure.

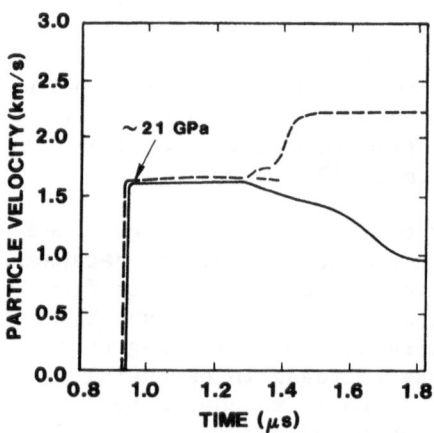

Fig. 2. Reshock and Unloading   Fig. 3. Dynamic Strength and Shear
Profiles in Aluminum (Asay &   Stress Behavior of Aluminum (Ref. 8)
Chhabildas, Ref. 8)

The measured variation of shear stress and strength as a
function of initial shock stress is summarized in Figure 3. For
unloading, the shear stress change from the shocked state is $(\tau_o + \tau_c)$, whereas for reloading it is $(\tau_c - \tau_o)$. Typical error estimates
are also shown. The two can be combined to estimate both the
shear stress, τ_o, and the critical shear strength, τ_c, also illustrated in Figure 3. Nominal times after shock compression are
about 1 μs.

There are several points to note. First, there is a large
increase in both $(\tau_o + \tau_c)$ and $(\tau_c - \tau_o)$ in the stress region
between 5 and 15 GPa. Second, a large departure is observed from
the normal elastic-plastic behavior which assumes that $\tau_o = \tau_c$.
Third, the shear stress, τ_o, remains essentially constant with shock

stress to within experimental error, whereas the critical strength increases with increasing shock stress. These observations will be considered in terms of metallurgical changes and thermal effects in the following sections.

Steady-Wave Risetime and Viscosity

An important aspect of the microstructure and material property changes which occur during passage of a shock wave is the time duration within which they must occur. Improved methods for measuring time-resolved profiles show that shock compression is not discontinuous but occurs within one to a few hundred ns over the stress range of about 1 to 10 GPa with the risetime decreasing rapidly with increasing stress amplitude. Irreversible derformation processes occur within the plastic portion of the wave. The plastic wave can change with time during the early evolution of the profile but achieves a steady shape after a short propagation distance due to a balance between the nonlinearity of the material compression behavior and rate-dependent dissipative processes which tend to disperse the wave.

Barker [10] has measured wave profiles in 6061-T6 aluminum from about 1 to 9 GPa, shown in Figure 4, which illustrate the finite risetime and the strong dependence on stress amplitude. Through measurements in samples of several thicknesses, Barker concluded that all the profiles with the exception of the lowest shock amplitude achieved near steady wave conditions.

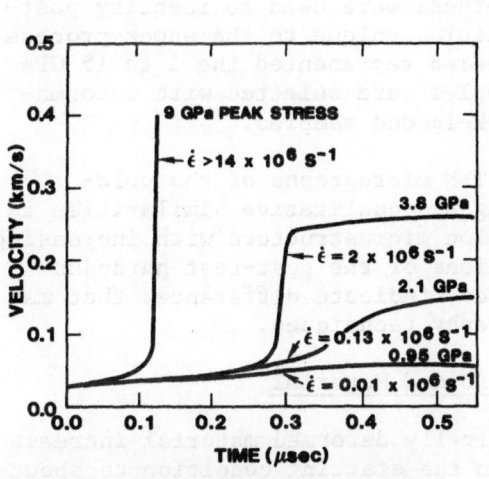

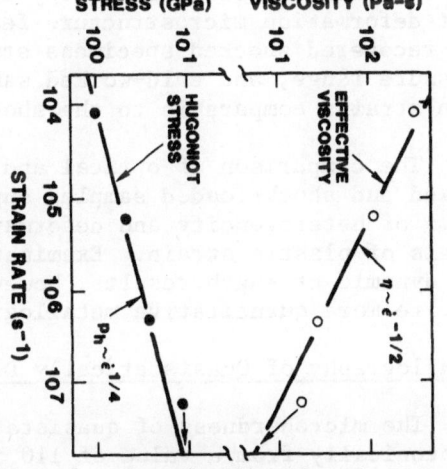

Fig. 4. Steady Wave Risetime   Fig. 5. Steady-Wave Strain Rate
         Measurements in Aluminum   and Viscosity in
         (Barker, ref. 10).   Aluminum.

In the hydrodynamic approximation of shock compression, a viscous relation has been found useful in characterizing material behavior. In more general elastic-plastic response, a more complicated viscous behavior is expected. It has been found useful, however, to classify the dissipation over a steady-wave shock compression process by an effective viscosity. The viscosity coefficient is quantified experimentally as the ratio of the maximum viscous stress, which is proportional to the maximum difference between the Rayleigh line and the Hugoniot, and the strain rate from the maximum slope of the wave profile.

Effective viscosity coefficients for aluminum from the wave profile measurements of Barker are shown in Figure 5. The increase is stress amplitude with steady-wave strain rate is well described by a one-fourth power relationship, while the viscosity shows an inverse square root dependence on strain rate.[9]

Gilman[11] has suggested underlying microscopic viscous shock compression processes. Although the actual processes responsible for shock viscosity have not yet been determined, the ideas broached by Gilman are critical to an understanding of the shock process. The heterogeneous deformation features observed in aluminum and the measured amplitude and trend of the shock viscosity coefficient are interrelated and a satisfactory understanding of the dynamics of the shock process must include both concepts.

QUASISTATIC AND SHOCK METALLOGRAPHY STUDIES

Microhardness measurements were used to compare the shock and quasistatically-deformed aluminum specimens, and both optical and transmission electron microscopy methods were used to identify post-test deformation microstructure features unique to the shock process. The recovered shocked specimens studied represented the 1 to 15 GPa pressure range, and cold-worked samples were selected with deformation strains comparable to the shock-loaded samples.

The comparison of optical and TEM micrographs of the cold-worked and shock-loaded samples suggests qualitative similarities in terms of heterogeneity and deformation microstructure with increasing levels of plastic strain. Examinations of the post-test hardness and dynamic strength results, however, indicate differences that may require more quantitative metallography techniques.

Metallography of Quasistatically Deformed Material

The microhardness of quasistatically deformed material increases monotonically from a value of 110 in the starting condition to about 140 for heavy cold working as shown in Figure 6. Saturation of the hardness is typical of cold deformation.

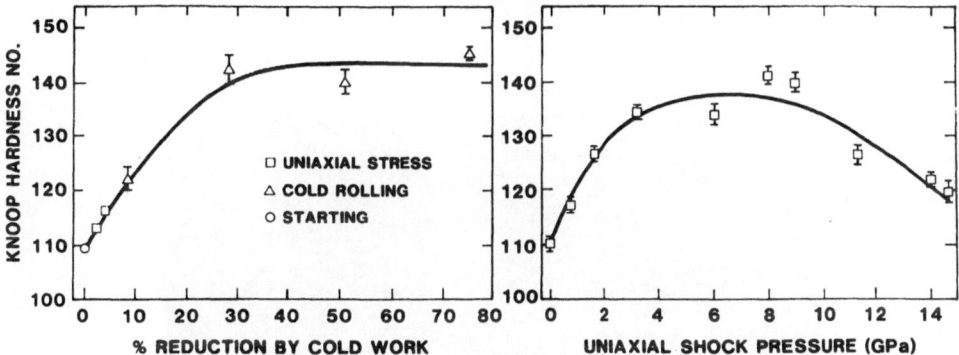

Fig. 6.   Post-Test Hardness of Shock-Deformed and Cold-Worked
          Aluminum

In optical metallography, the elongated grains and dark spots
are characteristic of the starting material. Elongated grains are
a consequence of bar stock preparation. The dark spots are precipitates of Mg_2Si and $\alpha A\ell$-Fe-Si commonly seen in commercial 6061
aluminum. With optical microscopy resolution no change is detected
until slip bands appear within the grains at 8% deformation. These
bands are regions of intense, localized deformation occurring on
normal dislocation slip planes and persist to the largest strains
examined. Figure 7a illustrates the optical microstructure observed
at 8% strain.

TEM examination showed planar slip with regions of intense
shear 1 to 2 μm apart for all cold work levels beyond 1.5%. The
density of dislocations between these slip bands increases with
deformation. At the largest deformation of 70%, the planar structure
begins to decompose into a more cellular arrangement. A TEM
micrograph of 52% cold-worked material is shown in Figure 7b.

Metallography of Shock Deformed Specimens

Microhardness, shown in Figure 6, for the shock-deformed
specimens rises with peak shock stress to a value of about 140,
identical to the saturation hardness found for the cold-worked
specimens. Unlike the cold-worked material, hardness decreases at
higher stresses, similar to related work in nickel and 304 stainless
steel.[12] In the latter, however, the hardness decrease occurring
at higher pressures was attributed to thermal recovery effects.
The bulk shock temperature rise in aluminum released from 11 GPa is
too low (< $100^\circ C$) to result in significant thermal recovery.

Optical metallography showed that shear banding similar to that
present in quasistatically deformed material is produced when samples
are subjected to shock compression to stresses as low as 0.95 GPa.

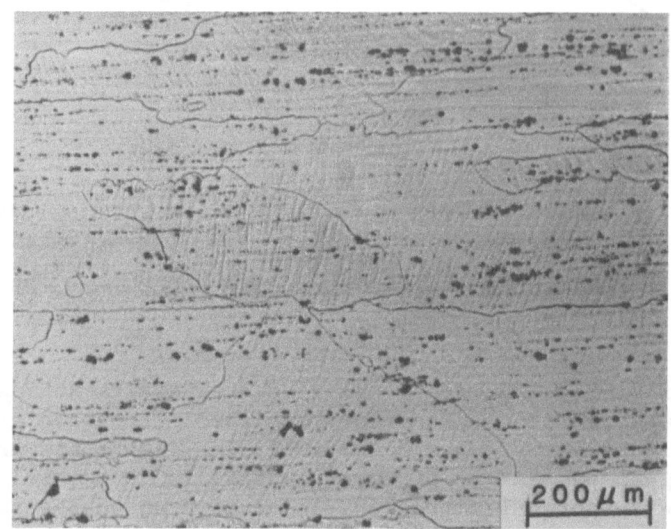

(a) 8% COLD WORK

(b) 52% COLD WORK

Fig. 7a,b. Optical and TEM Metallographs of
            Cold-Worked Aluminum.

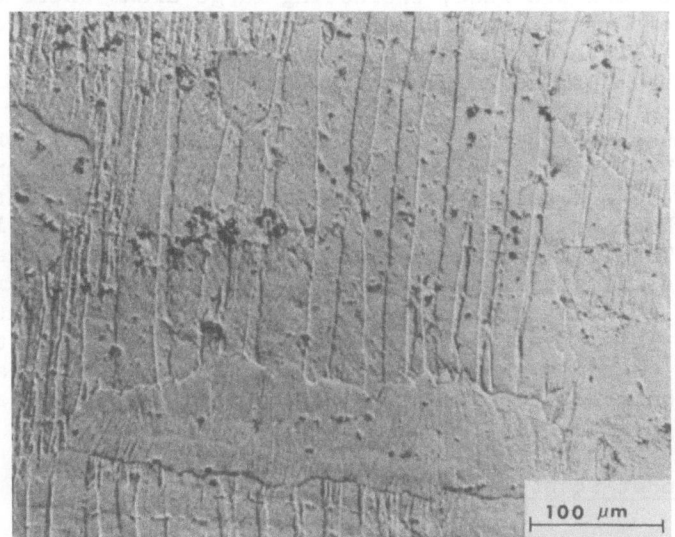

(c) 14.7 GPa Shock

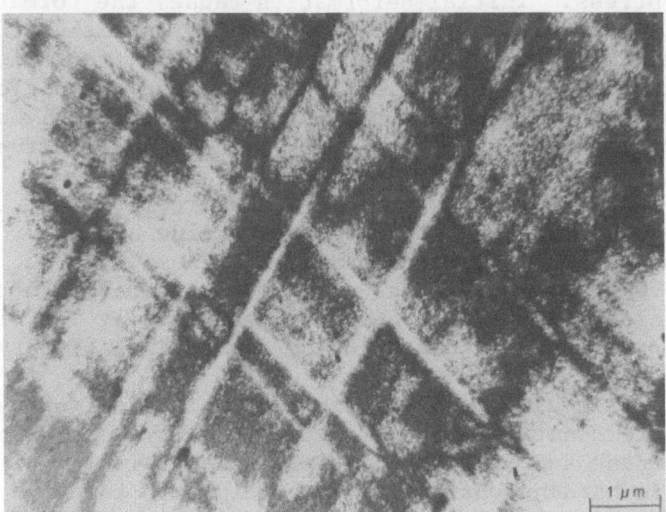

(d) 9.0 GPa SHOCK

Fig. 7c,d.   Optical and TEM Metallographs of
             Shock-Deformed Aluminum.

Bands become more evident at higher shock pressure, persisting to the highest shock stress of 14.7 GPa. The bands at 8.0 and 14.7 (Figure 7c) GPa are wide, indicating large areas where crystallite orientation within the individual grains has undergone changes or rotation. Width of the bands formed under shock appear somewhat greater than for the cold-worked material.

TEM observations at 0.95 GPa showed linear arrays of dislocations with slip band spacing of about 1 to 2 μm, similar to the microstructure formed by cold working 1.5%. At shock stresses of 3.2 GPa shear bands become well developed with spacing less than 1.0 μm. This well-developed shear band structure persists at shock stresses yielding a maximum hardness, (9.0 GPa, Figure 7d) where a high density of dislocations between linear features is evident. This planar microstructure exists even after the hardness begins to decrease at a shock pressure of 11.4 GPa. A similarity is noted between microstructure in the high shock stress samples and that developed at 70% cold work.

Comparison of Static and Shock Deformation

Comparison of the shocked and cold-worked material shows that the microstructure evolves in a similar way with increasing strain or shock stress. Initial deformation causes the formation of planar slip features with sufficiently low densities of dislocations to permit TEM imaging. At higher deformation, the density of dislocations in regions of localized slip and the dislocation density between slip bands both increase. At the highest strains, planar structure begins to give way to a cellular structure.

Post-test hardness evolves very differently, however. Hardness in cold worked material saturates at a value near 140, whereas shock-loaded samples achieve maximum hardness of about 140 between 6.0 and 9.0 GPa and decreases at higher stresses. Furthermore, hardnesses of samples with qualitatively similar TEM microstructures can be very different, depending on deformation history.

Since it is almost aphoristic that mechanical properties like hardness are functions of dislocation density, the disparity between hardness and microstructures is surprising. While 6061-T6 aluminum is hardened primarily by precipitate zones of Mg_2Si, generation of dislocations causes the work hardening. However, alterations in the extant precipitate zone structure can also result in hardening, and there is some evidence[13] that unusual precipitation events can occur under shock. Unfortunately, TEM observations are not sensitive to such subtle changes in precipitate zones. Point defects can also contribute to strength, and excess vacancy concentrations have been observed in some shock-deformed material.[12] At the very high dislocation densities in the samples, order-of-magnitude changes could occur and not be

observed by TEM. Thus, TEM can provide a good qualitative representation of the microstructural evolution but a poor quantitative
assessment of details. Consequently, dislocation densities,
precipitate zones, vacancies, or a combination of the three elements
could be changing sufficiently to account for the hardness behavior
of the shocked samples.

To counter this deficiency, X-ray diffraction and positron
annihilation techniques will be used to estimate dislocation
densities and search for changes in precipitate zones and excess
vacancies which might occur under shock deformation. Preliminary
X-ray diffraction results indicate that the dislocation density in
the shock-loaded material reaches a maximum at the maximum hardness
and then decreases at higher pressures where the hardness also
falls.[14]

Correlation of Hardness and Dynamic Strength Measurements

The evolution of shock microstructure parallels that of cold-
worked material; however, the hardness of the cold-worked material
saturates while the hardness of the shocked material achieves a
maximum, and then decreases -- an effect which may be accounted for
by thermal recovery. The dynamic strength of this alloy appears
to increase monotonically and saturate -- a behavior more like the
hardness of the cold-worked specimens.

This dynamic behavior could be explained in the following way.
The first shock results in a microstructure unique to that shock
stress. The microstructure is much harder (stronger) than the
virgin structure, and this is reflected in the strengths measured
by the second recompression or unloading wave. The shock strength
saturates rather than decreases as observed in post-test hardnesses
because presumably, the ~ 1 µs time interval between the development
of a structure from the first shock and the testing of its strength
from the second is insufficient for normal thermal recovery effects
to occur.

THERMAL TRAPPING CALCULATIONS IN SHOCKED ALUMINUM

Plastic flow in shock loading involves nonconservative microstructural processes which lead to the dissipation of energy,
primarily as heat. If deformation and dissipation is heterogeneous,
then local temperature nonuniformities may persist in the shock
state, which may alter plastic flow. Thermal activation barriers
governing dislocation processes are lowered and drag processes or
viscosities resisting flow are less. Equilibrium quantities of
defects will also increase if generating mechanisms exist.
Submicron-sized temperature heterogeneities persisting after passage
of the shock wave will be quenched with high cooling rates. Thus,
activation barriers will increase, locking in defects or substructure

created within zones of intense shear. Knowledge of the magnitude
and duration of local temperatures during shock loading would be
helpful in assessing these deformation and metallurgical processes.
Accordingly, we have attempted to model the nonequilibrium thermal
effects of heterogeneous shock deformation.[15]

Heterogeneous Deformation and Adiabatic Heating in Shock-Wave Loading

The quasistatic and shock compression experiments on aluminum
suggest that plastic deformation occurs through heterogeneous
microslip. The intensity of deformation features appears relatively
independent of loading rate for values approaching 10^7/s. At
higher strain rates, increased intensity is observed, but attention
will focus on the lower-rate region. Plastic work performed during
deformation is dissipated within the slip bands and distributed to
the cooler bulk of the material through thermal conduction.
Consequently, a local shear band temperature, T_ℓ, is expected to
depend on the competing processes of dissipation, thermal diffusion,
and the nominal spacing of the shear bands.

To model local heating effects, the deformation fabric is
considered a lattice of slip bands with equal spacing, d (approximately 10 μm for the intense shear bands observed optically in
aluminum). All deformation and dissipation is assumed to occur
within these bands. Isentropic heating is accounted for and
Fourier heat conduction is assumed. Within the model, plastic
shearing implies a relative velocity, v, between faces of the
shearing regions, and strain rate, $\dot{\varepsilon} = v/d$. The relative velocity
is expected to depend on the shear stress, τ, the displacement, and
the temperature, T_ℓ, suggesting a functional dependence, $\dot{\varepsilon} = f$
$(\tau,\varepsilon,T_\ell)$. A thermal diffusivity, χ, suggests a thermal relaxation
time of the order d^2/χ. Deformation to a strain ε defines a
characteristic strain rate, $\dot{\varepsilon}_c \sim \varepsilon\chi/d^2$, below which thermally
equilibrated deformation is expected above which heat trapping
within slip bands should occur. For a 5 GPa shock wave in aluminum,
$\dot{\varepsilon}_c \simeq 2.5 \times 10^4$/s. Present calculations will explore the case where
$\dot{\varepsilon} > \varepsilon_c$, which occurs in the moderate shock regime of aluminum where
possible thermal trapping effects may alter deformation processes
under rapid loading conditions.

Energy and Risetime Relations under Steady-Wave Shock Compression

As indicated in a previous section, a steady wave propagates
without change in shape and, for a particular material and stress
amplitude, is characterized by a specific risetime. The stress-
strain path is on a straight line (the Rayleigh line) from the
origin to the Hugoniot state. In the weak shock regime, the energy

dissipated is approximately the area enclosed by the Rayleigh line and the Hugoniot. The Hugoniot is adequately described by a linear shock velocity–particle velocity relation ($U_S = C_O + Su_p$), and it can be shown that a first-order expression for the energy dissipated is $E = \rho_o C_o^2 S\varepsilon/3$ where ρ_o is the initial density.[9]

The average strain rate for a steady shock is determined from the Hugoniot strain and the risetime through $\dot{\varepsilon} = \varepsilon/t$. Defining a property, A, as the product of the energy dissipated and the risetime, A = Et, then experimental data indicate that A is invariant for aluminum as well as several other metals in the range of measurement. This implies a fourth-power dependence on the Hugoniot pressure,

$$\dot{\varepsilon} = [S/3A(\rho_o C_o^2)^3]p^4 \ , \tag{1}$$

and an inverse square-root dependence of the effective viscosity,

$$\eta = \frac{3}{16} [3 A \rho_o C_o^2]^{1/2} \dot{\varepsilon} - 1/2 \ , \tag{2}$$

consistent with the data in Figure 4. Implications of this invariant energy feature are not yet fully understood; however, the energy relation and the experimental expression, $\varepsilon = 10^{-3} \dot{\varepsilon} 1/4$, for steady waves in aluminum will be used in the following calculation.

Heterogeneous Temperature Calculations in Aluminum

An approximate method has been developed[15] to calculate local temperatures within the slip bands which accounts for both steady-wave dissipation and thermal conduction. It is assumed that energy is dissipated at a constant rate, F_o, over the risetime, t. The temperature in the slip plane is then given by

$$T_\ell = T_s + \frac{2F_o}{\rho_o C_v} \sqrt{\frac{t}{\pi \chi}} \ , \tag{3}$$

where C_v is the specific heat and T_s is the isentropic temperature far from the slip band. Equation 3 is combined with the steady-wave relations

$$E = \frac{1}{3} \rho_o C_o^2 S\varepsilon^3, \quad \dot{\varepsilon} = \varepsilon/t, \quad \text{and} \quad F_o = Ed/t \ , \tag{4}$$

to account for energy dissipation throughout the slip band lattice with spacing, d, providing a local slip band temperature,

$$T_\ell = T_o e^{\gamma \epsilon} + \frac{SC_o^2 d}{3C_v \sqrt{\pi \chi}} \dot{\epsilon}^{5/2} \epsilon^{1/2} \quad , \tag{5}$$

in terms of steady-wave strain and strain rate.

Using shock wave properties for aluminum, calculated local temperatures have been compared with corresponding isentropic temperatures.[15]   The calculation, based on intuitively reasonable concepts of dissipation and thermal conduction, indicates a transition at approximately 5 GPa steady-wave shock compression from thermal equilibrated deformation to deformation where thermal trapping of dissipated shock energy leads to significantly higher local temperatures.[15]   The calculated temperatures are expected to occur immediately after passage of the shock wave but will persist only briefly beyond this time.   Slip band features observed in shock-recovered aluminum specimens are several hundred nanometers in thickness.   Thermal cooling rates for bodies of this thickness are calculated to be on the order of 10^{11} K/s; thus, local temperatures should only persist for a few nanoseconds after passage of the shock wave.

SUMMARY AND CONCLUSIONS

The present work describes an in-progress study focused on the shock behavior of metals.   Motivation was provided by recent observations of strength[8] and viscosity[9] behavior of metals in wave profile measurements.   Heterogeneous shear deformation appears to play a role in the dynamic deformation process.   Rational explanations seem to follow from this concept; and we are focusing our study on an understanding of heterogeneous deformation on the microscale.   Velocity interferometer measurements of wave profiles provide only bulk or average material response.   Consequently, a large effort is being directed toward metallographic methods of inferring the shock deformation process.   Substantial work of this type has been undertaken in the past. We are attempting to expand on this work by correlating microstructure with careful wave profile measurements.

The production and measurement of controlled one-dimensional compressive waves is now fairly advanced.   The soft recovery of specimens through stress and temperature states which do not complicate the microstructure left by the passing shock wave is a more difficult task.   We are addressing the problems both by improving recovery techniques and determining deformation effects associated with the shock stress release and thermal recovery process.

The post-shock metallographic studies, in addition to microhardness measurements, have focused on optical and transmission electron microscopy.   These techniques have provided a useful

qualitative tool for identifying trends of the deformation process
in both shocked and cold-worked specimens. Similarities in the
quasistatic and high-rate heterogeneous deformation process have
been observed. TEM methods have not identified reasons for
differences in cold-worked and post-shock hardness measurements or
microstructure features responsible for the strength and stress-
state behavior of aluminum on the Hugoniot. These effects, which
may be related to alterations in precipitate structure or vacancy
concentration as well as dislocation density variations, are
difficult to explore with TEM techniques, and, consequently, other
methods including x-ray diffraction and positron annihilation are
being pursued to better account for the shock strengthening
phenomenon.

Similar strain levels in shock-loaded and cold-worked samples
can be achieved and comparisons made. Stress states experienced
by the materials are markedly different, however. This is because
the shocked specimen must traverse the Rayleigh line during passage
of the steady wave and an additional viscous stress must be supported.
The viscous stress becomes substantial at Hugoniot pressures above
about 5 GPa where it can exceed the static flow stress. Additional
defect production or precipitate alteration due to the shock-
induced intensified shear stress does not therefore seem unreasonable
and may explain the increased shock strength on the Hugoniot.

It is difficult to explain the reduced shear stress state on the
Hugoniot without a thermal mechanism. Usual concepts of plastic
flow suggest that the stress state should reside on the yield surface.
The data show only a small residual state of shear stress relative
to the strength at Hugoniot states greater than about 5 GPa. This
is expected if the flow stress were small as the material approached
the Hugoniot state and if some rapid strength recovery mechanism
were operating during the microsecond or less before the strength is
tested with a release or reloading wave. Heterogeneous deformation
and thermal trapping during the high-rate deformation process could
cause the reduced flow stress, and microscale thermal equilibration
could provide the recovery process. This has yet to be verified;
however, appropriate model calculations suggest local high temperatures within slip bands during the shock process.

Factors governing shock viscosity and risetime of the plastic
wave have not yet been determined. Viscous flow should be associated
with the microscopic process of dislocation multiplication and motion,
vacancy production, precipitate alteration and so forth. There are
tentative indications, however, that shock wave risetime and
viscosity are governed by more fundamental, mechanism independent,
energy principles. If so, the microscopic shock processes would be
an effect rather than a cause, occuring in the energetically
favorable way, consistent with the time constraints. This idea is
speculative but it is clear that a better understanding is necessary

here before a comprehensive theory of the shock deformation process
will emerge.

To close, it should be iterated that effects relating to the
stress state and strength of metals during shock compression occur
which fall outside of the behavior in normal plastic processes.
The behavior of these dynamic processes still remains largely unexplained. We are currently pursuing mechanisms relating to
heterogeneous deformation processes and possible accompanying
thermal heterogeneity effects as responsible, in part, for the
observed dynamic results. Although the approach offers feasible
explanations for some aspects of the problem, considerable work
remains to demonstrate the relationship.

ACKNOWLEDGMENTS

The authors would like to express their gratitude to Rod Clifton,
Brown University, and Lalit Chhabildas, Sandia National Laboratories,
for helpful discussions during the course of this work.

This work was performed at Sandia National Laboratories and
supported by the U.S. Department of Energy under DE-AC04-76-DP00789.

REFERENCES

1. L. C. Chhabildas, in Shock Waves in Condensed Matter -- 1981,
    (AIP), edited by W. J. Nellis, L. Seaman, R. A. Graham,
    p.621 (1982).
2. L. M. Barker and R. E. Hollenbach, J. App. Phys. 43, 4669 (1979).
3. L. M. Barker and R. E. Hollenbach, J. Appl. Phys. 41, 4208 (1970).
4. J. L. Wise, to be published.
5. R. J. Lawrence and D. S. Munson, Sandia Laboratories Report No.
    SC-RR-710284 (1971).
6. A. L. Stevens and O. E. Jones, J. Appl. Mech. 39, 359 (1972).
7. L. Davison, D. M. Webb and R. A. Graham, in Shock Waves in
    Condensed Matter -- 1981, (AIP), edited by W. J. Nellis, L.
    Seaman, R. A. Graham, p. 67 (1982).
8. J. R. Asay and L. C. Chhabildas, in Shock Waves and High-Strain-
    Rate Phenomena in Metals, edited by M. A. Myer and L. E. Murr
    (Plenum Press) p. 47 (1981).
9. D. E. Grady, Appl. Phys. Lett. 38, 825 (1981).
10. L. M. Barker, in Behavior of Dense Media under High Dynamic
    Pressures, (Gordon and Breach, NY) p. 483 (1968).
11. J. J. Gilman, J. Appl. Phys. 50, 4959 (1979).
12. L. E. Murr, in Shock Waves and High-Strain-Rate Phenomena in
    Metals, edited by M. A. Myer and L. E. Murr (Plenum Press,NY)
    p. 607 (1981).
13. C. H. Hill and H. J. Rack, Mat. Sci. and Engr. 51, 231 (1981).
14. J. G. Byrne, University of Utah, to be published.
15. D. E. Grady and J. R. Asay, J. Appl. Phys. (in press).

ADIABATIC SHEARING – GENERAL NATURE AND MATERIAL ASPECTS

H.C. Rogers

Drexel University
Department of Materials Engineering
Philadelphia, PA 19104

INTRODUCTION

The phenomenon loosely termed "adiabatic shearing" is one aspect of high velocity deformation that has been receiving substantially greater attention in recent years because of the growing number of areas in which it is now recognized to play a significant role. This will become obvious from the details not only of the rest of the papers in this section, but from others in this conference as well. Although shearing deformation is well known and well understood from a solid mechanics point of view, the term adiabatic shearing will not be found in mechanics texts, the reason being that it is an aspect of real material behavior rather than of an idealized continuum. Moreover, the discussion throughout this paper will concentrate on the behavior of metals although some aspects of adiabatic shearing behavior are also applicable to polymers and other non-metals as well. In metals, plastic deformation takes place by the motion and interaction of dislocations. The work of plastic deformation is partly stored in the metal as elastic strain energy known as cold work while the remainder is converted to heat. It has been experimentally determined that at room temperature 90-95 percent of the work of deformation goes into heat. Ideally, if the deformation took place instantaneously, all would be retained locally and the deformation would be truly adiabatic. During dynamic deformation, however, a certain fraction of the heat generated is lost to the surrounding metal, the exact amount depending both on the rate of deformation and the thermal properties of the material. In any case, if sufficient heat is retained that the deformation process is thought to be significantly modified by it, the deformation is referred to as "adiabatic".

There will be no attempt here to be comprehensive in the coverage of adiabatic shearing but sufficient background and clarification will be given to put the discussion of the influence of material variables on this phenomenom in context. Additional information, if desired, can be obtained from recent more extended reviews of the subject (1-3).

A further point of clarification relates to the term "shear localization" or localized shear. Plastic deformation in metals by its very nature takes place principally by the motion of dislocations along planes or bands that are crystallographic in nature. The spatial orientation of these active slip planes varies from grain to grain in metals and changes with deformation. Although there is a tendency for deformation to derandomize the distribution of slip planes in the polycrystalline solid leading to the development of crystallographic texture, these shear bands are each confined to a single grain. These aspects of shear localization will be discussed by Asaro in another paper in this conference. The discussion herein will be concerned with macroscopic bands extending over many grains.

A recognition of the magnitudes of the local strains and strain rates involved in adiabatic shearing during dynamic deformation is necessary to appreciate the difference between it and the more familiar quasistatic deformation. Strains as large as 100 are developed in microseconds, giving rise to strain rates of 10^7-10^8 per second. Moreover, hypervelocity impact is not required to produce these strains and strain rates; they are the result of extreme strain localization during what might be a relatively low displacement rate.

This localization of deformation is exemplified by the deformed shear band in the aluminum alloy 2014-T6 shown in Fig. 1. Precipitate particles and grain boundaries provide the markers for shear strain delineation, which is obviously very high in an extremely thin zone of deformation. There is no evidence from the microstructure that this was an adiabatic shear band; only the fact that the deformation was caused by projectile impact would necessitate considering deformation heating as a contributing factor in strain localization. On the other hand, in the case of carbon steels, there is a "signature" frequently left by adiabatic shearing often described as a "white band" and shown clearly in Fig. 2. The "white" appearance of such a band is caused by its failure to provide significant strucural detail on etching with nital or other conventional steel etches. In addition to their peculiar etching characteristics, these white shear bands are also extremely hard. It was this hardness that led Zener and Hollomon (4) to the conclusion that they are bands of martensite resulting from the deformation heating of the steel locally to temperatures above that required for austenitization followed by a rapid quench by the surrounding metal on cessation of deformation. Although they estimated that the shear strain in the bands that they observed in a mild steel might be as high as 100, they also calculated that a shear

strain of only 5 would have been sufficient to heat the band to 1000°C
if the deformation was adiabatic. That austenitization has occurred
has not in fact been unequivocally established; nevertheless, the
evidence is sufficiently good and the character of the band sufficiently different from the normal deformed band that these bands are
referred to as "transformed" bands.

    Since the conclusion of Zener and Hollomon nearly forty years
ago that these bands resulted from deformation-induced heating and
subsequent quenching, there have been many investigations into their
actual structure with conflicting results. There is no doubt that
their hardness exceeds that obtainable by conventional quenching
techniques in the same steel. Rogers and Shastry (5) have shown that
the hardness of the transformed bands in a 1040 steel is independent

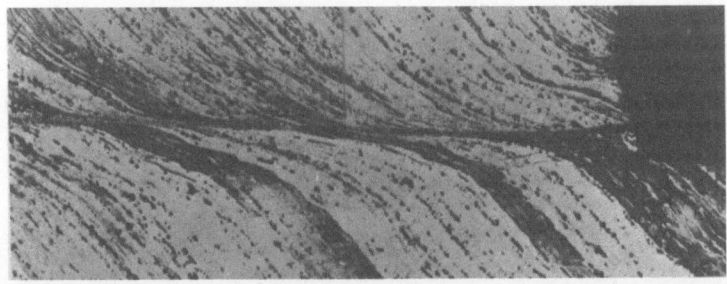

Figure 1. Deformed shear band produced below a flat-ended projectile in aluminum alloy 2014-T6, showing the high degree of shear in
the band (2).

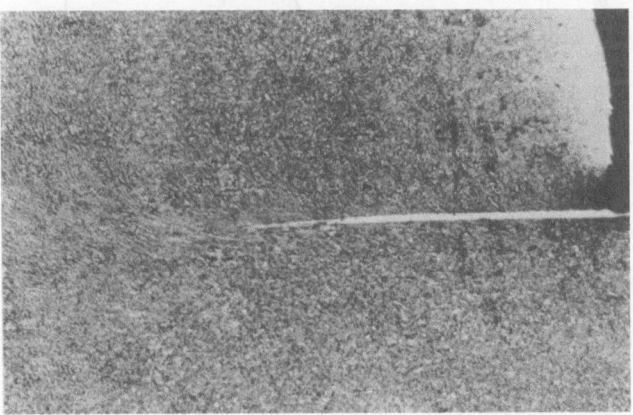

Figure 2. A sectioned plate clearly showing a shear band "propagating" ahead of a projectile in an impacted plate of AISI 1040 steel
quenched and tempered at 400°C (100X).

of the velocity of deformation within the range studied and is also
independent of the hardness level of a given target steel. They
further showed that the hardness of the band is linearly related to
the carbon level of the target material, Fig. 3. The substructure
in these bands consists of fine equiaxed grains with diameters in
the range of tenths of micrometers. Rogers and Shastry have shown
that the extraordinary hardness of these transformed bands can be
accounted for by considering the combined hardening effect of ultra-
fine grain size and that resulting from the carbon supersaturation
in the transformation product (i.e., martensite). The strong interest
in the nature of the transformed bands in steels, however, stems
more from a scientific desire for understanding than from practical
considerations since in most but not all instances it is the behavior
of the material during deformation and shear band development and
not subsequent to it that is of primary importance.

Adiabatic shearing is involved in a wide variety of processes or
situations where dynamic deformation occurs. Ordnance problems in
particular are heavily concerned with adiabatic shearing phenomenology; these include armor penetration, penetrator performance, and
fragmentation. Machining, even at conventional speeds, produces
high deformation rates in the workpiece ahead of the tool. Impact

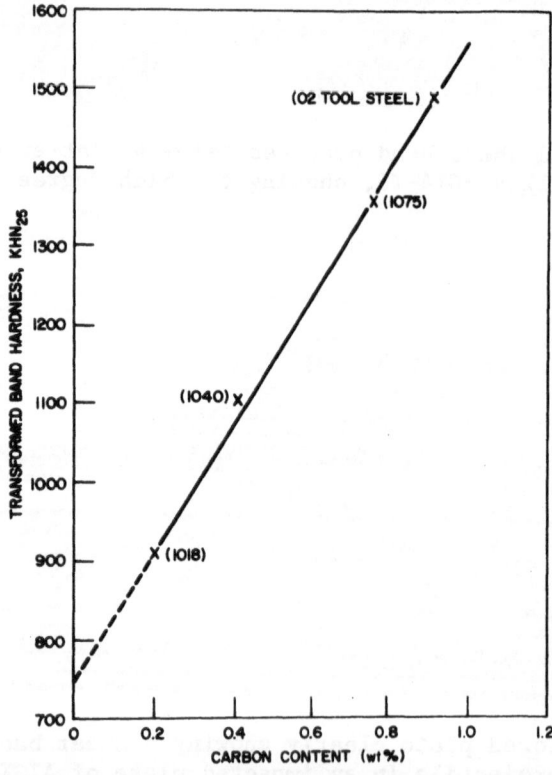

Figure 3. The effect of carbon content on the hardness of transformed shear bands in several steels (5).

erosion of metals has now been convincingly shown to involve adiabatic shearing in material removal mechanisms. Other areas of
commercial significance in which adiabatic shearing plays a significant role are ore crushing, impact tooling failure, and metal shaping
and forming processes. These will not be discussed here in detail;
additional discussion is contained in the review articles previously
mentioned. What should be noted, however, that in almost all these
instances there is an interaction between a tool or hard body and a
somewhat softer body or workpiece. The resulting deformation in
the work piece is not homogeneous at the outset and becomes increasingly inhomogeneous as the dynamic deformation continues. There
may also be other reasons for the initial deformation to be inhomogeneous. As Semiatin et al.(6,7) have demonstrated, when titanium
alloys are hot upset using cold tooling, thermal gradients are set
up in the hot metal as a result of tool-workpiece contact which in
turn produce regions of higher and lower flow stress. This temperature inhomogeneity combined with frictional effects leads to localization of deformation in narrow zones bordering the cooler regions.
Only in the case of fragmentation of explosively-expanded smooth
bore steel cylinders is there no identifiable reason for the deformation to be initially inhomogeneous although well defined transformed bands are observed at later stages in the deformation process.
Until further information is gained either about the explosive
process itself or of identifiable structural inhomogeneities in the
metal, the band formation in this instance must be treated as a
homogeneous nucleation and band growth process (8,9). Most other
cases must be considered as instances of inhomogeneous deformation
localizing catastrophically under the imposed dynamic displacements
primarily because of the thermal and mechanical properties of the
deforming material.

FRACTURE

    The state of stress under which the deformation is taking place
plays a critical role in determining whether or not adiabatic shear
bands are observed. Under tensile loading they are not observed
because as deformation tends to localize fracture ensues early by
void growth and rupture, preventing sufficient accumulation of local
deformation to raise the temperature significantly. As the hydrostatic component of the stress operative during shearing in any
dynamic loading situation changes from tension to compression,
fracture tends to be increasingly suppressed and greater local
strains required for significant heating can accumulate. Adiabatic
shearing has been observed under torsional loading but most commonly
arises during dynamic deformation conditions that tend to be highly
compressive, thus suppressing void nucleated fracture.
    From the point of view of understanding the formation of adia-
batic shear bands, their subsequent fracture is obviously of little
concern. From a practical point of view, however, the nature of the
fracture may yield information about the time of fracture in a
complex dynamic loading situation. Of even greater importance is

that the fracture of the shear bands during deformation may limit
certain operations or the fracture characteristics of the transformed
bands may weaken the metal when subjected to future loading. Consider the shear band in the U-2 Mo alloy shown in Fig. 4. This
band has partially failed by ductile rupture although there is little
evidence of void growth outside the band. This means that the band
was weaker than the matrix and the tensile stress producing the
fracture was imposed while the band was still hot. The fact that
the voids are more or less equiaxed indicates that the intense
shearing had stopped and the major stress direction at the time of
fracture was normal to the band. Such effects are commonplace as a
result of wave propagation and reflection during impact loading.
The sequence and timing of the stressing determines the observed
fracture behavior. Fracture in steels is somewhat more complex in
that the hard transformed bands are brittle as well. Since they
are fine grained with little or no crystallographic or fiber texture,
transformed bands fracture brittlely in directions normal to the
local tensile stress. As mentioned above, in complex loading
situations a tensile wave may impact an existing transformed band
from any direction depending on the particular geometry of loading.
Some examples are shown in Fig. 5. If the matrix is sufficiently
ductile, brittle fractures will be confined to the band; if not,
the fractures may extend into the matrix or the matrix itself may
fracture in a brittle manner. An important point that is frequently
overlooked when considering the effects of adiabatic shear banding in
steels is that once transformed bands have formed from one dynamic
loading, although these bands may not be fractured, they remain as
brittle fracture paths in what may be an otherwise ductile material.
If the part or structure is then subjected to repeated loading as
is the case with hammers, ore crushing equipment, and probably armor,
chunks of the part can spall off as brittle fracture takes place
through preexisting brittle adiabatic shear bands. This may also
occur during multiple-impact solid particle erosion. In any case,
the result may be catastrophic failure or rapid deterioration of
the part containing shear bands of this type.

ANALYSES

A number of analyses have been made of the adiabatic shearing
phenomenon, including among others those that have been referenced.
Most of the analyses accompany or are complementary to experimental
studies of dynamic deformation in which adiabatic shearing presumably
plays a part. Many of the analyses consist of two parts: a mechanical instability analysis and a thermal model assumed to be appropriate for the particular system under study. Some are quite simple;
others, more sophisticated. The majority of the analyses assume that
instability arises at the time when the change in the force, dP = 0;
since $P = \sigma A$, this is equivalent to

$$\sigma dA + A d\sigma = 0 \qquad [1]$$

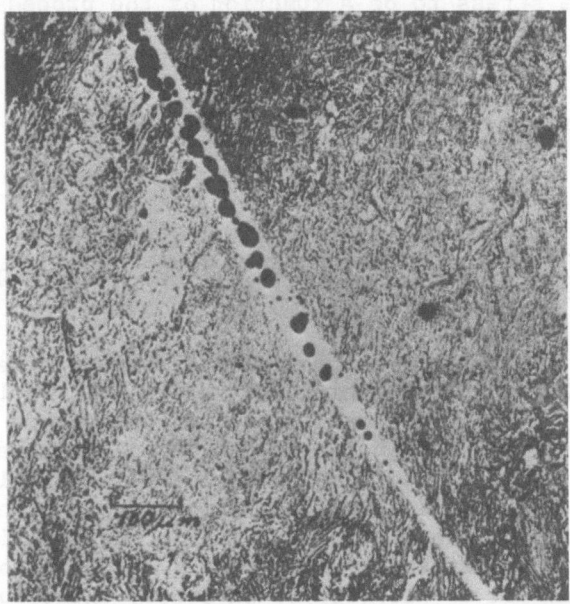

Figure 4.   Section of a fractured shear band in a U-2Mo alloy (10).

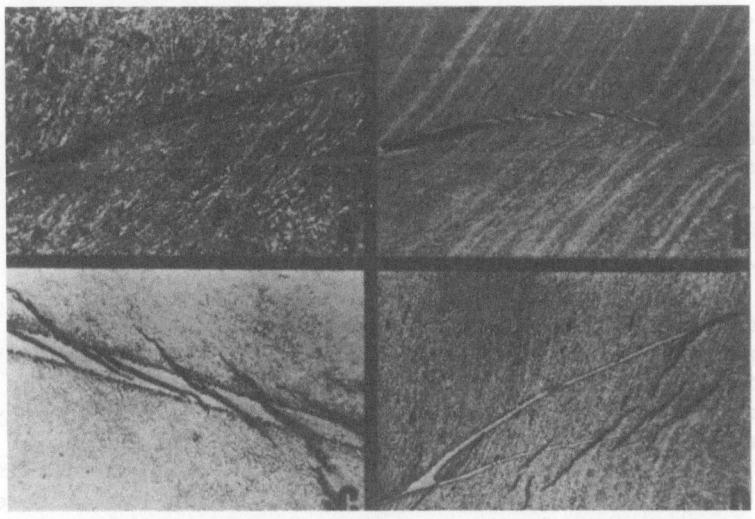

Figure 5.   Several types of brittle fractures of transformed adiabatic shear bands (11).

Considering the stress to be a function of the process variables: strain, strain rate, and temperature (ε, $\dot{\varepsilon}$, T), $d\sigma$ can be written as

$$d\sigma = \left(\frac{\partial\sigma}{\partial\varepsilon}\right)_{\dot{\varepsilon},T} d\varepsilon + \left(\frac{\partial\sigma}{\partial\dot{\varepsilon}}\right)_{\varepsilon,T} d\dot{\varepsilon} + \left(\frac{\partial\sigma}{\partial T}\right)_{\varepsilon,\dot{\varepsilon}} dT$$

and equation 1 can be rewritten as

$$\sigma dA + A\left[\left(\frac{\partial\sigma}{\partial\varepsilon}\right)_{\dot{\varepsilon},T} d\varepsilon + \left(\frac{\partial\sigma}{\partial\dot{\varepsilon}}\right)_{\varepsilon,T} d\dot{\varepsilon} + \left(\frac{\partial\sigma}{\partial T}\right)_{\varepsilon,\dot{\varepsilon}} dT\right] = 0 \qquad [2]$$

In tension the first term is negative and leads to the conventional plastic instability relationship. In shear, with no loss of cross-sectional area, instability must arise from the terms in the brackets alone. Under compressive loading, the second term must in addition be sufficiently negative to overcome a normally positive geometrical hardening that accompanies compressive loading. Moreover, it is highly unlikely that the instability will occur normal to the compression direction; rather it will occur by a shear localization, as observed by Semiatin et al. (6,7). In shear, then, substituting τ and γ for the general terms σ and ε,

$$\left(\frac{\partial\tau}{\partial\gamma}\right)_{\dot{\gamma},T} + \left(\frac{\partial\tau}{\partial\dot{\gamma}}\right)_{\gamma,T} \frac{d\dot{\gamma}}{d\gamma} + \left(\frac{\partial\tau}{\partial T}\right)_{\gamma,\dot{\gamma}} \frac{dT}{d\gamma} = 0 \qquad [3]$$

In many of the analyses such as those of Recht (12), Staker (9), and Olson et al. (13) the strain rate hardening term

$$\left(\frac{\partial\tau}{\partial\dot{\gamma}}\right)_{\gamma,T} \frac{d\dot{\gamma}}{d\gamma}$$

is considered stabilizing but negligible given the materials and strain rate range under study. Thus at instability, Eq. 2 becomes

$$\left(\frac{\partial\tau}{\partial\gamma}\right)_{T,\dot{\gamma}} + \left(\frac{\partial\tau}{\partial T}\right)_{\gamma,\dot{\gamma}} \frac{dT}{d\gamma} = 0 \qquad [4]$$

Recht used as a thermal model one for uniform, constant-rate heat generation at a plane in an infinite medium as he was investigating instability in the shear zone ahead of a cutting tool during orthogonal machining. His analysis also factored in the thermophysical properties of the material being sheared. The end result of this analysis was a quantitative comparison of the critical strain rates at which catastrophic localized shear occurred in two different materials. In terms of relative sensitivity to localized adiabatic shear, Recht showed that this model did order materials in the same way as observed experimentally. According to this model, the criti-

cal strain rate for catastrophic shear in mild steel in 1400 times
greater than that for titanium. Based on the data supplied by Recht
in his Table I, it can easily be shown that of the ratio 1400:1, the
difference in thermophysical properties contributes a factor of
approximately 6 and the ratio of the yield stresses, a factor of 4,
while the major contributor is the difference in mechanical behavior,
which contributes a factor of 58.   In the Recht analysis, this latter
factor is of the form:

$$\left[\left(\frac{\partial \tau}{\partial \gamma}\right) \Big/ \left(\frac{\partial \tau}{\partial T}\right)\right]^2 \tag{5}$$

where the upper term is the material's capacity to strain harden and
the denominator defines its tendency to thermally soften. Not only
does this factor contribute the major portion of the difference in
critical strain rate sensitivity of two different materials in the
Recht analysis but also the second power relationship magnifies the
influence of small differences in the strain hardening capacities
and/or thermal softening tendencies of various materials.

Staker (9) and Olson et al. (13) used a similar instability
analysis to model the appearance of adiabatic shear bands in the
controlled explosive expansion of steel cylinders, Staker argues,
however, that because of the short times involved in explosive
loading the deformation can be considered completely adiabatic with
no need to consider the thermophysical properties of materials.
Hence $dT/d\gamma = \tau/C$ where C is the volume specific heat in units of
KPa/°C. He also considers strain rate effects to be of second order
in the steels studied and therefore can be neglected. For a power
law hardening material with strain hardening exponent, n, Eq. 4 then
gives the critical strain at which instability occurs:

$$\gamma_C = -\frac{Cn}{\left(\frac{\partial \tau}{\partial T}\right)_{\gamma,\dot{\gamma}}} \tag{6}$$

If other strain hardening laws are assumed, the relation given in
Eq. 6 is modified accordingly. For the types of steels studied,
however, the above relation holds quite well as shown in Fig. 6.

Semiatin et al. (6,7) on the other hand have analyzed the conditions for instability during the hot upsetting of titanium alloys
both conventionally with cold tools and also isothermally. In both
instances there is a tendency toward localized flow because of die-
workpiece interface friction that leads to restricted flow zones
beneath the tools. For nonisothermal forging, chill zones set up
by contact with the cold tools enhance the tendency toward localization of the deformation. The analytical model developed by
Semiatin et al., in contrast to those previously discussed, does
not discount strain rate effects and must also account for the geometrical hardening that normally results from compressive loading,
although the same instability relation is utilized. A parameter, α,
that measures a material's tendency toward marked localization is
developed. This parameter is of the form

$$\alpha = \frac{\eta' - 1}{m} \qquad\qquad [7]$$

where η' contains both the strain hardening and thermal softening terms while m is the strain rate sensitivity [m = $(\partial \ln\sigma/\partial \ln\dot{\epsilon})_{\epsilon,T}$]. The -1 results from the geometrical hardening that occurs in axisymmetric compression. In plane strain upsetting, shearing is favored and α simply becomes η'/m. Their results indicate that this parameter is useful in identifying those combinations of materials and processing conditions that are likely to result in localized shearing during hot forging.

MATERIAL PROPERTIES

The most significant material characteristics that emerge from the above and other analyses are the temperature dependence of the flow stress and the strain hardening capacity, generally characterized by the strain hardening exponent, n. There is general agreement that the susceptibility to adiabatic shear localization increases as the rate of strain hardening decreases and/or the rate of thermal softening increases. On the other hand, the role of strain rate sensitivity is less clear. This may reflect differences in materials investigated or differences in the strain rate, temperature, and loading regime utilized, of the geometry of loading used in different studies. Another material characteristic that can have a significant

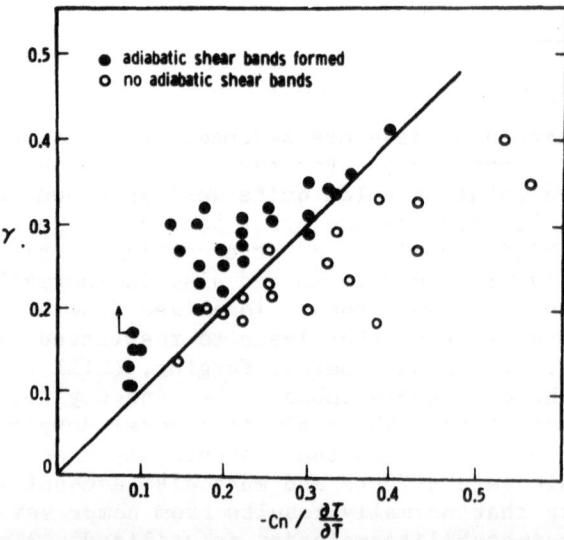

Figure 6. The relation between measured true shear strain γ and a material parameter $-Cn/(\partial\tau/\partial T)$ defining shear instability (9).

effect on the generation of adiabatic shear bands, however, is the basic strength level of the deforming material. Because the heat generated locally comprises the major part of the local specific work done by the deformation, it is obvious that for a given strain increment a higher flow stress causes more heat generation and a greater temperature rise. Thus the maximum temperature attained can be calculated from the following relation

$$T_{max} = \frac{F}{C} \int_0^{\gamma_f} \tau d\gamma + T_0 \qquad [8]$$

where T_0 is the initial temperature, γ_f is the final shear strain obtained, and the factor F takes into account both the fraction of work converted to heat and the fraction of the heat generated that is lost to the surroundings through thermal conduction, etc. For any set of alloys having the same base element, the heat capacity and thermophysical properties are usually similar so that for equivalent deformation states, the greatest amount of heat will be generated in the strongest alloy. This is at least one of the reasons for the generally observed enhanced susceptibility to adiabatic shear band formation of high strength quenched and tempered steels compared to their lower strength counterparts.

The influence of microstructure and changes in microstructure with strain rate and temperature on the susceptibility to adiabatic shearing is the least understood aspect of the influence of material characteristics on the adiabatic shearing phenomenom. By microstructure is meant here the number, size, shape and distribution of precipitates or second phases that influence the strength of metals in some degree. This means an interaction of the microstructure with the dislocation substructure to alter its character in terms of dislocation density and/or arrangement. The complexity that arises is that changes in the process variables: strain, strain rate, and temperature can alter both the microstructure and the dislocation substructure as well as modify the behavior of individual dislocations or groups of dislocations. Thus there is a "history effect" during any stage of the adiabatic deformation process including the time after deformation has ceased. This is clearly illustrated by the results from the ongoing studies of adiabatic shearing at Drexel University shown in Fig. 7, and discussed in more detail previously (5). Room temperature microhardness measurements transverse to a transformed band in an impacted 1040 steel target, quenched and tempered at 400°C, show the usual extreme hardness of such a band. On either side of the band are regions of matrix material that have suffered various degrees of deformation strengthening and thermal softening. Generally the hardness values in the untransformed material decrease with distance from the band center until the matrix hardness is reached. This is also seen for the deformed bands observed in 1018 steel and 304 stainless steel. Thermal softening effects are most clearly identifiable, however, in the zones adjacent to the transformed bands in similarly impacted but more lightly

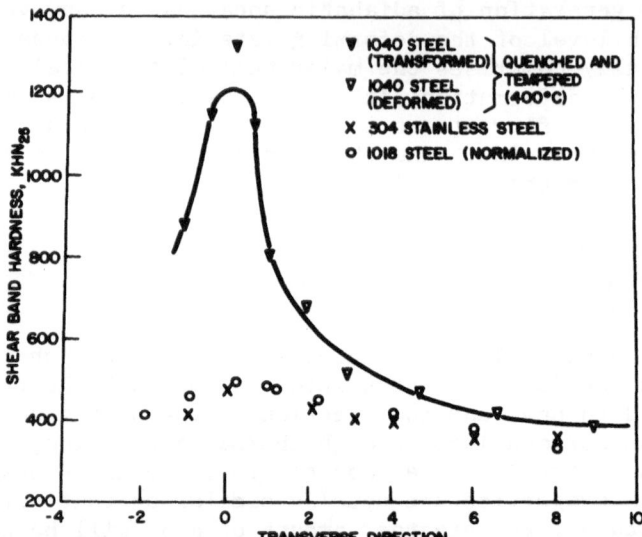

Figure 7. Profile of the microhardness across adiabatic shear bands in several steels (5).

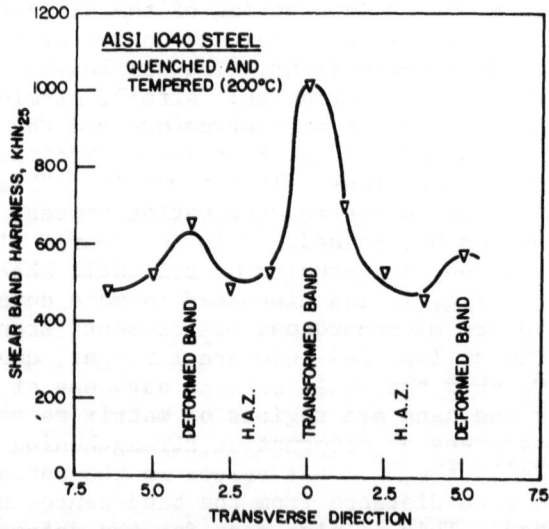

Figure 8. Profile of the microhardness across a transformed shear band in AISI 1040 quenched and tempered at 200°C (5).

tempered 1040 steel, Fig. 8. The deformation-induced heating, which, in the center of the shear band, was high enough to cause transformation, was also great enough in the adjacent heat affected zones to cause a hardness reduction. The fact that such a reduction is observed only in the steel tempered at 200°C is considered to be a reflection of its more thermally unstable microstructure. The major point to be made from a review of this data is that, in a deformed band, the material that is hardest and strongest with respect to post-deformation room temperature mechanical behavior – the material at the center of the band – is precisely the material that was weakest and had the greatest amount of deformation and highest strain rate during dynamic loading. Thus during deformation thermal softening is overcoming the strengthening effects of strain and strain rate on dislocation behavior while producing a dislocation substructure that is more resistant to dislocation motion at subsequent ambient temperature and quasistatic rates of deformation. Kunze et al. (14) also observed hardness values about 50% higher than the matrix in deformed adiabatic shear bands in several nonferrous metals.

EFFECTS OF COLD WORK

From the preceding discussion it appears that the materials most resistant to the development of adiabatic shear bands should have low strength, rapidly work harden, and have excellent resistance to thermal softening. Notwithstanding, the preponderance of situations where adiabatic shear localization is a problem require the use of hard and/or strong materials. Steels, particularly in the quenched and tempered condition, or other high strength ferrous alloys are most often utilized.   Unfortunately, as Hollomon and Zener (15) showed, high strength and a high rate of work hardening are essentially mutually exclusive properties of steel. For a wide variety of steels the product of the yield stress and the strain hardening exponent is found to be a constant, ϕ, having a value in the range of 8,000-16,000 psi. Rosenfield and Hahn (16) made a similar analysis, incorporating the results of their investigation of line-pipe steel with those of other studies including data on the very high strength maraging steels. These data, Fig. 9, confirm the general conclusion of Hollomon and Zener. There does not, therefore, appear to be an easy solution to the problem of obtaining both high strength and a high rate of work hardening in the same class of materials.

The general employment of thermal treatments to change the strength of a given composition of steel has caused some ambiguity in understanding of the material variables that control the development of transformed bands in carbon or low alloy steels. Heat treatments that lead to lower strength also normally produce coarse microstructures that tend to be thermally stable while strong, lightly tempered or patented steels contain fine, less stable carbides. It is not clear, therefore, whether it is the strength level per se or the lower stability of the microstructure in high strength steels that

is the most important factor in determining susceptibility to highly
localized adiabatic shear banding.

A series of experiments have been carried out at Drexel University in an attempt to clarify the problem. Two different materials,
a 1018 steel and a 70–30 brass, were used as target materials in
impact studies employing flat, stepped-nose projectiles. The
materials were each tested in the annealed condition and also after
67 per cent cold reduction by rolling. The cold working strengthens
the metals, reduces their rate of strain hardening but leaves the
non-dislocation microstructure essentially unaltered. Presumably
the cold worked targets would have a significantly higher susceptibility to adiabatic shearing instability than those that are annealed.
Some of the test results can be seen in Figs. 10 through 14. In the
annealed 1018 steel even the highest velocities available failed to
produce transformed bands although they far exceeded the velocities
that were capable of producing such bands in several different 0.40 C
quenched and tempered steels. The target that had been cold rolled
67%, however, did show transformed bands when impacted at 100 m/s,
although at lower impact velocities it too developed only deformed
bands. Plug formation occurred in all cold worked targets, while the
annealed targets exhibited much more general deformation except in
the immediate vicinity of the projectile step where the deformation
was concentrated in a deformed band. The brass behaved in a similar

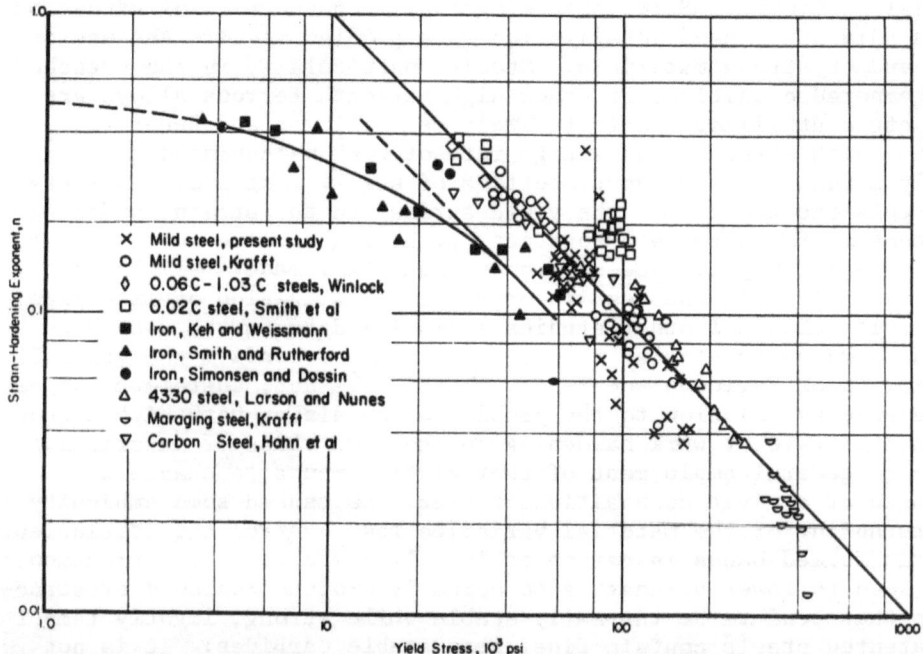

Figure 9. Effect of yield stress on strain-hardening exponent (16).

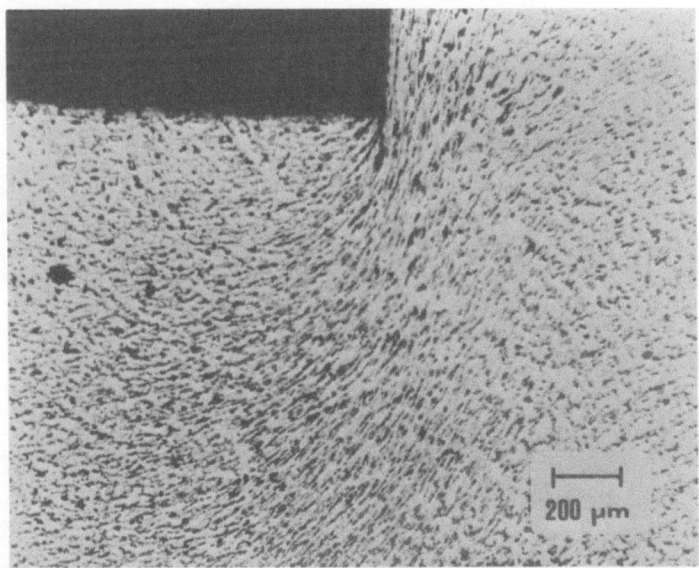

Figure 10. Diffuse deformed shear band at corner of impact crater in annealed 1018 steel impacted at 100 m/s. Etched in nital.

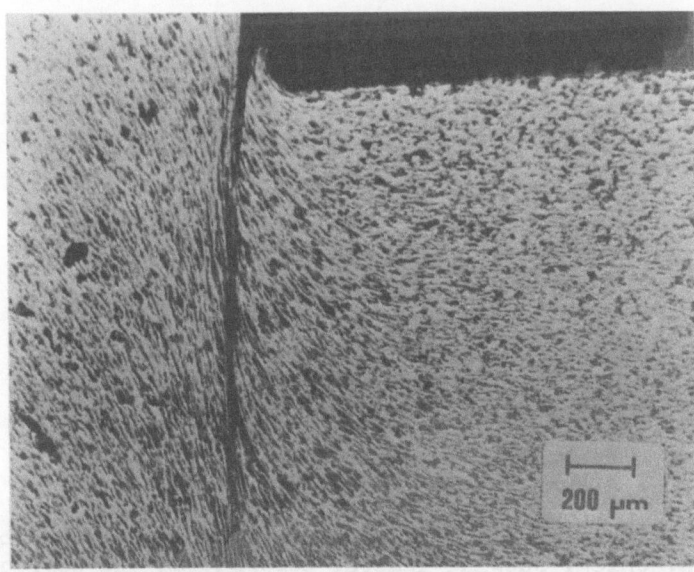

Figure 11. Deformed shear band at corner of impact crater in 1018 steel, cold rolled 67% and impacted at 94 m/s. Etched in nital.

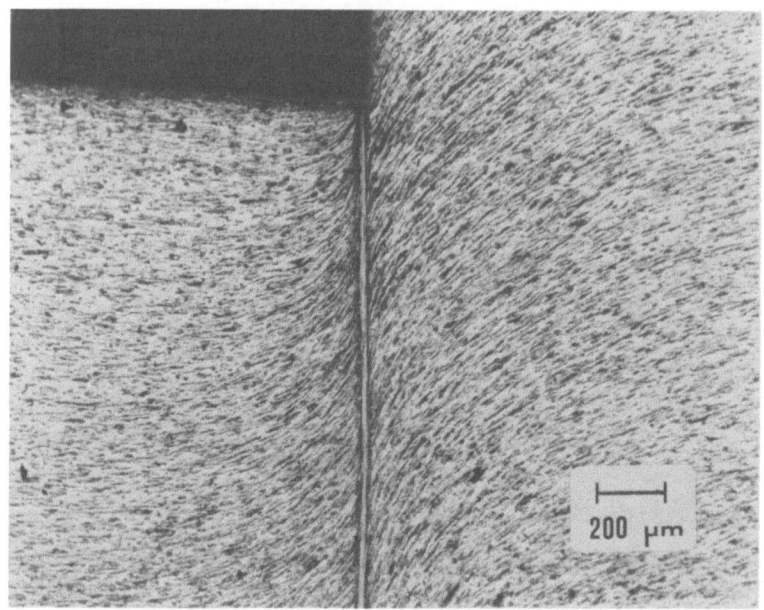

Figure 12.   Transformed shear band at corner of impact crater in
1018 steel, cold rolled 67 percent and impacted at 100 m/s.   Etched
in nital.

Figure 13.   Deformation of back side of an annealed 70-30 brass target, .250 in. thick, showing general deformation and bulging.   Target
was mounted on a backing plate with a central .500 in. diameter hole.

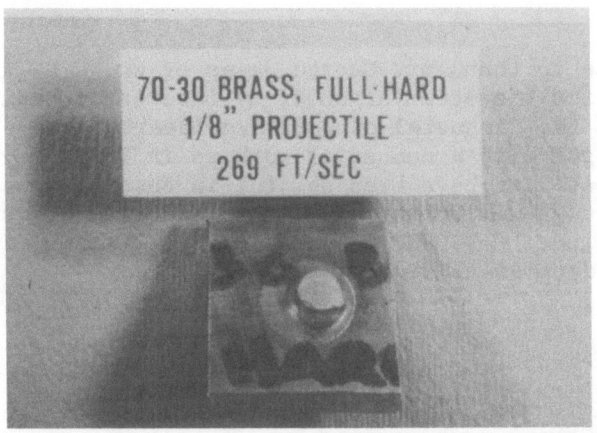

Figure 14.   Back side of an impacted 70-30 brass target, .250 in.
thick, cold rolled 67%. Deformation is highly localized resulting
in plug formation. Target was mounted on a backing plate with a
central .500 in. diameter hole.

manner including plug development in the cold worked target with the
exception that the bands obviously were all of the deformed type.

SUMMARY

     The appearance of adiabatic shear bands in materials during
dynamic loading depends both on the rate and other aspects of the
dynamic loading of materials and also certain material properties.
Most analyses agree, supported by the available experimental evidence,
that there is a significantly greater tendency for the formation of
adiabatic localized shear bands in materials that have a low rate of
work hardening and/or an increased tendency to thermally soften.
Under conditions where a certain displacement is imposed, high
strength materials will be more susceptible to adiabatic shearing
than those of lower strength whether or not the high strength arises
from heat treatment or from cold work. The importance of the strain
rate sensitivity of materials is much less clear. Equally uncertain
is the role of the thermal stability of the second phase microstructure. Because of the complexity of the problem and the general
unavailability of comparable experimental results for a wide range
of dynamic deformation conditions, stress states, geometries, etc.,
there is still much work to be done before there is sufficient
understanding of the adiabatic shearing problem that optimum conditions to resist such localized deformation can be stated from first
principles.

ACKNOWLEDGMENT

I would like to thank Dr. George Mayer of the U.S. Army Research Office for his stimulation of my interest in adiabatic shearing and for the financial support provided by that office. Continuing dialogue with a number of workers in U.S. Army and other government laboratories have been helpful in shaping the direction of the studies at Drexel University. Also appreciated is the help of my co-workers, Dr. C.V. Shastry and Mr. F.T. Zimone, in the experimental aspects of this work.

REFERENCES

1. H.C. Rogers: "Adiabatic Shearing: A Review", Drexel University Report prepared for U.S. Army Research Office, (1974).
2. H.C. Rogers: Ann. Rev. Mater. Sci. 9:283 (1979).
3. A.J. Bedford, A.L. Wingrove and K.R.L. Thompson, J. Aust. Inst. Metals 19:61 (1974).
4. C. Zener and J.H. Hollomon, J. Appl. Phys. 15:22 (1944).
5. H.C. Rogers and C.V. Shastry, in "Shock Waves and High-Strain-Rate Phenomena in Metals", M.A. Meyers and L.E. Murr, eds., Plenum Press, New York, 1981, p. 285.
6. S.L. Semiatin and G.D. Lahoti, Met. Trans. 12A:1705 (1981).
7. S.L. Semiatin, G.D. Lahoti, S.I. Oh, "The Occurrence of Shear Bands in Metalworking", in Material Behavior Under High Stress and Ultra High Loading Rates, Proc. 29th Sagamore Army Res. Conf., Lake Placid, New York, 1982 (to be published).
8. D.C. Erlich, D.R. Curran and L. Seaman, "Further Development of a Computational Shear Band Model", Rep. No. AMMRC TR 80-3, March 1980, SRI International Report prepared for Army Materials and Mechanics Research Center, Watertown, MA.
9. M.R. Staker, Acta Met. 29:683 (1981).
10. C.J. Irwin, "Metallographic Interpretation of Impacted Ogive Penetrators", DREV-R-652/72, Canada, 1972, 46 pp.
11. M.E. Backman and S.A. Finnegan in "Metallurgical Effects at High Strain Rates", R.W. Rohde et al., eds., Plenum Press, New York, p. 531.
12. R.F. Recht, J. Appl. Mech., Trans. ASME 31E:189 (1964).
13. G.B. Olson, J.F. Mescall and M. Azrin, in "Shock Waves and High-Strain-Rate Phenomena in Metals", M.A. Meyers and L.E. Murr, eds., Plenum Press, New York, 1981, p. 221.
14. H.-D. Kunze, K.-H. Hartmann and J. Rickel, Pract. Metallog. 18:261 (1981).
15. J.H. Hollomon and C. Zener, Trans. AIME 158:283 (1944).
16. A.R. Rosenfield and G.T. Hahn, Trans. ASM 59:962 (1966).

THE OCCURRENCE OF SHEAR BANDS
IN METALWORKING

S. L. Semiatin,[*] G. D. Lahoti,[+] S. I. Oh[*]

[*]Metalworking Section
Battelle's Columbus Laboratories
Columbus, Ohio 43201

[+]Timken Research
The Timken Company
Canton, Ohio 44706

ABSTRACT

Shear bands, or regions of localized plastic flow crossing
many grains, are not uncommon during the deformation processing of
metals. They may occur under nominally isothermal conditions
(tooling and workpiece at the same initial temperature) as well as
non-isothermal conditions (tooling and workpiece at different temperatures). Under isothermal conditions, the localization of plastic
flow is a function of the geometry involved in the metalworking operation, the deformation rate, and material properties such as the
work-hardening or flow softening rate of the material and its strain
rate sensitivity. For non-isothermal metalworking operations, these
as well as other process and material parameters controlling temperature, and hence flow stress and strain, gradients must be considered
to predict the occurrence of shear bands. Methods of predicting the
occurrence and severity of shear bands are presented. The power of
the analytical tools are demonstrated with results from studies of
the hot forging behavior of several titanium alloys. Other observations of flow localization in a hot-work tool steel, a uranium alloy,
and a superalloy are used to illustrate the generality of the concepts presented. In addition, shear band phenomena in metalcutting
operations are reviewed in the context of the metalworking studies.

119

INTRODUCTION

      The occurrence of flow instability and flow localization is
well known in the deformation processing of ductile metals. Under
tensile modes of loading, flow instability leads to the formation
of flow localizations in the form of necks.[1-4] These necks can
be either rather symmetric relative to the applied loading as in
the case of tension testing of round bars or somewhat asymmetric
relative to the applied loading as in the formation of localized
(through-thickness) necks in sheet metal forming. The understanding
of flow localization in these cases is important because the occurrence of necking often limits the amount of uniform deformation that
can be imposed and hence determines the overall formability of
the metal.[5]

      In contrast to flow localization in tensile modes of loading,
the formability of metals under compressive loading conditions, or
workability as it is more commonly known, is not controlled by
necking. In the so-called primary or bulk forming operations, defects such as free surface fractures,[6] grain-boundary and triple-
point cracks,[7] and central bursts[8-10] often limit workability.
Yet another kind of defect, the shear band, can appear, however.
This defect arises from a flow localization process as do necks in
tensile modes of loading. In the present context, the primary
attribute of a shear band will be taken to be a localization of
plastic flow crossing many grains. Hence, shear bands are macroscopic internal defects which should not be confused with deformation
bands or slip bands which are microscopic regions of localized flow
occurring in the individual grains of a polycrystalline aggregate
subjected to monotonic or fatigue loads.[11]

      Shear band occurrence has been documented in a variety of bulk
forming operations.[12] In cold rolling, for instance, heavy reductions have been observed to lead to shear bands in pure aluminum,[11] aluminum alloys,[11] and low carbon steel[13]. Similarly,
Kobayashi and his coworkers[14,15] have studied the formation of
shear bands and shear fractures in aluminum alloy 7075-T6 during
the cold forging operations of indentation and lateral sidepressing.
In this work, shear bands were shown to initiate along directions
coinciding with the slip lines or velocity discontinuities of
plasticity theory.[16] Osina[17] also noted the development of
shear bands in medium carbon steels in the cold heading operation.
Shear bands have been noted to develop during hot metalworking
processes as well. For example, Semiatin and Lahoti[18-20] have
established the flow localization process that leads to shear bands
in isothermal, hot forging and isothermal, hot torsion of Ti-6Al-
2Sn-4Zr-2Mo-0.1Si.

      In all of the above metalworking operations, the occurrence of
shear bands has been attributed to a loss or exhaustion of strain-

hardening capacity of the metal being deformed (Figure 1). During cold working, this may occur because of such effects as the achievement of stable deformation textures,[11,13,21] dissolution of precipitates, or thermal softening arising from deformation heating.[12] During hot working, a flow stress maximum and exhaustion of strain hardening may occur because of these effects also as well as other phenomena such as dynamic recovery, dynamic recrystallization, and coarsening of Widmanstatten and martensitic microstructures.[22] When shear bands occur as a result of deformation heating, they are often referred to as adiabatic shear bands. A number of simple criteria have been developed to predict the strain at which the flow stress passes through a maximum under these conditions and thus the onset of the flow localization process.[12,23-27] Recently, it has been shown,[18-20] however, that a flow stress maximum is a necessary but not sufficient condition for shear band occurrence under nominally isothermal deformation conditions, or conditions in which the workpiece and tooling are at the same temperature at the beginning of the deformation. It has been shown that the workpiece must also undergo a certain degree of flow softening, or a decrease of flow stress with increasing deformation.

The effects of non-isothermal working conditions (workpiece and tooling at different temperatures) on the localization of plastic flow and the occurrence of shear bands has also recently come under investigation.[28] In these cases, the heat transfer between workpiece and forming tools may give rise to sharp temperature gradients in the workpiece which may tend to localize flow if the workpiece metal has a flow stress which is very temperature dependent.

The objectives of this paper will be to review the conditions under which flow may localize into shear bands during bulk forming processes and metalcutting operations. The applicability of the analyses to be presented will be illustrated with observarions of flow localization and shear band formation for a number of alloy systems.

SHEAR BAND FORMATION UNDER ISOTHERMAL METALWORKING CONDITIONS

By far, the largest number of observations of shear bands in worked metals has been in processes done under nominally isothermal conditions, or conditions in which the workpiece and tooling have the same initial temperature. At low strain rates, any deformation heat that is developed dissipates into the tooling and thus temperature changes in the workpiece are negligible. At high strain rates, deformation, or adiabatic, heating raises the workpiece temperature, but limited heat transfer between the workpiece and tooling occurs because of short deformation times. For purposes of discussion, therefore, flow localization and shear band formation under nominally isothermal conditions focuses for the most part on the workpiece material and its flow properties.

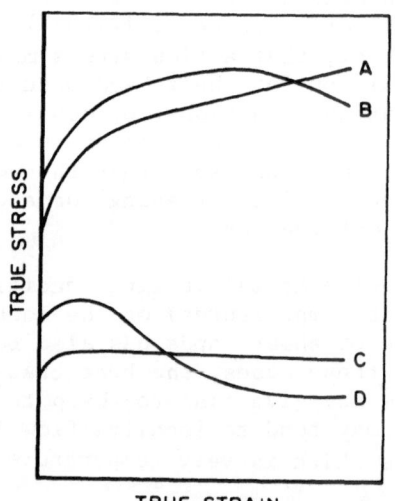

TRUE STRAIN

Figure 1.   Typical Flow Curves for Metals Deformed at Cold-Working
            Temperatures (A - Low Strain Rate, B - High Strain Rate)
            and at Hot Working Temperatures (C,D).   Strain Hardening
            Persists to Large Strains for Curve A.   The Flow Stress
            Maximum and Flow Softening in Curve B Arise from Deforma-
            tion Heating.   The Steady State Flow Stress Exhibited by
            Curve C is Typical of Metals which Dynamically Recover. [22]
            The Flow Stress Maximum and Flow Softening in Curve D
            May Result from a Number of Metallurgical Processes. [22]

Conditions for Flow Localization under Isothermal Conditions

Insight into the effect of material properties on the tendency for flow localization under isothermal conditions may be gotten by analyzing the axisymmetric problem of upsetting of a cylindrical workpiece. The analysis that follows is similar to that employed by Semiatin and Lahoti,[18] Jonas, et al.,[29] and Dadras and Thomas[30].

Under equilibrium conditions, the axial force F that is transmitted through a cylindrical workpiece under compression must be constant. Denoting the axial coordinate by x, this condition is

$$\frac{dF(x)}{dx} = 0 \quad ,$$

or simply

$$\frac{dF}{dx} = 0 \quad . \tag{1}$$

Assuming uniaxial stress conditions, at any point in the sample, the axial force is simply

$$F = \sigma A \quad ,$$

in which $A \equiv$ cross-sectional area of the sample and $\sigma \equiv$ the axial stress,* and Equation (1) becomes

$$\sigma \frac{dA}{dx} + A \frac{d\sigma}{dx} = 0 \quad . \tag{2}$$

For the time being, we shall write this equation simply as

$$\sigma dA + A d\sigma = 0 \quad , \tag{3}$$

but we should not forget that the σ and A in it are the values of the stress and cross-sectional area at particular points in the sample.

Denoting the axial strain, axial strain rate, and temperature (all field quantities whose values depend of position as well) by

---

* In this and subsequent equations, normal compressive stresses, strains, and strain rates have negative signs.

ϵ, $\dot{\epsilon}$, and T, respectively, Equation (3) becomes

$$0 = \sigma dA + A\left\{\left(\frac{\partial \sigma}{\partial \epsilon}\right)\bigg|_{\dot{\epsilon},T} d\epsilon + \left(\frac{\partial \sigma}{\partial \dot{\epsilon}}\right)\bigg|_{\epsilon,T} d\dot{\epsilon} + \left(\frac{\partial \sigma}{\partial T}\right)\bigg|_{\epsilon,\dot{\epsilon}} dT\right\} \qquad . \quad (4)$$

Assuming that the stress field may be taken to be uniaxial, the stress σ is equal to the compressive flow stress $\sigma = \sigma(\epsilon,\dot{\epsilon},T)$ of the metal and the various material property influences on flow stability may be identified through Equation (4). In upsetting, σdA or the geometric hardening term is a stabilizing influence. For the terms in the braces, the first (strain hardening) is typically stabilizing at cold working temperatures and neutral or destabilizing because of flow softening at hot working temperatures. The second term here, the strain rate hardening term, is usually stabilizing, and the third term, the temperature dependence of the flow stress, is typically destabilizing because of thermal softening. As an aside, one may also note that the form of the above equilibrium equation is identical to the maximum load condition for instability in homogeneous compression (or tension).

For homogeneous compression $dA/A = -d\epsilon$. Thus, after dividing by $A\sigma d\epsilon$ and rearrangement, Equation (4) becomes

$$0 = -1 + \left[\left\{\left(\frac{\partial \sigma}{\partial \epsilon}\right)\bigg|_{\dot{\epsilon},T} d\epsilon + \left(\frac{\partial \sigma}{\partial T}\right)\bigg|_{\epsilon,\dot{\epsilon}} dT\right\}/\sigma d\epsilon\right] + \frac{m}{\dot{\epsilon}}\frac{d\dot{\epsilon}}{d\epsilon} \qquad , \quad (5)$$

in which $m = (\partial \ln\sigma/\partial \ln\dot{\epsilon})_{\epsilon,T}$, the strain rate sensitivity. The term in brackets on the right-hand side of Equation (5) is the normalized strain-hardening (or flow softening) rate determined at constant strain rate, γ':

$$\gamma' \equiv \left(\frac{1}{\sigma}\frac{d\sigma}{d\epsilon}\right)\bigg|_{\dot{\epsilon}} = \left\{\left(\frac{\partial \sigma}{\partial \epsilon}\right)\bigg|_{\dot{\epsilon},T} d\epsilon + \left(\frac{\partial \sigma}{\partial T}\right)\bigg|_{\epsilon,\dot{\epsilon}} dT\right\}/\sigma d\epsilon \qquad . \quad (6)$$

Note that γ' includes the effects of deformation heating on flow curves determined at constant strain rate. With this definition, Equation (5) becomes

$$\alpha \equiv -\frac{1}{\dot{\epsilon}}\frac{d\dot{\epsilon}}{d\epsilon} = \frac{\gamma' - 1}{m} \qquad . \quad (7)$$

The factor α, commonly called the flow localization parameter, is used as a measure of a material's tendency to form catastrophic or marked flow localizations under conditions of force equilibrium. Note that it is a function solely of the material properties, γ' and m, either or both of which may be functions of ϵ. Thus, to determine the propensity to form flow localizations, flow curves from constant rate compression tests and estimates of m (from "step-strain rate change" tests, for example) are required, and α should be examined as a function of strain. By definition, the onset of flow localization corresponds to $\alpha = 0$. This corresponds to materials which exhibit $\gamma' = 1$ or a fairly large flow softening rate (since σ and ε are both negative). In actuality, Jonas, et al.[29], have suggested and Semiatin and Lahoti[18,20] have verified that materials will show noticable flow localization when $\alpha \geq 5$, approximately. In uniaxial compression, such flow localization ordinarily corresponds to local bulges. Bulging has been observed in upset tests on both uranium and the titanium alloy Ti-6Al-2Sn-4Zr-2Mo-0.1Si (Ti-6242Si).[22,31]

Another mode of flow localization is possible though. This is shear band formation. Since shear bands are planar or two-dimensional defects, gross shear bands are usually not observed in axisymmetric deformations such as upsetting unless the bulging deformation (or perhaps friction) can cause a shift in strain state. Such shifts have indeed been observed for Ti-6242Si.[31] Other examples will be cited in a subsequent section.

In contrast to axixymmetric deformation modes, in plane strain compressive modes of deformation (e.g., lateral sidepressing, rolling) shear band formation is favored over the bulging mode of flow. This may be deduced as follows. First, since shear bands involve pure shearing deformation, the variation with position of shear stress along the directions (or, more precisely, the planes) of pure shear should be examined. In plane strain, these directions are the slip line directions. Denoting the net force on these planes by F_τ, the shear stress on these planes by τ, the displacement coordinate perpendicular to these planes by x' (which lies parallel to the conjugate set of slip line directions), and the cross-sectional area of these planes by A_τ, equilibrium requires that

$$\frac{dF_\tau}{dx'} = 0 \quad , \tag{8}$$

or

$$A_\tau \frac{d\tau}{dx'} + \tau \frac{dA_\tau}{dx'} = 0 \quad . \tag{9}$$

Since the slip line directions are directions of zero extension, dA_τ/dx' is identically equal to zero. Thus, flow localization tendencies may be determined by examining conditions under which $d\tau/dx'$ is equal to zero. If we assume $\tau = \tau(\Gamma, \dot{\Gamma}, T)$, where Γ and $\dot{\Gamma}$ are shear strain and shear strain rate, it is easy to show in a manner analogous to that used to derive Equation (7) that the rate at which shear strain localizes under these conditions is α_τ defined by

$$\frac{1}{\dot{\Gamma}}\frac{d\dot{\Gamma}}{d\Gamma} = - \frac{\frac{1}{\tau}\left(\frac{\partial\tau}{\partial\Gamma}\right)\Big|_{\dot{\Gamma}}}{(\partial\ln\tau/\partial\ln\dot{\Gamma})\big|_{\Gamma,T}} \quad . \tag{10}$$

Assuming that the von Mises criterion holds so that $\tau = \bar{\sigma}/\sqrt{3}$ $\Gamma = \sqrt{3}\,\bar{\epsilon}$, and $\dot{\Gamma} = \sqrt{3}\,\dot{\bar{\epsilon}}$, where $\bar{\sigma} = \bar{\sigma}(\bar{\epsilon},\dot{\bar{\epsilon}},T)$ is the effective stress-strain curve, Equation (10) becomes

$$\alpha = \frac{1}{\dot{\bar{\epsilon}}}\frac{d\dot{\bar{\epsilon}}}{d\bar{\epsilon}} = - \frac{\frac{1}{\bar{\sigma}}\left(\frac{\partial\bar{\sigma}}{\partial\bar{\epsilon}}\right)\Big|_{\dot{\bar{\epsilon}}}}{m} = \frac{\gamma'}{m} \quad , \tag{11}$$

where γ' and m have the same definitions as before, i.e., the normalized flow softening rate in compression and the strain rate sensitivity parameter. This equation tells us that the rate at which effective strain rate concentrations may develop in plane strain is proportional to the ratio of the normalized flow softening rate to the rate sensitivity. When it is compared to Equation (7), it is seen that the strain localization tendencies in plane strain are greater for shear band formation than they are for bulging. Thus, it is not surprising that shear band development is most common in plane strain, and that the best idea of the effect of material properties on shear band occurrence in metalworking is obtained from deformation studies done under plane strain conditions.

Shear Band Occurrence in Isothermal, Hot Forging of Titanium Alloys

The correlation between α (Equation (11)) and the occurrence of shear bands may be illustrated by examining data on the hot forging behavior of Ti-6Al-2Sn-4Zr-2Mo-0.1Si (Ti-6242Si) and Ti-10V-2Fe-3Al (Ti-10-2-3) under plane strain conditions. The Ti-6242Si alloy, whose beta transus temperature is 988 C (1810 F), is an α/β alloy that is widely used in jet engines. A newer titanium alloy, Ti-10-2-3 is a near β alloy which is finding increasing application in structural applications which require a good blend of strength and fracture toughness. Its beta transus temperature is 799 C (1470 F).

Flow curves for Ti-6242Si at hot working temperatures show a range of behaviors which are determined largely by starting microstructure.[18] Most of the different possibilities are illustrated with data from isothermal, hot compression tests run at a constant strain rate of 10 sec.$^{-1}$ (Figure 2). When the alloy has a starting microstructure of equiaxed alpha in a transformed beta matrix ("$\alpha+\beta$" microstructure), the flow curves show moderate to small amounts of flow softening ("small" γ''s) following the flow stress

maxima. These levels of softening can be attributed to deformation
heating and the general dependence of flow stress on temperature.[31]
On the other hand, when the starting microstructure is an acicular,
or Widmanstatten, alpha one("transformed β", or simply, "β" microstructure), much more marked flow softening ("large" γ''s) is observed. A small portion of the softening in these cases arises from
deformation heating. Most of it, however, comes from changes in the
morphology of the alpha and beta phases during a process which is

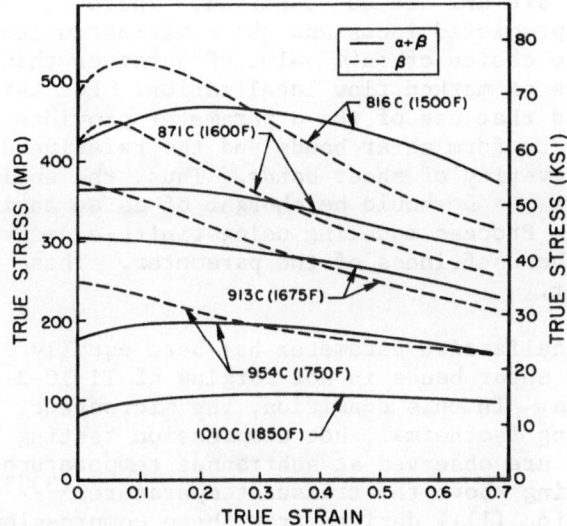

Figure 2.   True Stress-Strain Curves for Ti-6242Si of α+β and β
            Microstructures Determined at a Constant True Strain
            Rate = 10 Sec.$^{-1}$.

similar to dynamic spheroidization in eutectoid steels.[32]   The m
values, or strain rate sensitivity parameters, for the two microstructures are similar and fall generally in the range of 0.1 to 0.3
with the general trend being that rate sensitivity increases with decreasing strain rate and increasing hot working temperature.

      As can be expected, forging temperature and strain rate through
their effect on material properties and thus α (Equation (11)) play a

strong role in determining the occurrence of shear bands. Observations of shear bands in plane strain lateral sidepressing of Ti-6242Si
cylinders (Figures 3,4) show a strong correlation between the occurrence of shear bands and the α parameters. Such observations are
best summarized in defect maps, or workability diagrams, similar to
those used in selecting working conditions to avoid grain boundary
and triple-point cracking during hot working.[7] A diagram of this
type for Ti-6242Si is shown in Figure 5. The curves in these diagrams are the loci of strain rates and temperatures at which the maximum value of α over the strain interval of 0 to 0.7 is equal to 5,
and have been derived from the flow stress data determined in isothermal, constant rate compression tests.

In general, the defect map loci for α = 5 separate regions in
which shear bands are and are not observed. There is slight disagreement between the predicted locus and the β microstructure data, but
this may be due to choice of an α value of 5 rather than 4 perhaps
for the occurrence of marked flow localization. For this reason, it
should be realized that use of the α parameter provides an insight
into the tendency to form shear bands and the relative degree of
localization or severity of shear bands. Thus, the application of a
criterion based on α = 5 should be thought of as an engineering tool
or rule of thumb. Process modeling using finite-element mehtods has
shown, however, the usefulness of the parameter. These results will
be discussed later.

The flow localization parameter has been equally successful in
the prediction of shear bands in hot forging of Ti-10-2-3 in the beta
annealed condition. In this condition, the microstructure is totally
beta phase. During isothermal, hot compression testing, large amounts
of flow softening are observed at subtransus temperatures and negligible flow softening above the transus temperature.[33,34] The α
parameters (Equation (11)) derived from these compression data have
also been found to correlate well with the occurrence of shear bands
during the isothermal, hot forging operation of plane-strain lateral
sidepressing (Figure 6).

Shear Band Occurrence in Isothermal Deformation of Other Alloys

In order to demonstrate further the success of this phenomen-
ological description of the effect of material properties on flow
localization and the occurrence of shear bands under isothermal conditions, observations for several other alloy systems will be cited.

No discussion of flow localization could be complete without
some mention of shear banding in ferritic steels. In an extensive
investigation of the flow properties of several steels, Doraivelu[35]
noticed shear bands at certain temperatures and strain rates in
uniaxial compression for one of the steels. For example, he found
shear bands in upset samples of an alloy similar to H26 deformed

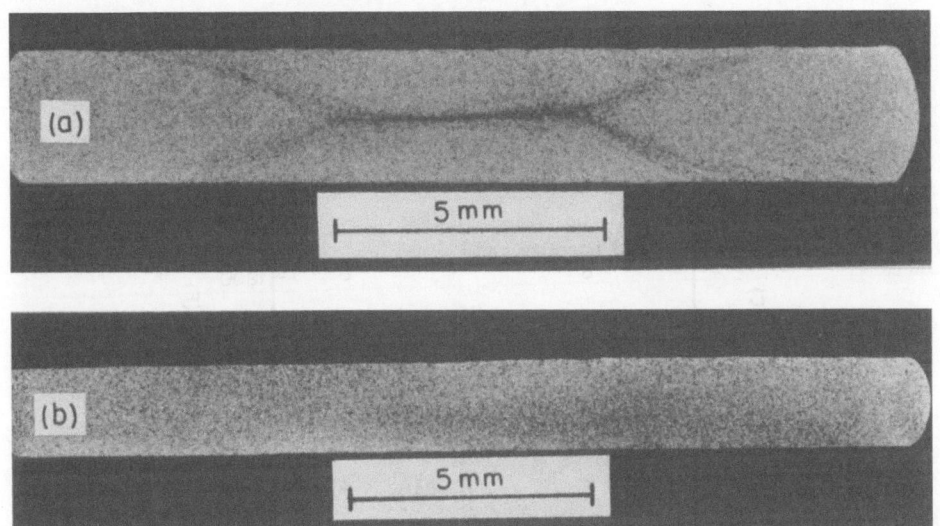

Figure 3.   Transverse Sections of Isothermally Sidepressed Cylinders
            of $\alpha+\beta$ Microstructure Ti-6242Si: (a) Sidepressed at
            843 C (1550 F), $\alpha_{max} \approx 7.0$, (b) Sidepressed at 913 C
            (1675 F), $\alpha_{max} \approx 3.5$. $\dot{\varepsilon} \approx 10$ Sec.$^{-1}$.

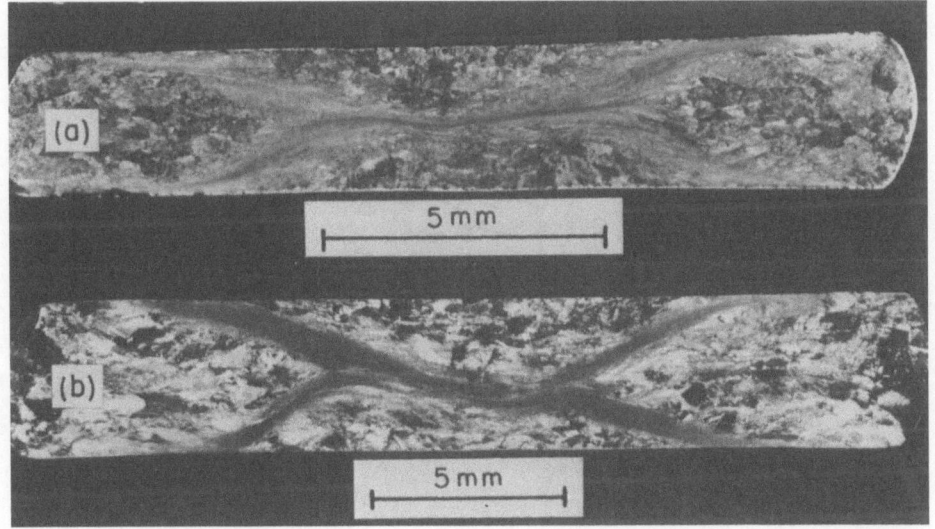

Figure 4.   Transverse Sections of Isothermally Sidepressed Cylinders
            of β Microstructure Ti-6242Si: (a) Sidepressed at 843 C
            (1550 F), $\alpha_{max} \approx 11$, (b) Sidepressed at 913 C (1675 F),
            $\alpha_{max} \approx 5$. $\dot{\varepsilon} \approx 10$ Sec.$^{-1}$.

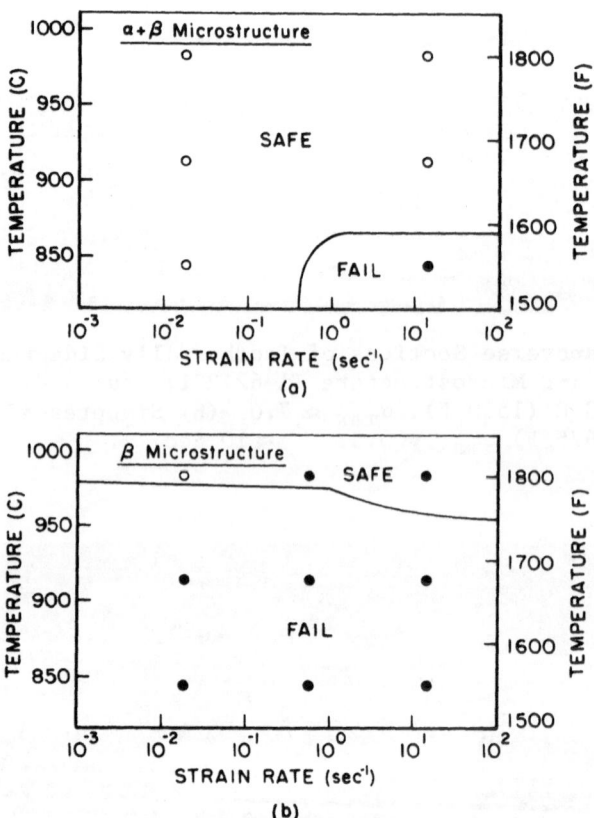

Figure 5.   Workability Maps for Occurrence of Shear Bands in Hot
            Forging of Ti-6242Si with (a) α+β Microstructure and
            (b) β Microstructure.   Workability Predictions (—) and
            Forging Conditions in Which Shear Bands Were (●) and
            Were Not (o) Observed are Noted.

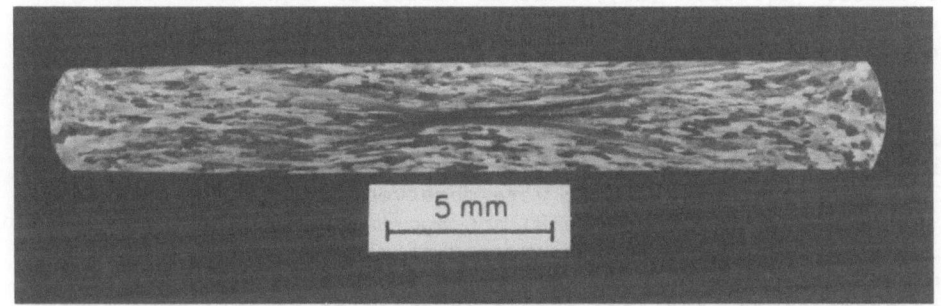

Figure 6.   Transverse Section of Cylinder of Beta Annealed Ti-10-2-3
             Isothermally Sidepressed at 704 C (1300 F), $\dot{\varepsilon} \approx 10$ Sec.$^{-1}$,
             $\alpha_{max} \approx 7.7$.   Specimen Sidepressed at 816 C (1500 F),
             $\dot{\varepsilon} \approx 10$ Sec.$^{-1}$, for which $\alpha_{max} \approx 1.8$, Exhibited no Shear
             Bands.

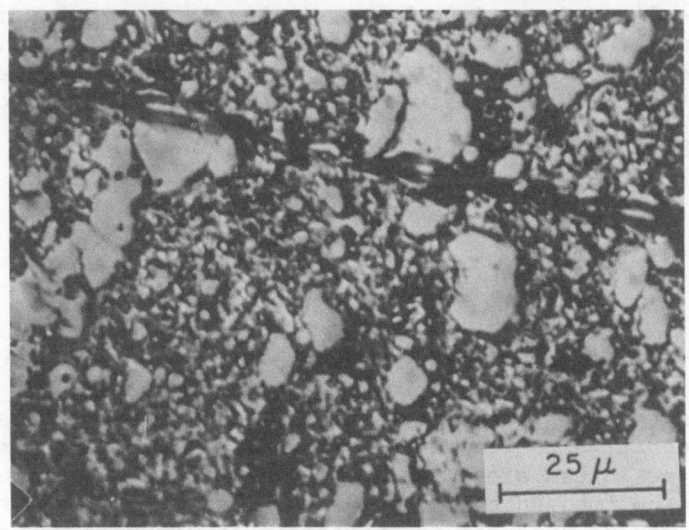

Figure 7.   Micrograph off an Axial Section of an H26 Upset Specimen
             (Deformed in a Drop Hammer at 100 C (212 F)) Which Ex-
             hibits Shear Bands. $\dot{\varepsilon} \approx 470$ Sec.$^{-1}$.

at 100 C (212 F) and a strain rate of 470 S^{-1} (Figure 7). If one
assumes that some effect arising from friction or bulging led to a
change in strain state to plane strain, one can calculate that the
α parameter (Equation (11)) reached values as high as 16 during the
upsetting operation (Table 1), and it is not surprising that shear
bands were observed.

In two other non-ferrous alloys, U-0.75Ti and JBK-75 (an alloy
similar to A286), shear band observations were found to correlate
with large degrees of flow softening and thus high α values as well
(Table 1). These shear bands were observed in isothermal upsetting
of the U-0.75Ti alloy (Figure 8)[36] and isothermal, hot sidepressing
of JBK-75 (Figure 9). The occurrence of shear bands in U-0.75Ti is
important with regard to its use as a material for kinetic energy
armor penetrators, and care should be taken to control processing
conditions which produce products not prone to formation of such
defects and eventual failure.

Shear Band Occurrence under Adiabatic Conditions

For yet other alloys for which no shear band observations have
been made, the propensity for flow localization can be gaged from
calculations based on published material property data. These
calculations have been performed for room temperature deformation
at $\dot{\varepsilon}$ = 10 sec.$^{-1}$ to determine the strain levels at which adiabatic
shear bands may initiate because of thermal softening and flow
then localize. According to Staker[26], the critical compressive
strain ε_c at which the flow stress attains a maximum in a power-

law hardening material is given by*:

$$\varepsilon_c = - \frac{\rho c n}{0.95 \left(\frac{\partial \sigma}{\partial T} \right)_{\varepsilon, \dot{\varepsilon}}} \tag{12}$$

In Equation (12), ρ is the density of the metal, c its specific
heat, and n the strain hardening exponent. It has been assumed
that 95 percent of the deformation work is transformed into heat.
Table 2 shows that for ductile (large n) metals whose flow stress
is not too sensitive to temperature (e.g., normalized 1006 steel,
annealed 304 stainless steel, and 6061-0 aluminum), large amounts
of strain may be imposed before shear bands or other modes of flow
localization may be expected to initiate. In contrast, quenched

---

* Since σ is negative for compressive conditions, $(\partial \sigma / \partial T)$ is
  positive and ε_c is negative.

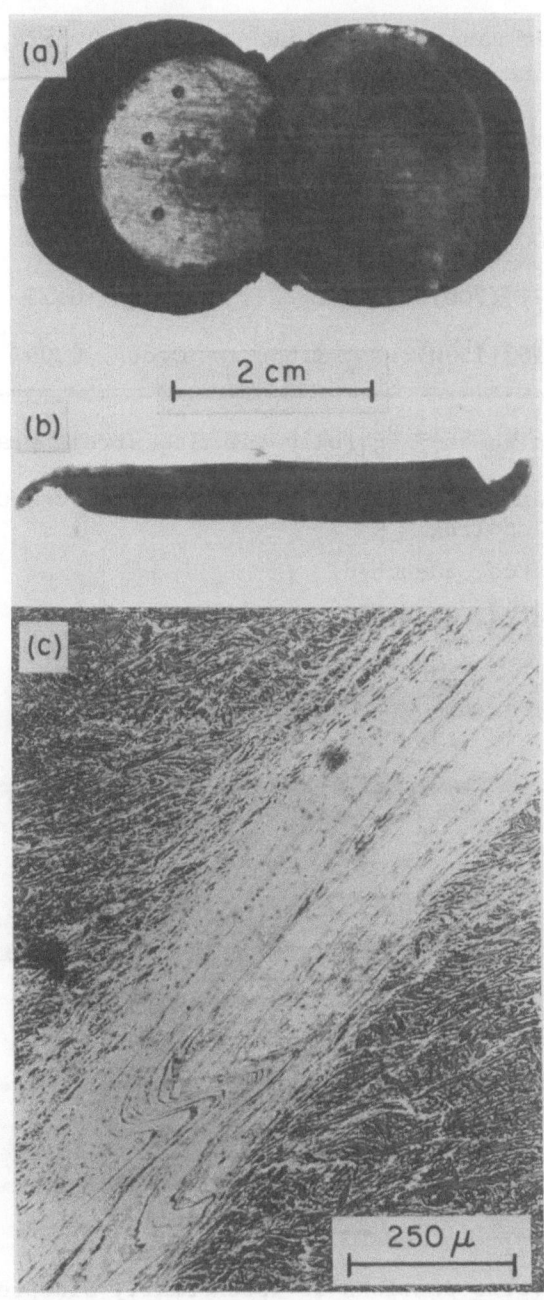

Figure 8.   (a) Top View, (b) Side View, and (c) Micrograph off an
Axial Section of a Solution Treated U-0.75Ti Upset
Specimen, Which Underwent Shear Localization During
Testing at 371 C (700 F), $\dot{\varepsilon} \approx 1$ Sec.$^{-1}$.

Table 1.   Shear Band Observations In H26, U-0.75Ti, and JBK-75

Alloy	Temperature C (F)	Strain Rate Sec.$^{-1}$	γ'_{max}*	m	$\alpha_{max} = \dfrac{\gamma'_{max}}{m}$
H26[a]	100 (212)	470	0.80	0.05	16
U-0.75Ti[b]	371 (700)	1	1.30	0.15	8.7
JBK-75[c]	816 (1500)	2.5	0.86	0.05	17.2

* γ'_{max} = max value of $\left|\dfrac{1}{\sigma}\dfrac{d\sigma}{d\varepsilon}\right|$ following flow stress peak

[a] Quenched & Tempered followed by annealing at 840 C (1545 F) plus furnace cooling

[b] Solution treated, quenched

[c] Solution treated, quenched

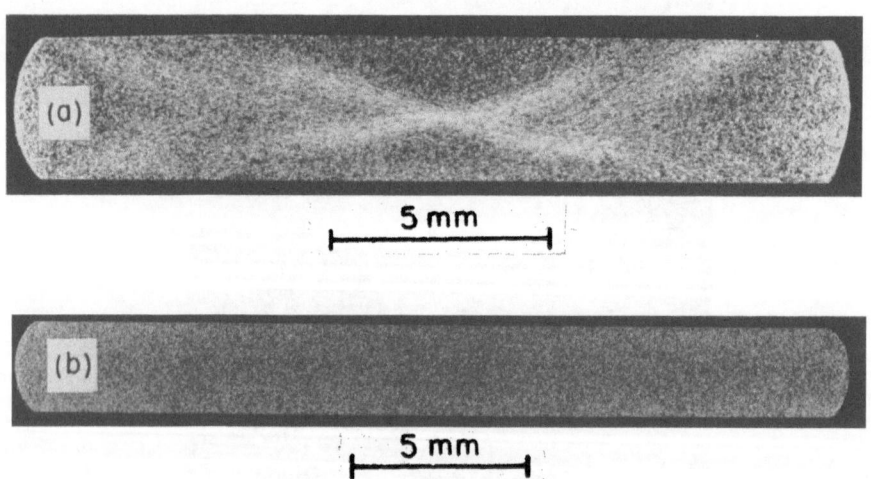

Figure 9.   Transverse Sections of Isothermally Sidepressed Cylinders of Solution Treated JBK-75 Alloy:
(a) Sidepressed at 816°C (1500°F), α_{max} = 17.2
(b) Sidepressed at 982°C (1800°F), α_{max} = 3.5
$\dot{\varepsilon} \approx 2.5$ Sec.$^{-1}$.

Table 2. Flow Localization Tendencies of Various Alloys under Adiabatic Shear Band Conditions*

Alloy (Condition)	$\rho\left(\dfrac{gr.}{cc}\right)$	$c\left(\dfrac{cal.}{gr.C}\right)$	$n^{(a)}$	$\dfrac{\partial\sigma}{\partial T}^{(a)}$ (kPa/C)	$m^{(a)}$	$\lvert\dot{\varepsilon}_c\rvert$	$\dfrac{\varepsilon(\alpha=5)}{\varepsilon_c}$	$\dfrac{\varepsilon(\alpha=10)}{\varepsilon_c}$
1006 Steel (normalized)	7.86	0.11	0.24	625	0.01	1.45	1.43	2.53
1043 Steel (normalized)	7.83	0.11	0.12	750	0.01	0.61	1.34	2.03
4340 Steel (Q & T, R_c = 26)	7.83	0.11	0.092‡	1075‡	0.01	0.33	1.22	1.56
4340 Steel (Q & T, R_c = 39)	7.83	0.11	0.055‡	1925‡	0.01	0.11	1.11	1.25
4340 Steel (Q & T, R_c = 52)	7.83	0.11	0.043‡	3100‡	0.01	0.053	1.07	1.14
304 Stainless Steel (annealed)	8.03	0.09	0.50	1975	0.02	0.80	1.19	1.47
316 Stainless Steel (annealed)	8.03	0.11	0.35	2975	0.03	0.46	1.25	1.65
6061-0 Aluminum	2.71	0.23	0.24	500	0.002	1.33	1.06	1.12
Pure Titanium (mill annealed)	4.51	0.125	0.30	1250	0.025	0.60	1.33	2.00
Ti-6Al-4V (α/β Processed)	4.46	0.13	0.02	2300	0.015	0.022	1.09	1.32

* Mechanical Property Data from References 26,37.

‡ Mechanical Property Data from Quasistatic Tests.

(a) Data for Room Temperature Deformation at $\dot{\varepsilon}$ = 10 Sec.$^{-1}$.

and tempered steels such as 4340 can be deformed only small amounts
before flow starts to localize. This is particularly striking in
conditions with the highest hardnesses.

It is also instructive to determine the amount of quasi-stable
post-uniform deformation beyond ε_c that can be expected. This may
be done by examing the compression flow curves to determine the
strains at which the α parameter reaches values of say 5 or 10.
For a power-hardening material, γ' is given by

$$\gamma' = \left(\frac{1}{\sigma}\frac{\partial\sigma}{\partial\varepsilon}\right)\Bigg|_{\dot{\varepsilon}} = \left(\frac{\partial\ln\sigma}{\partial\varepsilon}\right)\Bigg|_{\dot{\varepsilon},T} + \left(\frac{\partial\sigma}{\partial T}\right)\Bigg|_{\dot{\varepsilon},\varepsilon}\frac{dT}{\sigma d\varepsilon} \quad ,$$

or,

$$\gamma' = \frac{n}{\varepsilon} + \frac{0.95}{\rho c}\frac{\partial\sigma}{\partial T} \quad .$$

Using Equation (12), this may be rewritten

$$\gamma' = \frac{n}{\varepsilon} - \frac{n}{\varepsilon_c} \tag{13}$$

and

$$\alpha = \frac{\gamma'}{m} = \left(\frac{n}{\varepsilon} - \frac{n}{\varepsilon_c}\right)/m \quad . \tag{14}$$

Solving Equation (14) for $\varepsilon(\alpha)$ and dividing by ε_c, the amount of
quasi-stable deformation relative to ε_c is given by

$$\frac{\varepsilon}{\varepsilon_c} = \left(1 - \frac{\alpha m|\varepsilon_c|}{n}\right)^{-1} \quad . \tag{15}$$

It appears that large m and large ε_c tend to increase the amount of
quasi-stable deformation, and large n tends to decrease it. This
result is plausible for m, but not for n. This is because n affects
strain hardening as well as the amount of deformation heating. A
better way of looking at the influence of n and ε_c is through
Equation (12) which demonstrates that increasing $\frac{|\varepsilon_c|}{n}$ is equivalent
to decreasing flow stress dependence on temperature, which should
tend to extend quasi-stable deformation. This trend is that predicted by Equation (15).

Estimates of $\varepsilon/\varepsilon_c$ for various materials are also given in
Table 2. The estimates illustrate the above result. For example,
1006 (with small $\partial\sigma/\partial T$) can be expected to undergo large quasi-
stable deformations prior to noticable flow localization (at $\alpha \approx 5$),
whereas the quenched and tempered 4340 steel (large $\partial\sigma/\partial T$) can be
expected to form shear bands almost immediately following ε_c.

Computer Simulation of Shear Band Occurrence under Isothermal Conditions

In order to obtain a comprehensive idea of the process by
which flow localizes and shear bands form during isothermal deformation, as well as the effect of material properties on the phenomenon, a series of computer simulations of the lateral sidepressing
operation were performed. This was done using an advanced finite
element code called ALPID (Analysis of Large Plastic Incremental
Deformation) developed by Oh.[38] Input material properties to
the program included hypothetical stress-strain curves (Figure 10)
(which, it was assumed, implicitly include the effects, if any, of
deformation heating) as well as various m values (0.0, 0.125, and
0.30). The flow curves exhibited different γ'''s (0.0 or 1.25) at
ε's > 0.10. Since materials whose flow curves show monotonically
increasing flow stress do not exhibit flow localization under
typical conditions, flow curves without maxima were not included
in this study. With the material properties assumed, however,
α's (Equation (11)) between 0 and ∞ were studied, however.

Concentrations of effective strain rate predicted from the
various simulations (Figure 11) indicate a trend which is qualitatively quite similar to that which would be expected based on
consideration of the α parameters alone. (Iso-strain rate contours
for only one-fourth of the sidepressing are shown in Figure 11 because of symmetry considerations.) For the two cases for which α
is less than 5 (Case I, $\alpha = 0$ and Case II, $\alpha = 4.2$) deformation is
relatively uniform, with only a minor flow localization observed
for Case II, $\alpha = 4.2$. Similarly, the flow pattern for Case III,
in which α is undefined, is only moderately nonuniform. In this
case, $\gamma' = 0$, $m = 0$ is equivalent to a rigid, perfectly plastic
material, or the kind of material postulated in slip-line field
theory. Thus, it is not surprising that the solution for Case III
is very similar to slip-line field solutions for various height-
to-width ratios proposed by Kobayashi, et al.[14] For Cases IV
and V ($\alpha = 10$ and $\alpha = \infty$, respectively), marked flow localization
may be noticed. For both of these cases (as well as Case II), the
direction of flow localization forms an angle of 45° with the
ordinate at low reductions. At higher reductions, the direction
of the flow localization rotates away from the ordinate, in agreement with experimental observations (Figures 3,4,6,9; Reference 20).

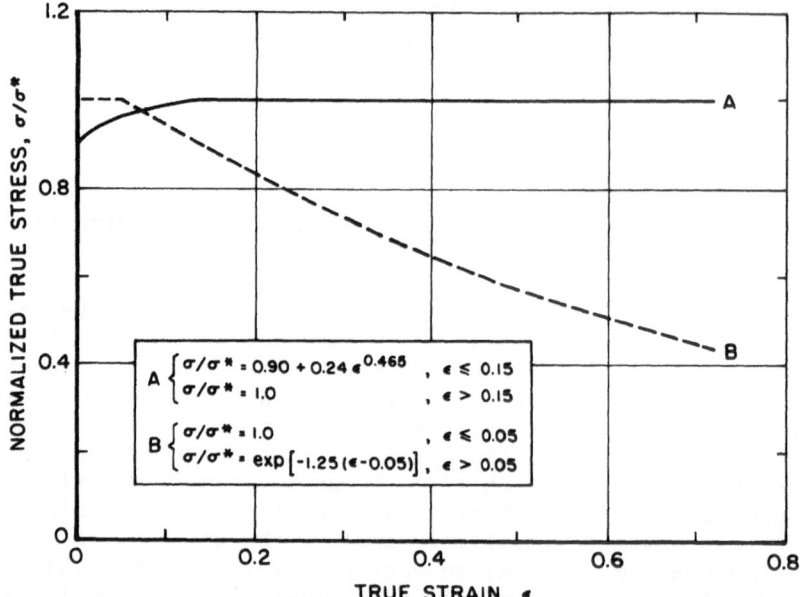

Figure 10. Hypothetical Flow Curves for Metals which Show
            Extremes of No Flow Softening (A) and Large Amount
            of Flow Softening (B).

Another interesting observation is the tendency of the flow localizations to form "flats" near the center of the specimen, much like
those actually observed.

      The simulation results allow a more complete evaluation of the
accuracy of the α-parameter method of predicting shear bands in
isothermal forging. As shown by the simulation, intense shear
bands should definitely be expected when α is equal to 10 or more.
Furthermore, modest localizations may be expected for α's between
4 and 10. Thus, the use of α = 5 as a criterion for shear band
formation should be thought of as an approximation. For the most
part, it indicates the degree of tendency towards flow localization
and not the occurrence per se. Another feature illustrated by
process simulation results is the fact that flow localization is a
process and not an event. Strain and strain rate concentrations
do not occur instantaneously. This conclusion is very similar to
that regarding the necking-controlled failure process in sheet
forming[4] in which flow localizes gradually and not at the point
of instability.

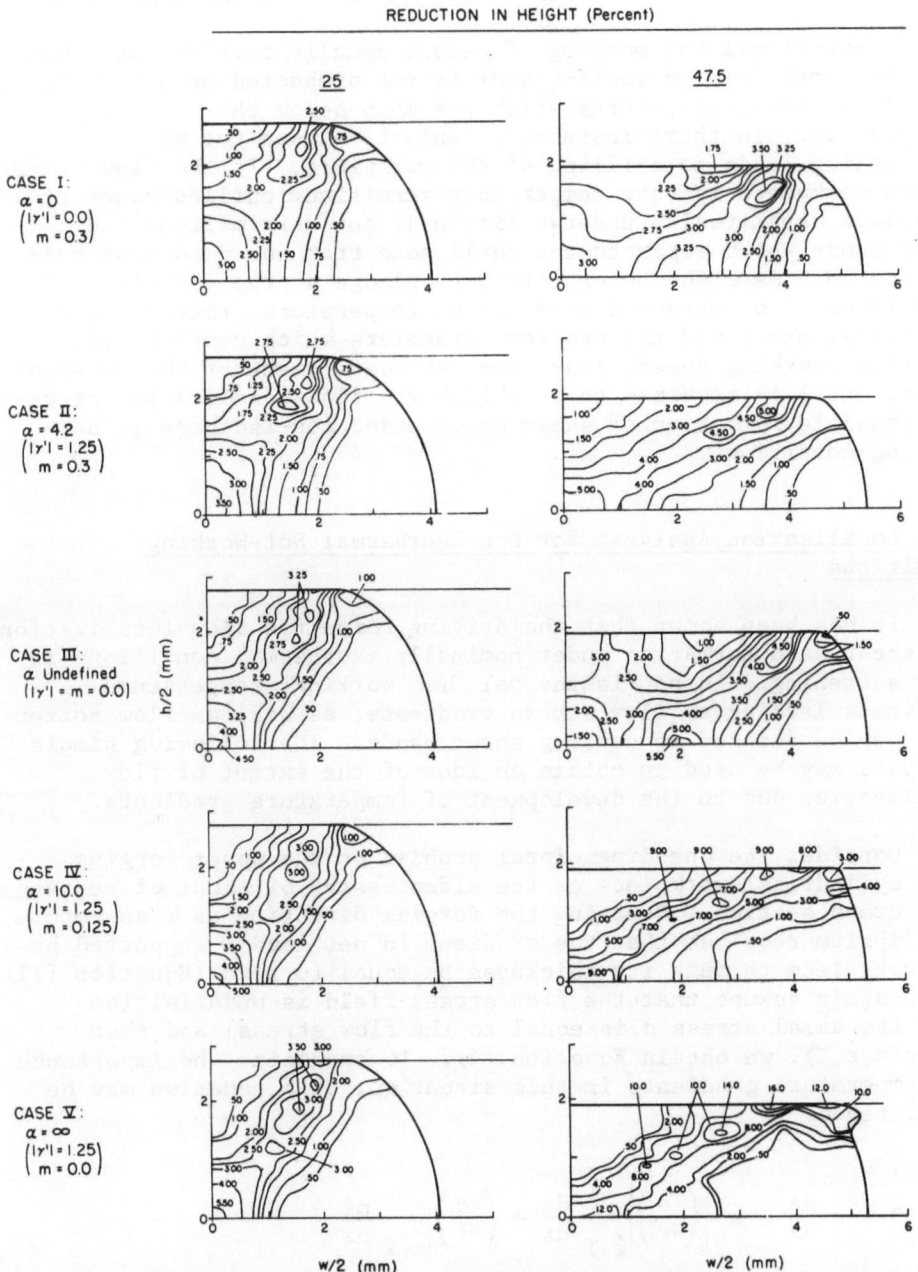

Figure 11. Predicted Effective Strain Rate Contours as a Function
            of Reduction for Metals with Varying Degrees of Flow
            Softening Rate γ' (Figure 10) and Strain Rate Sensitivity
            m. Strain Rates are in Units of Sec.$^{-1}$.

SHEAR BAND FORMATION UNDER NON-ISOTHERMAL HOT-WORKING CONDITIONS

 Conventional hot working of metals usually involves deformation of a hot workpiece by tooling that is not preheated or heated to relatively low temperatures which are much below the workpiece temperature. In these instances, contact between the workpiece and tooling leads to chilling of the workpiece. If the flow stress of the workpiece is very temperature sensitive, chilled zones of workpiece metal (which undergo little if any deformation) and shear bands which separate the chill zone from the deforming bulk are formed. As might be expected, knowledge of the workpiece properties (flow stress dependence on temperature, thermal conductivity, etc.) and the process parameters which control heat transfer (working speed, heat transfer coefficient of the lubricant layer, etc.) is required to establish the flow localization process and possible formation of shear bands under non-isothermal, hot-working conditions.

Flow Localization Analysis for Non-Isothermal Hot-Working Conditions

 It has been shown that the driving force for flow localization and shear band formation under nominally isothermal conditions is flow softening. In non-isothermal hot working, temperature gradients leading to flow stress gradients, as well as flow softening, may be deduced as causing shear bands. The following simple analysis may be used to obtain an idea of the extent of flow localization due to the development of temperature gradients.

 Consider the one-dimensional problem of the upset forging of a cylindrical workpiece or the sidepressing of a bar of rectangular cross section. Denoting the forging direction as x as before, equilibrium requires that the gradient in net load F supported by the workpiece through its thickness be equal to zero (Equation (1)). If we again assume that the flow stress field is uniaxial (so that the axial stress σ is equal to the flow stress) and that $\sigma = \sigma(\varepsilon, \dot{\varepsilon}, T)$, we obtain Equation (4). To emphasize the importance of temperature gradients in this situation, this equation may be rewritten as

$$0 = \sigma \frac{dA}{dx} + A \left\{ \left(\frac{\partial \sigma}{\partial \varepsilon} \right) \bigg|_{\dot{\varepsilon}, T} \frac{d\varepsilon}{dx} + \left(\frac{\partial \sigma}{\partial \dot{\varepsilon}} \right) \bigg|_{\varepsilon, T} \frac{d\dot{\varepsilon}}{dx} \right.$$

$$\left. + \left(\frac{\partial \sigma}{\partial T} \right) \bigg|_{\varepsilon, \dot{\varepsilon}} \frac{dT}{dx} \right\} .$$

(16)

If we restrict our attention to the early stages of flow localization in which $dA/dx \approx 0$, Equation (16) reduces to

$$0 = \left(\frac{\partial \sigma}{\partial \epsilon}\right)\bigg|_{\dot{\epsilon},T} \frac{d\epsilon}{dx} + \left(\frac{\partial \sigma}{\partial \dot{\epsilon}}\right)\bigg|_{\epsilon,T} \frac{d\dot{\epsilon}}{dx} + \left(\frac{\partial \sigma}{\partial T}\right)\bigg|_{\epsilon,\dot{\epsilon}} \frac{dT}{dx} \quad . \tag{17}$$

It is seen that strain and strain rate gradients are controlled
by material properties (strain hardening, or flow softening, strain-
rate hardening, and flow stress dependence on temperature) and
temperature gradients which are established by a variety of material
properties and process variables. Strain hardening and strain rate
hardening tend to minimize flow localization under conditions involving temperature gradients. On the other hand, flow softening
and negative dependence of flow stress on temperature both tend
to act in concert with temperature gradients to produce flow
localization.

In order to separate the effects of thermal gradients and
flow softening on flow localization, let us assume that the flow
stress is not a function of strain at fixed strain rate and temperature. Many metals behave this way at large strains during hot
working. With this assumption, Equation (17) becomes after
rearrangement

$$\frac{d\dot{\epsilon}/\dot{\epsilon}}{dx} = - \frac{\frac{1}{\sigma}\left(\frac{\partial \sigma}{\partial T}\right)\bigg|_{\epsilon,\dot{\epsilon}}}{m} \frac{dT}{dx} \equiv \eta \frac{dT}{dx} \quad . \tag{18}$$

One can note the marked similarity between Equation (11) and
Equation (18). In the former, the "destabilizing" influence is
flow softening, whereas, for the latter, it is a thermal gradient
in conjunction with a negative dependence of flow stress on temperature. In addition, in both instances large strain-rate
sensitivity tends to act as a stabilizing influence.

To get a handle on the temperature gradients that can be expected in upsetting of cylinders or sidepressing of rectangular bars,
the classical solution for one dimensional heat conduction between
two semi-infinite bodies may be employed. If we assume perfect
heat transfer between the two bodies, one at an in initial temperature T_D and one at an initial temperature T_S, the temperature of the
body whose initial temperature is T_S as a function of x and time t
($T(x,t)$) is the following: (39)

$$\frac{T(x,t) - T_S}{T_D - T_S} = \frac{\sqrt{\rho_D c_D k_D}}{\sqrt{\rho_D c_D k_D} + \sqrt{\rho_S c_S k_S}} \cdot \text{erfc}\left(x/2\sqrt{k_S t/\rho_S c_S}\right) . \tag{19}$$

In this expression, ρ, c, and k are density, specific heat, and thermal conductivity, respectively, and the subscripts D and S refer to the two bodies. Taking the partial derivative with respect to x of Equation (19), we obtain

$$\frac{\partial T}{\partial x} = \frac{\left(T_S - T_D\right)\sqrt{\rho_D c_D k_D}}{\left(\sqrt{\rho_D c_D k_D} + \sqrt{\rho_S c_S k_S}\right)\left(\sqrt{\pi k_S / \rho_S c_S}\right)} \cdot \exp\left(-\rho_S c_S x^2 / 4 k_S t\right) \quad . (20)$$

Identifying S with the workpiece in the forming operation and D with the tooling, Equation (20) in conjunction with Equation (18) may be used to get an idea of strain-rate gradients due to chilling if one nelgects heat generation due to deformation. Also, it should be kept in mind that Equation (20) is only valid for times in which the chilling on one end of the workpiece is not "seen" (i.e., influenced) by that on the other end.

Flow Localization During Non-Isothermal Forging of Ti-6242Si

The application of Equations (18) and (20) will be illustrated by examining the non-isothermal compression deformation of Ti-6242Si alloy. For the moment let us focus our attention on non-isothermal upsetting of $\alpha+\beta$ microstructure Ti-6242Si cylindrical specimens. In particular, let us examine the deformation of 1.02 cm (0.40 in.) diameter X 1.52 cm (0.60 in.) high cylinders of this material which have been preheated to 954 C (1750 F), and upset between tool steel dies heated to 191 C (375 F) in a mechanical press running at a nominal strain rate of 10 sec.$^{-1}$. The observed flow pattern for a fifty percent reduction under such conditions exhibits chill zones of limited deformation (Figure 12a). Shear bands separating the chill zones from the deformed bulk may be noted in optical micrographs of specimens which have undergone this deformation.[28]

To obtain a liberal estimate of the temperature and hence strain rate gradients for this deformation it will be assumed that the deformation and heat transfer occur sequentially. That is to say, it shall be assumed that the deformation is imposed during some time Δt, and this is followed then by heat transfer (without deformation) for an identical time increment Δt as well. In a mechanical press, the strain rate drops off during the last third of an upset test.[40] The time required for a 50 percent reduction ($\varepsilon = 0.70$) is therefore approximately $\Delta t = 0.09$ secs. Using this for the time t in Equation (20), an upper limit on the thermal gradients at this reduction may be predicted using the physical properties listed in Table 3. The predicted thermal gradients using these physical properties are given in Table 4.

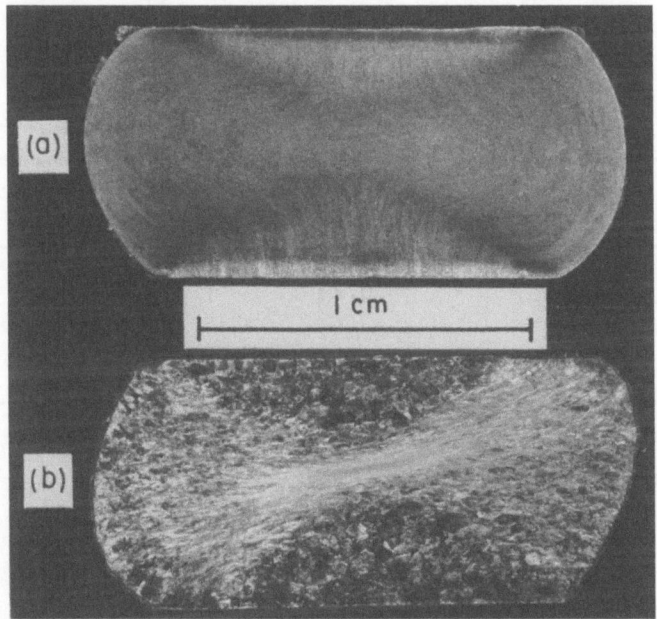

Figure 12.   Axial Cross-Sections of Lubricated Compression Specimens
             of (a) α+β Microstructure Ti-6242Si and (b) β Micro-
             structure Ti-6242Si Non-Isothermally Deformed to 50
             Percent Reduction in Height in a Mechanical Press
             ($\dot{\varepsilon} \approx 10$ Sec.$^{-1}$). Specimen Preheat Temperature 954 C
             (1750 F), Die Temperature 191 C (375 F), Dwell Time
             0 Secs.

     The other information needed to apply Equation (18) is the
value of η, or the ratio of the normalized thermal softening rate
to the strain rate sensitivity. The Ti-6242Si alloy shows a very
strong dependence of flow stress on temperature. Plots of flow
stress versus temperature have been obtained from isothermal, hot
compression data (such as those in Figure 2). When such plots are
made, however, the flow stresses should be plotted versus instantaneous test temperature. This temperature is equal to the nominal
or initial test temperature plus increases in temperature due to
deformation heating.[31] For Ti-6242Si deformed at a strain rate
of 10 sec.$^{-1}$, the plots shown in Figure 13 are obtained for the two
different microstructures. It should be noted that the α+β data
exhibit a single σ vs. T trend line and the β data a series of lines
for different strain levels. Thus, these plots demonstrate that
flow softening in the α+β microstructure arises principally from
deformation heating whereas that for the β microstructure comes
from deformation heating as well as microstructural effects.

Table 3. Physical Properties of Ti-6242Si and H11 Tool Steel

	Density, ρ (gr./cc)	Specific Heat, c (cal./gr. C)	Thermal Conductivity, k (cal./sec. cm. C)
Ti-6242Si*	4.4	0.23	0.040
H11‡	7.8	0.11	0.080

* Properties for T ∼ 900 C (1650 F)
‡ Properties for T ∼ 20 C (70 F)

Table 4. Estimated Temperature and Strain Rate
Gradients at Fifty Percent Reduction
of α+β Microstructure Ti-6242Si*

$x^{\ddagger}$ (mm)	$\dfrac{dT}{dx}\left(\dfrac{C}{mm}\right)$	$\dfrac{d\log_{10}\dot{\varepsilon}}{dx}(mm^{-1})$	$\dfrac{\dot{\varepsilon}(x)}{\dot{\varepsilon}(x = 3.505\ mm)}$
0.5	342.7	7.44	<0.01
1.0	202.3	4.39	<0.01
1.5	84.0	1.82	<0.01
2.0	24.6	0.53	0.05
2.5	5.1	0.11	0.47
3.0	0.7	0.02	0.88
3.5	0.1	∼0.00	1.00

* Initial Specimen Temperature: 954 C (1750 F)
   Initial Die Temperature   : 191 C (375 F)
   Time of Deformation   : 0.09 Secs.

‡ x ≡ Distance from Die-Workpiece Interface

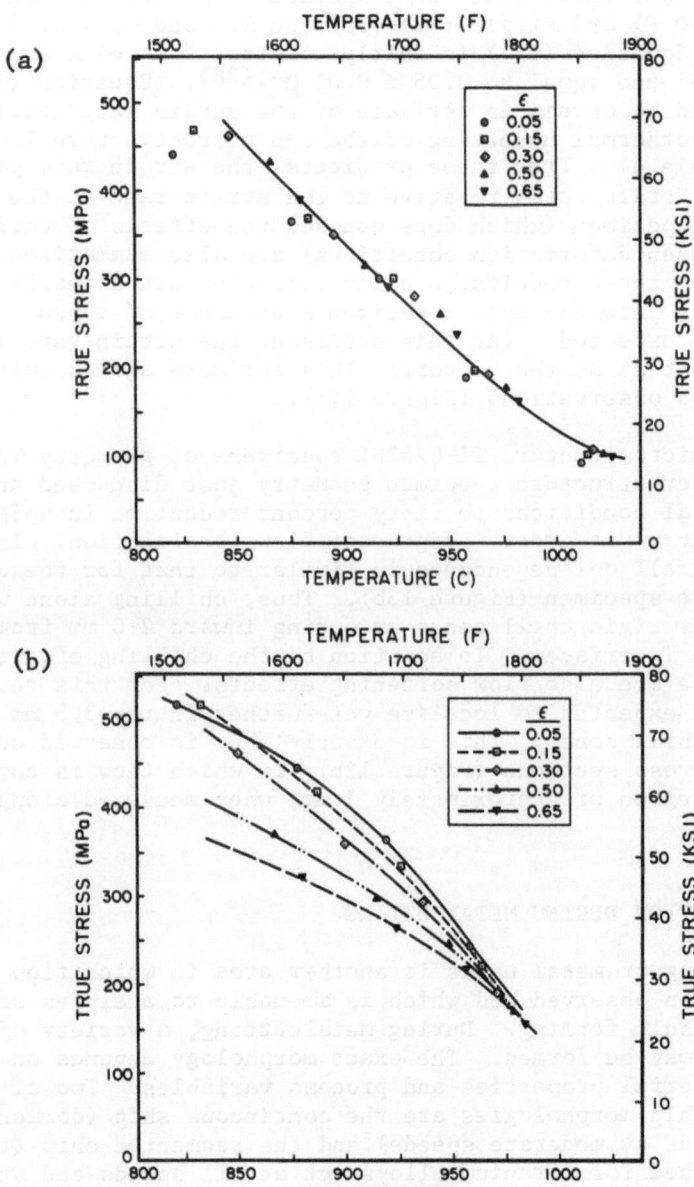

Figure 13.   Flow Stress vs. Temperature Plots for (a) α+β Micro-
             structure Ti-6242Si and (b) β Microstructure Ti-6242Si
             Deformed at a Strain Rate of 10 Sec.$^{-1}$.   Data Have
             Been Corrected for the Effects of Deformation
             Heating.[18,31]

Analysis of flow stress data such as that in Figure 13 and measured m values shows that for temperatures between 816 and 982 C (1500 and 1800 F) and strain rates between 0.1 and 10 sec.$^{-1}$, the η parameter (Equation (18)) is fairly constant for α+β microstructure Ti-6242Si and equal to 0.05 ± 0.01 C^{-1}[28]. Equation (18) may now be applied to obtain an estimate of the strain rate gradients in the non-isothermal upsetting of the α+β microstructure Ti-6242Si specimen (Table 4). From these gradients, the strain rate profile is determined. Strain rates relative to the strain rate at the center of the upset specimen (which does not see the effects of chilling under the chosen deformation conditions) are also summarized in Table 4. From these results, one may conclude that a nearly rigid zone extending into the upset specimen a distance of approximately 2.0 mm may be expected. (At this distance, the strain rate is only one-twentieth that at the center.) This estimate agrees quite favorably with observations (Figure 12a).

For β microstructure Ti-6242Si specimens of geometry identical to the α+β microstructure specimen geometry just discussed and upset under identical conditions to fifty percent reduction in height, one may expect a yet greater degree of flow localization. In this case, the overall σ-T dependence is similar to that for the α+β microstructure specimen (Figure 13b). Thus, chilling alone would lead to nearly rigid chill zones extending inward 2.0 mm from the die-workpiece interfaces. In addition to the chilling effects, however, there are also flow softening effects. For this reason, flow would be expected to localize yet further in the 3.5 mm region between the chill zones. This is exactly what is observed on metallographic cross sections (Figure 12b), in which flow is confined to a narrow region of approximately 1 mm, when measured along the specimen axis.

FLOW LOCALIZATION DURING METALCUTTING

Machining of metal parts is another area in which flow localization has been observed and which is amenable to analyses similar to those for bulk forming. During metalcutting, a variety of chip morphologies may be formed. The exact morphology depends on a number of material properties and process variables. Two of the most common ship morphologies are the continuous ship (common for many steels cut at moderate speeds) and the segmented chip (frequently obtained for titanium alloys cut at all speeds and steels at high speeds).[41,42,43] Examples of these two types of chips are shown in Figure 14. The formation of segmented chips often results from a flow localization process in which strain hardening tendencies are outweighed by softening due to deformation

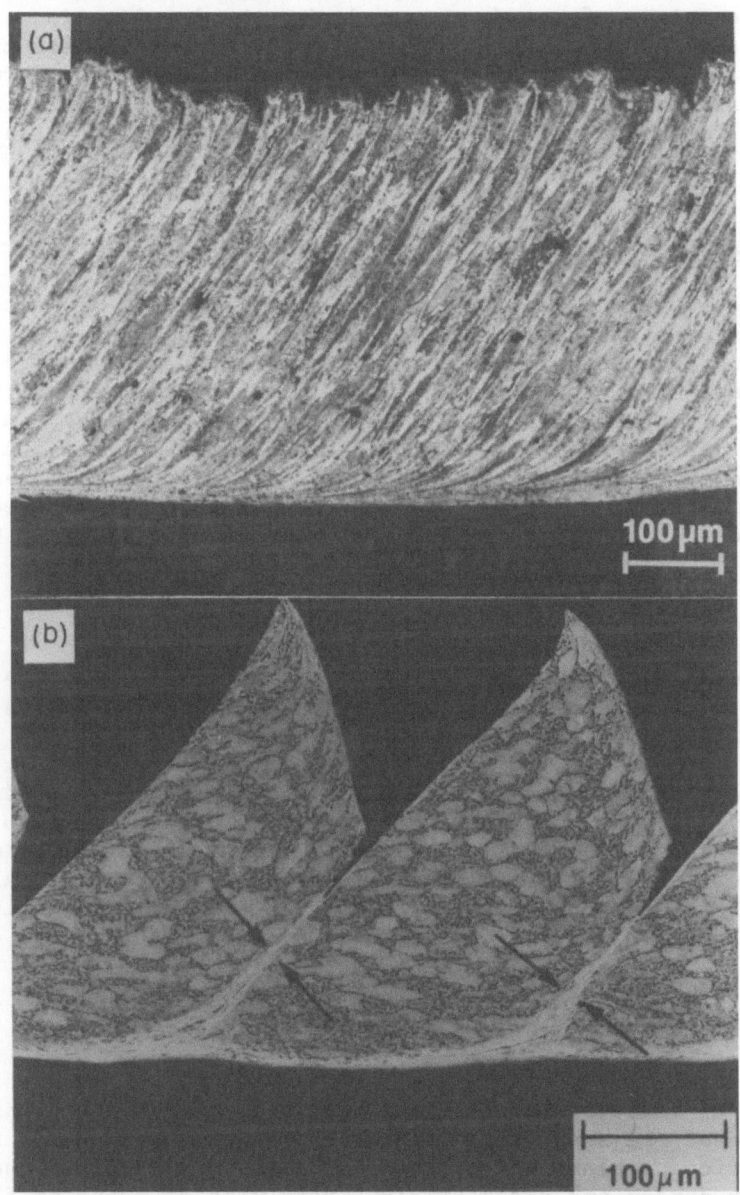

Figure 14.   Micrographs of Chips Formed During Metalcutting: (a)
             Continuous AISI 4340 Chip and (b) Segmented Ti-6Al-4V
             Chip.

heating.[23]   In addition, it has been suggested that the rate at
which the cutting force decreases during the localization process
must be compatible with the rate at which the machine structure
itself can unload.[44]   An estimate of the tendency for formation
of segmented chips which is based solely on the flow localization
analysis is therefore very approximate. This flow localization
analysis will be discussed next.

Flow Localization Analysis for Orthogonal Cutting

     The flow localization analysis which shall be discussed is
for the simple case of orthogonal cutting (Figure 15). In this
operation the cutting tool of rake angle $\alpha*$ causes metal to deform
under plane strain conditions. In particular, the deformation consists of a certain thickness of metal, d ($\equiv$ the depth of cut) that
is acted upon by the cutting tool being sheared so that the final
thickness is t_c ($\equiv$ the chip thickness). If we denote the angle of
the shear zone by ϕ, it may be shown[41] that the total shear strain
Γ_{max} imposed upon the metal for a continuous chip is given by

$$\Gamma_{max} = \tan(\phi - \alpha*) + \cot \phi \quad . \tag{21}$$

For large shear angles ($\phi \approx 30 - 40°$), the shear strains are on the
order of 2. For small shear angles ($\phi \approx 10°$), shear strains in
excess of 5 are obtained.

     Under the plane strain conditions of orthogonal cutting,
segmented chips may be expected to initiate alsong the zero extension directions which lie along the y' direction at angles of ϕ
to the cutting direction. Assuming that the cross-sectional area
in the shear zone does not vary with position, force equilibrium
requires the gradient of shear stress τ acting along these directions to vanish, or

$$\frac{d\tau}{dx'} = 0 \quad , \tag{22}$$

in which x' denotes the coordinate perpendicular to the long
direction of the shear zone (Figure 15). Unlike the other flow
localization problems which have been discussed thus far, however,
the shear stress and shear strain Γ vary with x' prior to flow
localization. As a first approximation, let us assume that the
shear strain varies linearly with x'

$$\Gamma = \Gamma_{max} x' \quad . \tag{23}$$

Since $\tau = \tau(\Gamma, \dot{\Gamma}, \text{ and } T)$, Equation (22) becomes then

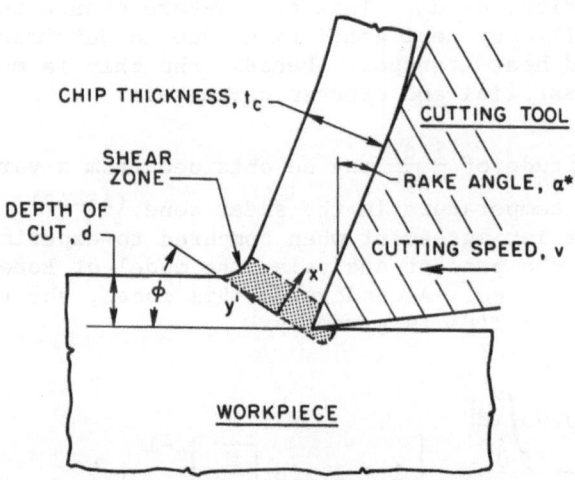

CHIP THICKNESS, t_c

SHEAR
ZONE

DEPTH OF
CUT, d

CUTTING TOOL

RAKE ANGLE, α^*

CUTTING SPEED, v

ϕ

x'

y'

WORKPIECE

Figure 15.   Schematic of Orthogonal Cutting Operation

$$0 = \left(\frac{\partial \tau}{\partial \Gamma}\right)\bigg|_{\dot{\Gamma},T} \Gamma_{max} + \left(\frac{\partial \tau}{\partial \dot{\Gamma}}\right)\bigg|_{\Gamma,T} \frac{d\dot{\Gamma}}{d\Gamma} \Gamma_{max}$$

$$+ \left(\frac{\partial \tau}{\partial T}\right)\bigg|_{\Gamma,\dot{\Gamma}} \frac{dT}{d\Gamma} \Gamma_{max} \quad . \tag{24}$$

After multiplying by $(1/\tau\Gamma_{max})$ and rearrangement, this equation reduces to

$$\frac{1}{\dot{\Gamma}} \frac{d\dot{\Gamma}}{d\Gamma} = \frac{\alpha}{\sqrt{3}} = -\left\{\left(\frac{1}{\tau}\right)\left(\frac{\partial \tau}{\partial \Gamma}\right)\bigg|_{\dot{\Gamma},T} + \left(\frac{\partial \tau}{\partial T}\right)\bigg|_{\Gamma,\dot{\Gamma}} \left(\frac{1}{\tau} \frac{dT}{d\Gamma}\right)\right\}/m \quad , \tag{25}$$

which is identical to Equation (10). All of the terms in this last expression except $(1/\tau)$ $(dT/d\Gamma)$ are material properties which are functions of $\Gamma,\dot{\Gamma}$ and T and thus, must be measured in conventional mechanical property tests. Moreover, inspection of Equation (25) demonstrates that the flow localization tendency increases with Γ, and thus flow localization is most likely to be initiated in the material elements just <u>leaving</u> the shear zone. To apply Equation (25), though, we need to estimate the rate of temperature change during deformation, $dT/d\Gamma$. This temperature change in the shear zone is controlled by heat generation (due to deformation), heat conduction, and heat transport (because the chip is moving). Thus it depends on material <u>and</u> process parameters.

The magnitude of $\frac{1}{\tau}\frac{dT}{d\Gamma}$ may be obtained from a variety of models for the temperature in the shear zone.[45-47] Most of the models give similar agreement when compared to experimental measurements.[48] In the present analysis, the model of Loewen and Shaw[45] will be used. According to this model, the temperature in the shear zone $T = T(\Gamma)$ is given by

$$T(\Gamma) = \frac{0.95\int_{0}^{\Gamma}\tau d\Gamma}{\rho c}\left[1 + 1.328\sqrt{\frac{\Delta\Gamma}{vd}}\right]^{-1} + T_{amb}. \qquad . \qquad (26)$$

In this equation, Δ is the thermal differsivity ($= k/\rho c$), v is the cutting speed, T_{amb} is the ambient temperature of the workpiece, and all other symbols are the same as defined previously. Again it has been assumed that 95% of the deformation work is transformed into heat. Differentiating Equation (26) with respect to Γ and dividing by τ, one obtains for a power hardening material $\tau = G\Gamma^{n}$

$$\frac{1}{\tau}\frac{dT}{d\Gamma} = \frac{0.95}{\rho c}\left\{\left[1 + 1.328\sqrt{\frac{\Delta\Gamma}{vd}}\right]^{-1}\right.$$
$$\left. - \frac{0.664}{(1 + n)}\sqrt{\frac{\Delta\Gamma}{vd}}\left[1 + 1.328\sqrt{\frac{\Delta\Gamma}{vd}}\right]^{-2}\right\} . \qquad (27)$$

Equation (27) in conjunction with Equation (25) may be employed to obtain estimates of the flow localization parameter.

Formation of Segmented Chips During Cutting of AISI 4340

The applicability of the analysis just presented will be illustrated with data obtained by Komanduri, et al.[43], for cutting of AISI 4340 steel tempered to a hardness of $R_{c}35$. At cutting speeds of 30 m/min to 60 m/min, continuous chips were formed (Figure 16). Above 60 m/min, segmented chips which formed apparently as a result of a flow localization process were obtained. The other cutting parameters were d = 0.05 cm and $\alpha^* = -5°$.

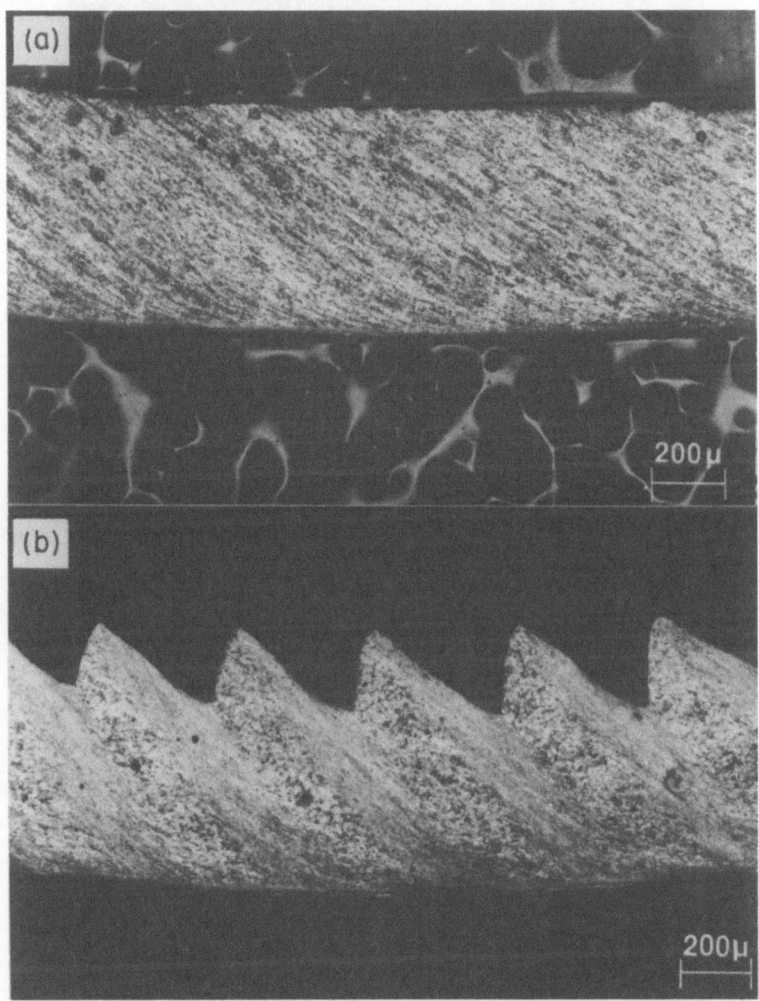

Figure 16. Micrographs of Chips of AISI 4340 (R_c = 35) Formed During Metalcutting. Chips Show Transition from Continuous Chip to Chip with Flow Localization to Chips which are Segmented. Cutting Speeds (m/min) were (a) 60, (b) 120, (c) 240, (d) 480 and (e) 960.

(continued)

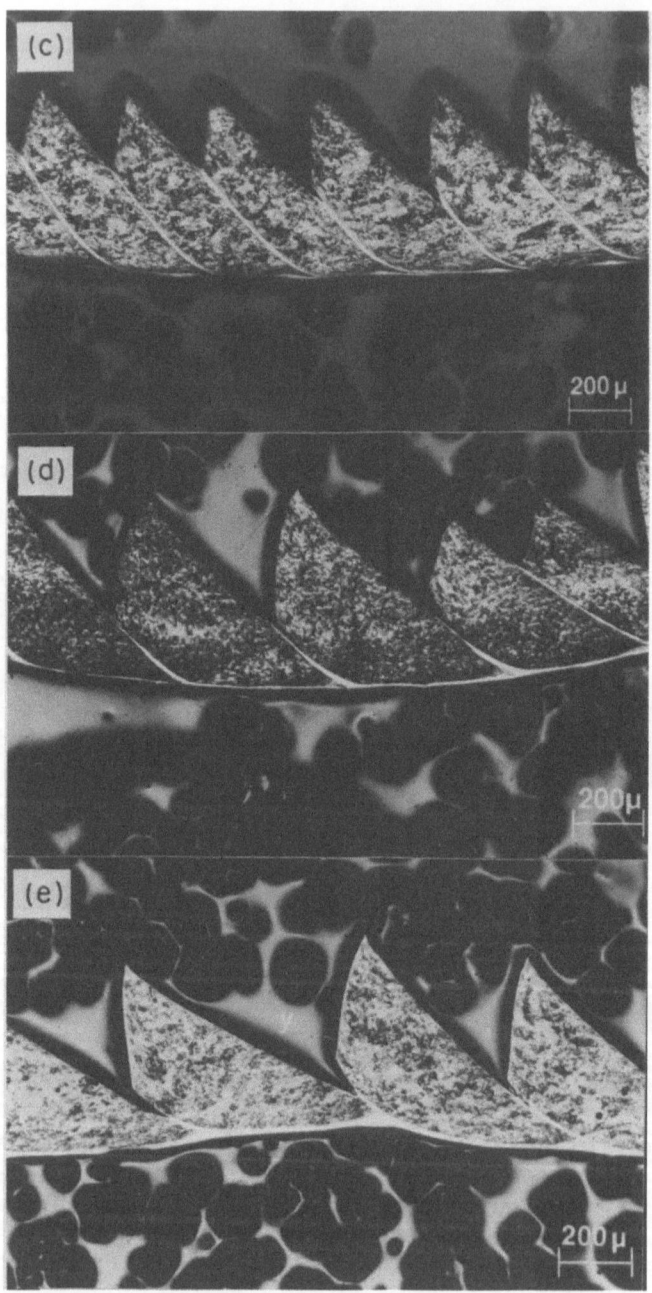

Figure 16.   (Continued)

From chip thickness measurements, it was found that $\phi \approx 30°$ and $\gamma \approx 2.5$. Physical properties for 4340 (Table 5) were obtained from References 49 and 50. Pertinent mechanical properties (Table 5) were obtained from References 51 (m values at $\dot{\varepsilon} \approx 10^3$ sec.$^{-1}$ $\approx$ approx. $\dot{\varepsilon}$ in metalcutting), 26 and 52 $\left(\frac{1}{\tau} \left(\frac{\partial \tau}{\partial T} \right) \right)$, and 26 and 53 $\left(\frac{\partial \tau}{\partial T} \right)$. Also, an approximate value of n (for use in Equation (27)) was obtained from Reference 26. All of these data except $\frac{\partial \tau}{\partial T}$ were for room temperature deformation. Despite the fact that Equation (26) shows that chip temperatures reach as high as 350 C (662 F), it is not thought that using room temperature data for these quantities introduce significant errors for the quenched and tempered steel.

Inserting the data in Table 5 (plus the previously mentioned cutting parameters) into Equations (25) and (27), estimates of the flow localization parameter α as a function of cutting speed were obtained (Table 6). It should be emphasized that these α's are for material points just leaving the shear zone i.e., at the elastic-plastic boundary. From the data in Table 6, it is seen that α reaches a value of 5 at a cutting speed of 60 m/min and higher values at higher cutting speeds. Since it is reasonable to expect that flow localization may be expected to occur only when $\alpha \geq 5$ over a non-negligible portion of the shear zone, segmented chips should start to form only at cutting speeds exceeding 60 m/min. This is indeed what has been observed for this material (Figure 16). Thus the flow localization analysis may be applied to the phenomenon of segmented-chip formation in metalcutting.

SUMMARY

The modes by which plastic deformation may localize during bulk deformation processing have been described. During bulk forming operations in which heat transfer between the workpiece and tooling are negligible (isothermal metalworking), flow localizes as a result of flow softening, and the tendency towards such localizations increases with increased flow softening rate and decreased strain rate sensitivity. Under axisymmetric deformation conditions, the flow localizations may appear as non-uniform bulges in the workpiece cross-section. On the other hand, in plane-strain modes of metal forming, shear bands are favored. Workability maps based on the value of the flow localization parameter α (= the ratio of the normalized flow-softening rate to the strain-rate sensitivity parameter) have been found to be very useful in delineating the strain rate and temperature regimes in which shear bands can be avoided.

Table 5.   Physical and Mechanical Properties of AISI 4340
           (R_C = 35)

Density, ρ $\left(\text{gr./cm}^3\right)$	:	7.83
Specific Heat, c (cal./gr. C)	:	0.11
Thermal Diffusivity, $\Delta\left(\text{cm}^2/\text{sec.}\right)$	:	0.0925
Rate Sensitivity at $\dot{\Gamma}$ = 2 x 10^3 sec.$^{-1}$	:	0.040
$\left\{\left(\partial\tau/\partial\Gamma\right)\big\|_{\dot{\Gamma},T}\right\}/\tau$ at Γ = 2.5	:	0.062
$\left(\partial\tau/\partial T\right)\big\|_{\Gamma,\dot{\Gamma}}$ (kPa/C)	:	968
n	:	0.07

Table 6. α Parameter Predictions for AISI
           4340 Steel (R_C = 35)

Cutting Speed, v		α
m/min.	(ft./min.)	
12	(40)	2.8
30	(100)	4.1
60	(200)	5.0
90	(300)	5.5
120	(400)	5.8
150	(500)	6.0
180	(600)	6.2
240	(800)	6.4
300	(1000)	6.6
∞	(∞)	8.4

Under non-isothermal bulk metalworking conditions (workpiece and tooling at different initial temperatures), flow may localize as a result of temperature gradients that are set up by heat transfer between the workpiece and tooling (i.e., chilling). In these instances, materials whose flow stress exhibits large sensitivity to temperature and low strain rate sensitivity are most likely to develop regions of localized flow and shear bands. To predict these defects, detailed knowledge of the material properties and a model to predict heat transfer and thus the workpiece temperature field are required.

Flow localization ideas may also be employed to evaluate the tendency to form segmented chips during metalcutting. As in bulk processing under nominally isothermal conditions, flow localization in metalcutting is controlled by the flow softening rate, which in turn is determined by the competitive forces of work hardening and thermal softening. The temperature field in the shear zone in metalcutting is affected by heat generation (due to deformation), heat conduction, and transport of heat with the moving chip. These influences must be considered to determine the rate of thermal softening and thus the overall hardening or softening rate at various points in the shear zone. With an understanding and model for the heat transfer phenomenon, flow localization parameters as a function of material properties and process variables, such as the cutting speed, may be estimated and the tendency to form segmented chips determined.

ACKNOWLEDGEMENTS

The authors acknowledge the gracious support of the Air Force Office of Scientific Research, AFSC, under Grant No. AFOSR-79-0048 (Dr. A. Rosenstein, Program Manager) for the bulk of this work. Technical discussions with Drs. S. B. Rao, A. Hoffmanner, and T. Altan of Battelle's Columbus Laboratories and Dr. J. J. Jonas of McGill Univeristy were helpful in formulating many of the ideas contained herein. The work on shear band occurrence in forging of JBK-75 was part of a joint effort between one of the authors (SLS) and Dr. M. Mataya of Rockwell International. In addition, the authors would like to express their appreciation to Dr. S. Doraivelu, Air Force Wright Aeronautical Laboratories (AFWAL), who supplied the photograph in Figure 7, and Dr. R. Komanduri, General Electric Corporate Research and Development, who gave us the photographs of the metalcutting chips in Figures 14 and 16. These chip photographs were taken as part of a DARPA sponsored program on Advanced Machining Research, managed by AFWAL under contract No. F33615-79-C-5119. Finally, the authors gratefully acknowledge the contributions of Mrs. Susan Tamplin who had the difficult task of typing the manuscript, and Messers. W. W. Sunderland and C. R. Thompson who prepared the line drawings and photographs.

REFERENCES

1. W. A. Backofen, Deformation Processing, Addison-Wesley, Reading, MA. (1967).
2. R. Hill, "On Discontinuous Plastic States with Special Reference to Localized Necking in Thin Sheets", J. Mech. Phys. Solids, 1, 19 (1952).
3. H. W. Swift, "Plastic Instability under Plane Stress", J. Mech. Phys. Solids, 1, 1 (1952).
4. Z. Marciniak and K. Kuczynski, "Limit Strains in the Process of Stretch Forming Sheet Metal", Inter. J. Mech. Sci., 9, 609 (1967).
5. S. P. Keeler, "Forming Limit Criteria - Sheets", in Advances in Deformation Processing, J. J. Burke and V. Weiss, eds., Plenum Press, New York (1978).
6. P. W. Lee and H. Kuhn, "Fracture in Cold Upset Forging - A Criterion and Model", Met. Trans., 4, 969 (1973).
7. R. C. Koeller and R. Raj, "Diffusional Relaxation of Stress Concentration at Second Phase Particles", Acta Met., 26, 1551 (1978).
8. A. L. Hoffmanner, " The Use of Workability Test Results to Predict Processing Limits", in Metal Forming - Interrelation Between Theory and Practice, A. L. Hoffmanner, ed., Plenum Press, New York (1971).
9. B. Avitzur, Metal Forming: Processes and Analysis, McGraw-Hill, New York (1968).
10. T. Miki, T. Tamano, and S. Yanagimoto, "Factors Causing Internal Cracks in Multistage Extrusion", in Proc. Sixth North American Metalworking Research Conference, R. S. Hahn and W. B. Rice, eds., Society of Manufacturing Engineers, Dearborn, MI. (1978).
11. K. Brown, "Role of Deformation and Shear Banding in the Stability of the Rolling Textures of Aluminum and Al-0.8% Mg Alloy", J. Inst. Metals, 100, 341 (1972).
12. H. C. Rogers, "Adiabatic Plastic Deformation", Ann. Rev. Mat. Sci., 9, 283 (1979).
13. P. S. Mathur and W. A. Backofen, "Mechanical Contributions to the Plane-Strain Deformation and Recrystallization Textures of Aluminum-Killed Steel", Met. Trans., 4, 643 (1973).
14. S. C. Jain and S. Kobayashi, "Deformation and Fracture of an Aluminum Alloy in Plane-Strain Sidepressing", Technical Report AFML-TR-70-90, University of California, Berkeley, CA. (July, 1970).
15. S. Kobayashi, C. H. Lee, S. Sohrabpour, F. Kanacri, and L. R. Beck, "Study of Deformation and Defects Occurrence in Advanced Forging Techniques", Technical Report AFML-TR-72-3, University of California, Berkeley, CA. (March, 1972).
16. W. Johnson and P. B. Mellor, Engineering Plasticity, Van Nostrand Rheinhold Company, London (1973).
17. V. Osina, "Forming of Metals at High Rates and Energies", Metal Treatment, 33, 193 (May, 1966).

18. S. L. Semiatin and G. D. Lahoti, "Deformation and Unstable
     Flow in Hot Forging of Ti-6Al-2Sn-4Zr-2Mo-0.1Si", Met. Trans.
     A, 12A, 1705 (1981).
19. S. L. Semiatin and G. D. Lahoti, "Deformation and Unstable Flow
     in Hot Torsion of Ti-6Al-2Sn-4Zr-2Mo-0.1Si", Met. Trans. A,
     12A, 1719 (1981).
20. S. L. Semiatin and G. D. Lahoti, "The Occurrence of Shear Bands
     in Isothermal, Hot Forging", Met Trans. A, 13A, 275 (1982).
21. I. L. Dillamore, J. G. Roberts, and A. C. Bush, "Occurrence
     of Shear Bands in Heavily Rolled Cubic Metals", Metal Sci.,
     13, 73 (1979).
22. J. J. Jonas and M. J. Luton, "Flow Softening at Elevated
     Temperatures", in Advances in Deformation Processing, J. J.
     Burke and V. Weiss, eds., Plenum Press, New York (1978).
23. R. F. Recht, "Catastrophic Thermoplastic Shear", J. Appl. Mech.,
     Trans. ASME, 31E, 189 (1964).
24. R. S. Culver, "Thermal Instability Strain in Dynamic Plastic
     Deformation", in Metallurgical Effects at High Strain Rates,
     R. W. Rhode, et al., eds., Plunum Press, New York (1973).
25. U. S. Lindholm, A. Nagy, G. R. Johnson, and J. M. Hoegfeldt,
     "Large Strain, High Strain Rate Testing of Copper", J. Eng. Mat.
     Techn., Trans. ASME, 102, 376 (1980).
26. M. R. Staker, "The Relation Between Adiabatic Shear Instability
     Strain and Material Properties", Acta Met., 29, 683 (1981).
27. G. B. Olson, J. F. Mescall, and M. Azrin, "Adiabatic Deformation
     and Strain Localization", in Shock Waves and High-Strain-Rate
     Phenomena in Metals, M. A. Meyers and L. E. Murr, eds., Plenum
     Press, New York (1981).
28. S. L. Semiatin and G. D. Lahoti, "The Occurrence of Shear Bands
     in Non-Isothermal, Hot Forging", Met. Trans. A (in press).
29. J. J. Jonas, R. A. Holt, and C. E. Coleman, "Plastic Stability
     in Tension and Compression", Acta Met., 24, 911 (1976).
30. P. Dadras and J. F. Thomas, Jr., "Compressive Plastic Instability
     and Flow Localization in Ti-6242", Res. Mechanica Letters, 1,
     97 (1981).
31. S. L. Semiatin, G. D. Lahoti, and T. Altan, "Determination and
     Analysis of Flow Stress Data for Ti-6242 at Hot Working Tempera-
     tures", in Process Modeling - Fundamentals and Applications to
     Metals, T. Altan, H. Burte, H. Gegel, and A. Male, eds.,
     American Society for Metals, Metals Park, OH. (1980).
32. J. L. Robbins, O. C. Shepard, and O. D. Sherby, "Accelerated
     Spheroidization of Eutectoid Steels by Concurrent Deformation",
     J. Iron Steel Inst., 202, 804 (1964).
33. I. A. Martorell, "Effects of Isothermal Forging Conditions on
     the Properties and Microstructures of Ti-10V-2Fe-3Al", Technical
     Report AFML-TR-78-114, Air Force Materials Laboratory, Air Force
     Wright Aeronautical Laboratories, AFSC, Wright-Patterson Air
     Force Base, OH. (December, 1978).
34. S. L. Semiatin, Battelle's Columbus Laboratories, Columbus, OH.,
     unpublished Ti-10-2-3 data (1982).

35. S. M. Doraivelu, "Studies on the Influence of Temperature and Mean Strain Rate on the Flow Stress of Alloy Steels", Ph.D. Thesis, Department of Metallurgy, Indian Institute of Technology, Madras, India (February, 1979).

36. S. L. Semiatin and A. L. Hoffmanner, Battelle's Columbus Laboratories, Columbus, OH., unpublished U-0.75Ti data (1979).

37. H. Meyer-Nolkemper, "Flow Curves of Metallic Materials", Report No. 4, Hannover Institute of Manufacturing Research, Hannover, Germany (1978).

38. S. I. Oh, "Finite Element Analysis of Metal Forming Processes with Arbitrarily Shaped Dies", Inter. J. Mech. Sci., 24 (1982), in press.

39. G. D. Lahoti and T. Altan, "Research to Develop Process Models for Producing a Dual Property Titanium Alloy Compressor Disk", Technical Report AFWAL-TR-80-4162, Battelle's Columbus Laboratories, Columbus, OH. (October, 1980).

40. T. Altan, F. W. Boulger, J. R. Becker, N. Akgerman, and H. J. Henning, Forging Equipment, Materials and Practices, Handbook MCIC-HB-03, Metals and Ceramics Information Center, Battelle's Columbus Laboratories, Columbus, OH. (October, 1973).

41. M. E. Merchant, "Mechanics of the Metal Cutting Process", J. Appl. Phys., 16, 267 and 318 (1945).

42. R. Komanduri and B. F. von Turkovich, "New Observations on the Mechanism of Chip Formation when Machining Titanium Alloys", Wear, 69, 179 (1981).

43. R. Komanduri, T. Schroeder, J. Hazra, B. F. von Turkovich, and D. G. Flom, "On the Catastrophic Shear Instability in High-Speed Machining of an AISI 4340 Steel", J. Eng. for Ind., Trans. ASME, 104, 121 (1982).

44. J. C. Lemaire and W. A. Backofen, "Adiabatic Instability in the Orthogonal Cutting of Steel", Met. Trans., 3, 477 (1972).

45. E. G. Loewen and M. C. Shaw, "On the Analysis of Cutting-Tool Temperatures", Trans. ASME, 76, 217 (1954).

46. J. H. Weiner, "Shear-Plane Temperature Distribution in Orthogonal Cutting", Trans. ASME, 77, 1331 (1955).

47. B. T. Chao and K. J. Trigger, "The Significance of the Thermal Number in Metal Machining", Trans. ASME, 75, 109 (1953).

48. G. Boothroyd, "Temperatures in Orthogonal Metal Cutting", Proc. Inst. Mech. Engrs., 177, 789 (1963).

49. Y. S. Touloukian, Thermophysical Properties of High Temperature Solid Materials, Volume 3: Ferrous Alloys, Macmillan Company, New York (1967).

50. T. Lyman, Metals Handbook, Vol. 1: Properties and Selection of Metals, Eighth Edition, American Society for Metals, Metals Park, OH. (1961).

51. T. Nicholas, "Tensile Testing at High Rates of Strain", Experimental Mechanics, 21, 77 (May, 1981).

52. T. B. Cox and J. R. Low, Jr., "An Investigation of the Plastic Fracture of High Strength Steels", NASA Technical Report No. 5 under Research Grant NGR 39-087-003, Carnegie-Mellon University,

Pittsburgh, PA. (May, 1973).

53. Anon., <u>Aerospace Structural Metals Handbook</u>, Mechanical
     Properties Data Center, Battelle's Columbus Laboratories,
     Columbus, OH., data for AISI 4340, Code 1206 (1982).

THE BEHAVIOUR OF METALS UNDER BALLISTIC IMPACT AT SUB-ORDNANCE VELOCITIES

I. M. Hutchings

University of Cambridge
Department of Metallurgy and Materials Science
Pembroke Street
Cambridge, England

INTRODUCTION

The behaviour of a material when struck by a projectile depends strongly on the velocity of impact. A useful method of characterizing the damage regime is by means of a dimensionless "damage number" (1), D:

$$D = \rho_T v^2 / Y_T \tag{1}$$

Here ρ_T is the density of the target material and Y_T is its flow stress: v is the impact velocity. The damage number is closely related to the Best or Metz number, B, defined by (2):

$$B = \rho_T v^2 / H_T \tag{2}$$

where H_T is the Brinell hardness of the target, expressed in suitable units. The damage number, or the Metz number, can be thought of as the ratio between the forces on the projectile due to the inertia of the target material, and those due to its strength. For $D \gg 1$, the strength of the material would be expected to be unimportant compared with its density, while for $D \ll 1$, the impact response of the target should depend almost entirely on its strength properties.

The regimes of behaviour expected over a wide range of damage number and impact velocity have been described by several authors (3-7). At very low velocities ($D \lesssim 10^{-5}$), the deformation of the projectile and target is purely elastic. No permanent deformation

161

results from the collision, and the projectile and target separate
under elastic restoring forces. For high values of damage number
($10^3 \gtrsim D \gtrsim 10$), extensive permanent deformation is to be expected,
and the inertial forces resisting penetration by the projectile
become more important as the velocity is increased. This regime may
be termed "hydrodynamic", and is encountered in typical terminal
ballistic applications. The majority of the many reviews of projectile impact have concentrated on velocities within this regime.

The intermediate regime ($10^{-4} \gtrsim D \gtrsim 1$), which corresponds
roughly to impact velocities on steel in the range from 5 to 500
ms^{-1}, is also of interest, for several reasons. The industrially
important processes of abrasive grit and bead blasting, shot
peening and peen forming (8) all involve impacts within this range
of velocities, as does erosive wear by solid particle impingement
(9) which occurs in situations as diverse as coal-fired power
stations and aircraft gas-turbine engines. The behaviour of materials in this regime is also relevant to the problems of foreign
object impact damage and blade containment in gas turbines (10),
nuclear reactor safety, and the performance of armour some way
below its ballistic limit (11).

In this review, we shall consider the behaviour of metals when
subjected to ballistic impact within the intermediate regime of
damage number ($10^{-4} < D < 1$), which for structural metals involves
impact at sub-ordnance velocities. The review will concentrate on
the problem of projectile impact on a half-space, that is, on an
effectively infinitely thick target, rather than on the penetration
of sheets, and will examine the mechanics of impact of projectiles
of simple geometry. Impact of a projectile on a half-space has been
employed by several workers as an experimental method for investigating the yielding of metals at high strain-rates, and work in
this area will be discussed. Finally, features of the deformation
in the target around an indentation resulting from impact will
be discussed, particularly the geometry and nature of sub-surface
shear bands and their role in the removal of material from the
target.

MECHANICS OF IMPACT AT SUB-ORDNANCE VELOCITIES

Recent advances in the capacity of computational codes, which
can now handle fully three-dimensional impact problems, have led to
excellent agreement between computed and experimentally observed
behaviour in the sub-ordnance and ordnance velocity regimes
(5,12,13). But since the use of these codes currently involves substantial computing power for times of the order of hours, and since
they are not widely accessible, there is still a place for simpler,
if less accurate, methods of analysing projectile impact. Fortu- -
nately, in the sub-ordnance regime, results accurate enough for many

purposes may be obtained from very simple models, which can provide valuable insight into the impact process.

Impacts on metals over the range of velocities of interest here lead to plastic deformation. In many practical applications, the plastic deformation is confined either to the target or to the projectile. Most theoretical analyses of impact within the velocity range make the assumption that either the target or the projectile remains rigid, and this leads to considerable simplification of the problem.

If the projectile does not deform, then the problem is one of dynamic indentation: this will be discussed, with the main emphasis on spherical indenters, in the next section. For the case where the target remains rigid, and deformation is confined to the projectile, most studies have been concerned with the mushrooming of cylindrical projectiles. This geometry, often termed "Taylor impact", will be discussed later.

IMPACT OF A RIGID PROJECTILE ON A DEFORMING TARGET

Simple Model for Impact

Most experimental and theoretical studies have examined the normal impact of spherical or conical projectiles. If the projectile is sufficiently strong, it will remain undamaged and produce a permanent indentation in the target which, in this velocity regime, conforms quite closely with the shape of the rigid indenter.

The most simple theoretical approach (1,4,14) is to assume that during impact the forces between projectile and target may be represented by a constant indentation pressure, or "dynamic hardness", acting uniformly over the area of contact. This pressure leads, for the case of normal impact, to a force directly opposing the motion of the projectile, which is proportional to the area of contact projected parallel to the plane of the target surface. For the case of a sphere (Figure 1), the retarding force is $P.\pi a^2$ where P is the indentation pressure and a is the chordal radius of the indentation as defined in Figure 1. Pile-up of material displaced by the indentation is ignored. From the geometry of the diagram,

$$a^2 = x(2r - x) \tag{3}$$

where r is the radius of the sphere and x is the indentation depth. For indentation such that $a < 0.63\ r$, an error of less than 10% in the force results from approximating equation (3) as

$$a^2 \simeq 2rx \tag{4}$$

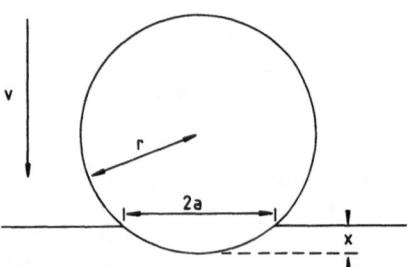

Figure 1.   The Geometry of Normal Indentation by a Sphere, in
             which Pile-Up is Neglected.   The Instantaneous
             Velocity is v.

Use of the approximate equation (4) results in a particularly simple
equation of motion:

$$\frac{4}{3} \pi r^3 \rho \ddot{x} = -2\pi r P x \tag{5}$$

where ρ is the density of the sphere material.   Solving this for
appropriate boundary conditions ($\ddot{x} = 0$, $\dot{x} = v_0$ at $t = 0$) yields the
expressions

$$x = r v_0 (2\rho/3P)^{\frac{1}{2}} \sin \omega_p t \tag{6}$$

and

$$a = \sqrt{2}\, r v_0^{\frac{1}{2}} (2\rho/3P)^{\frac{1}{4}} \sin^{\frac{1}{2}} \omega_p t \tag{7}$$

where

$$\omega_p = (3P/2\rho)^{\frac{1}{2}}/r \tag{8}$$

The timescale of the impact may conveniently be described by the "loading time", the time at which the sphere comes to rest, t_p :

$$t_p = \pi/2\omega_p = \frac{1}{2} \pi r (2\rho/3P)^{\frac{1}{2}} \qquad\qquad (9)$$

Elastic Effects

Perhaps the most serious criticisms of this analysis are that it totally ignores elastic deformation in the target, and is therefore unable to account for rebound of the sphere and a non-zero coefficient of restitution, and that no allowance is made for wave effects.

Neglect of elastic forces is certainly justifiable in the loading phase of the impact. At very low impact velocities, the collision would be purely elastic; the equations due to Hertz may be used to calculate the impact velocity at which plastic flow initiates (14,15). If P_e is the maximum mean contact pressure developed during purely elastic impact of a sphere on a half-space, then Hertz's equations predict that

$$P_e^5 = \frac{1280}{243\pi^4} \rho v_o^2 (1/f(E))^4 \qquad\qquad (10)$$

Here $f(E)$ is a function of the elastic constants of the sphere and target materials, given by

$$f(E) = (1 - v_1^2)/E_1 + (1 - v_2^2)/E_2 \qquad\qquad (11)$$

where v and E are Poisson's ratio and Young's modulus respectively. Plastic flow due to indentation is found to initiate in a metal, below the surface, when the mean pressure over the contact area is about 1.1 times its uniaxial yield stress. Fully developed plasticity, leading to a plastic zone extending to the surface around the indentation, as observed in a conventional hardness test, occurs when the mean pressure is $\sim$ 2.8 to 3 times the yield stress (14). The impact velocities needed to produce fully plastic indentations in metals, even in hard steels, may be shown from equation (10) to be less than 1 ms^{-1}, and therefore below the range under consideration here. Similarly, the proportion of the loading duration, t_p, during which the deformation of the target is purely elastic may be shown to be very small. Neglect of elastic forces during the loading phase is therefore reasonable.

During the unloading phase of the contact, however, elastic forces play a significant role. Although a small amount of reverse plastic flow is to be expected during the unloading (16), this flow, and the acceleration of the sphere away from the surface responsible for its rebound velocity, are driven by elastic strain energy stored in the target material during the formation of the indent. Hertz's equations enable a good estimate to be made of the rebound velocity

(14), and of the elastic force acting on the sphere during the
unloading phase. Unfortunately, an explicit analytical expression
cannot be obtained for the unloading force history, but as Hunter
(17) has demonstrated, a sinusoidal variation of force with time
approximates closely to the solution obtained numerically (with a
maximum discrepancy of $\sim$ 5%). The complete history of the force on
the sphere may therefore be described by two sinusoidal quarter-
cycles: one of angular frequency ω_p defined by equation (8), representing the plastic loading phase, and one of angular frequency ω_e
representing the elastic unloading phase. The value of ω_e may be
deduced from the velocity of rebound of the sphere. By equating
the rebound momentum to the time integral of the elastic unloading
force, it may be shown (18) that:

$$\omega_e = \omega_p/e \tag{12}$$

where e is the coefficient of restitution (rebound velocity/impact
velocity). The force-time relationship derived in this way for an
impact with e = 0.2 is shown in Figure 2.

The neglect of wave effects in this model must also be justified. Elastic waves will propagate within both the projectile and
the

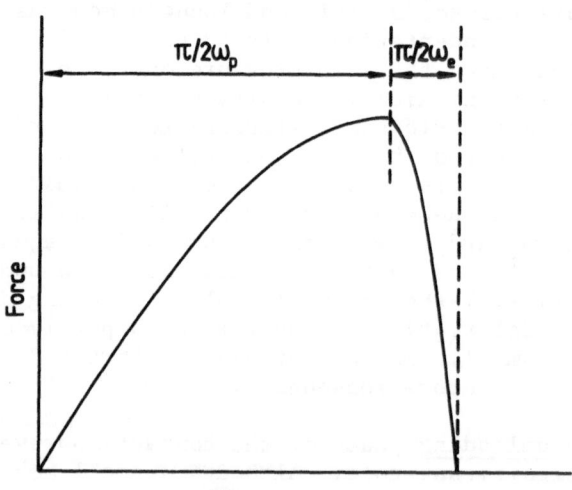

Figure 2.   Force-time history for a sphere striking a plane, with
            plastic loading and elastic unloading. The coefficient
            of restitution is 0.2 .
            (From ref. 18)

target, originating in the contact area. Davis and Hunter (19) have
shown that a quasi-static analysis, as used here, would be expected
to be valid at times long compared with the time taken for a longitudinal elastic wave to traverse the impinging sphere. The plastic
loading time derived from this analysis is indeed long compared
with the transit time for elastic waves in the sphere, as may be
seen from the following argument. The plastic loading time, t_p is
given by equation (9). For this time to be much greater than the
time for an elastic wave to cross the sphere, the inequality

$$t_p >> 2r/C_o \tag{13}$$

must hold, where C_o is the velocity of rod waves in the sphere
material:

$$C_o = (E/\rho)^{\frac{1}{2}} \tag{14}$$

where E is Young's modulus for the sphere. Hence,

$$\frac{1}{2} \pi (2\rho/3P)^{\frac{1}{2}} >> 2(\rho/E)^{\frac{1}{2}} \tag{15}$$

or,

$$P/E << 0.41 \tag{16}$$

For the sphere to indent the target, rather than to deform plastically itself, it has been found experimentally for quasi-static
loading (20,21) that the hardness of the sphere must be greater than
approximately 1.4 times that of the target. (This result is similar
to that found by Tabor (22) for scratching by sharp points. A
stylus dragged across a metal surface will produce a plastic groove
only if its hardness is greater than $\sim$ 1.2 times that of the surface.) If we assume that the ratio of 1.4 applies in the case of
dynamic indentation, then condition (16) implies that

$$H_S/E << 0.6 \tag{17}$$

where H_S is the Brinell or Vickers hardness of the sphere. This
inequality will certainly hold for any real material, unless its
strength approaches its theoretical strength (23), and we may therefore safely conclude that neglect of elastic waves in the sphere is
justified during nearly the whole of the impact period.

In the very first stages of loading, however, we would expect
wave propagation to be important. The pressure, P', generated when
two elastic half-spaces collide at velocity v may be derived from
one-dimensional elastic wave theory (e.g. ref. 1) as

$$P' = \rho_1 C_1 \rho_2 C_2 v / (\rho_1 C_1 + \rho_2 C_2) \qquad\qquad (18)$$

where ρ_1, ρ_2 and C_1, C_2 are the densities and elastic wave speeds in the two materials. A collision pressure P' given by equation (18) would be expected to result from a sphere striking a plane target, but only until it is relieved by tensile elastic waves propagating into the contact area from a free surface. The time for which the pressure will have this value may be simply estimated (24): the pressure should start to drop at a time $rv/2C^2$ after impact, and relief waves will have reached all parts of the contact zone at a time $4rv/C^2$, where C is the velocity of dilatational waves in the sphere material. For the impact velocities of interest here, these times will be substantially less than the time $(2r/C)$ for the transit of one-dimensional waves across the diameter of the sphere, and hence very much less than the plastic loading time t_p. It would therefore be extremely difficult to measure the duration of these pressures experimentally. Comparison may, however, be made with the results of the computations published by Evans et al. (25), who used a Lagrangian finite-difference code to analyse the elastic-plastic impact of a tungsten carbide sphere (r = 200 μm, v = 850 ms^{-1}) on a zinc sulphide target. The mean contact pressure derived from their calculations was indeed equal to P' at very short times, and fell rapidly with time, becoming sensibly constant at about twice the time predicted by the above simple theory.

It remains to justify neglecting the energy radiated into the target in the form of elastic waves. Hunter (17) obtained an expression for the energy absorbed by elastic waves during the purely elastic impact and rebound of a sphere. Hutchings (18) extended this theory for impact in which plastic flow occurred, assuming perfectly plastic loading and elastic unloading, leading to a force-time curve of the form illustrated in Figure 2. Within the subor-dnance velocity regime, elastic waves take up no more than a few percent of the initial kinetic energy of the sphere (18), and their neglect is fully justified in using the simple model presented above. The relatively small proportion of the total energy of the projectile which is finally partitioned into elastic waves and re-bound energy for impacts at these velocities is illustrated in the energy balance shown in Figure 3.

Results from the Simple Model

The simple theory of plastic impact presented above has been used by many investigators, and in the sub-ordnance velocity range gives quite good agreement with experimental observations. Since elastic effects are ignored, the volume of the indentation, V, would be expected to be directly proportional to the impact kinetic energy:

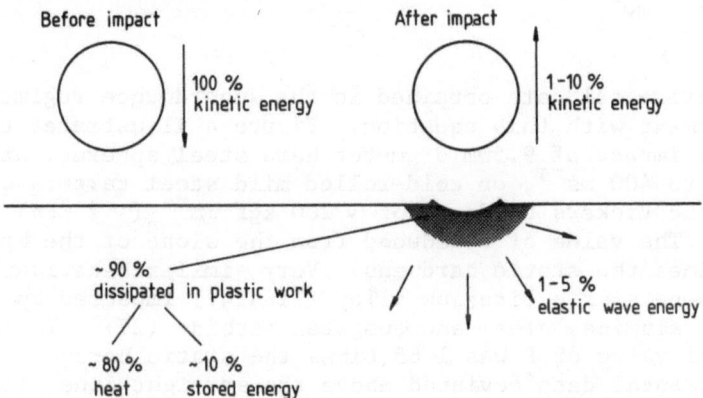

Figure 3. Typical Energy Balance for Normal Impact of a Rigid
           Sphere on a Metal Target at Subordinate Velocity.

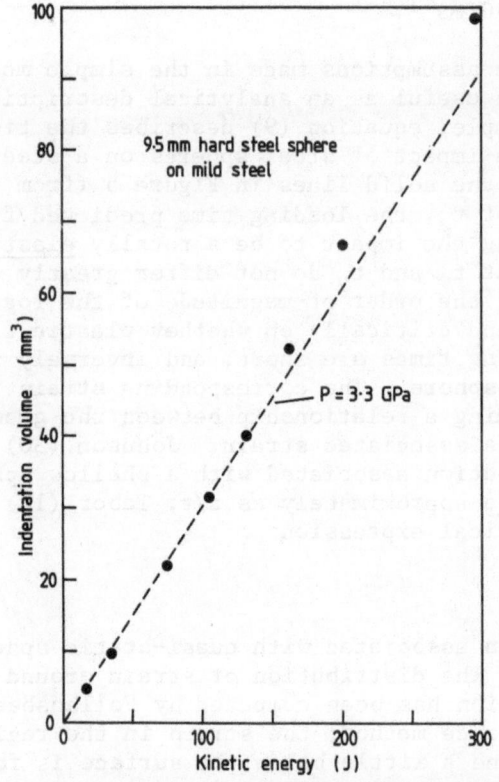

Figure 4. Indentation Volume vs. Projectile Kinetic Energy for the
           Normal Impact of 9.5 mm Hard Steel Spheres on to Mild
           Steel. (Data from ref. 26)

$$PV = \frac{1}{2} mv_o^2 \qquad\qquad\qquad (19)$$

Experimental data obtained in the subordnance regime shows fair agreement with this equation. Figure 4 illustrates this for the normal impact of 9.5mm diameter hard steel spheres, at velocities up to 400 ms^{-1}, on cold-rolled mild steel targets with a quasi-static Vickers hardness of $\sim$ 200 kgf mm^{-2} ($\sim$ 2 GPa) (data from ref. 26). The value of P deduced from the slope of the broken line is $\sim$1.7 times the static hardness. Very similar behaviour has been seen in targets of a titanium alloy (Ti6Al4V) impacted by 3.175mm spheres of alumina, steel and tungsten carbide (27). In that case the initial value of P was 1.65 times the static hardness, and again the experimental data deviated above the straight line, indicating an apparent reduction in P, at high velocities. A similar reduction in P as impact velocity increased was also reported by Goldsmith and Lyman (28), who measured the force directly with a Hopkinson pressure bar for the indentation of several metals by spheres at velocities up to 90 ms^{-1}.

Provided the assumptions made in the simple model are appreciated, it can prove useful as an analytical description of the impact process. For example, equation (9) describes the timescale of the event, and for the impact of steel spheres on a steel target provides the data shown by the solid lines in Figure 5 (from ref. 29). Also shown are values of t_e, the loading time predicted from Hertz's equations, assuming the impact to be a totally elastic process. It is significant that t_e and t_p do not differ greatly over the velocity range of interest: the order of magnitude of the loading time does not therefore depend critically on whether plastic flow occurs during impact. The loading times are short, and inversely proportional to the radius of the sphere. The corresponding strain rates may be estimated by assuming a relationship between the geometry of the indentation and its associated strain. Johnson (30) has shown that the strain distribution associated with a shallow spherical indentation should scale approximately as a/r; Tabor (14) had earlier proposed the empirical expression

$$\varepsilon \simeq 0.2 \ a/r \qquad\qquad\qquad (20)$$

for the mean strain associated with quasi-static spherical indentation of metals. The distribution of strain around a static spherical indentation has been computed by Follansbee et al. (31) by a finite difference method; the strain in the region just outside the contact area and a little below the surface is found to be represented quite accurately by Tabor's empirical relationship (equation 20) (32).

Although the numerical factor in equation (20) may well need modification for impact indentation (see below), it nevertheless

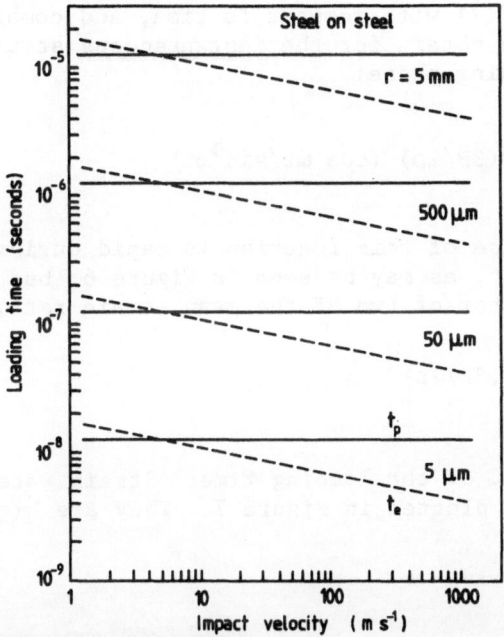

Figure 5.   Loading Time vs. Impact Velocity, for the Impact of Steel
            Spheres of Various Radii on to Mild Steel. Solid Lines:
            Perfectly Plastic Behaviour. Broken Lines: Elastic.

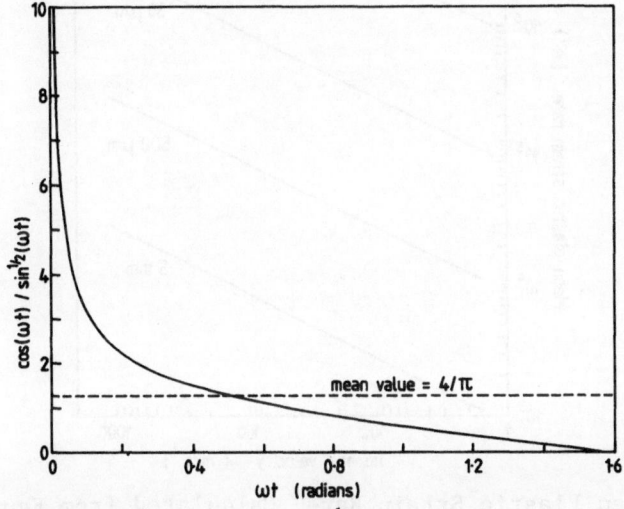

Figure 6.   The Function $\cos(\omega t)/\sin^{\frac{1}{2}}(\omega t)$, which is Proportional to
            the Instantaneous Strain Rate (equation 21).

provides a useful estimate of the strain involved. By differentiating equation (7) with respect to time, and combining it with
equation (20), we obtain for the <u>instantaneous</u> strain rate at time
t during the loading phase:

$$\dot{\varepsilon} \simeq 0.14 \; \frac{v_o^{\frac{1}{2}}}{r}(3P/2\rho)^{\frac{1}{4}}(\cos \omega t/\sin^{\frac{1}{2}}\omega t) \tag{21}$$

The time dependence of this function is rapid during the initial
part of the impact, as may be seen in Figure 6, but the strain rate
lies within a factor of two of the mean strain rate, given by (29):

$$\bar{\dot{\varepsilon}} \simeq 0.18 \; \frac{v_o^{\frac{1}{2}}}{r}(3P/2\rho)^{\frac{1}{4}} \tag{22}$$

over at least half of the loading time. Strain rates derived from
equation (22) are plotted in Figure 7. They are high, even for

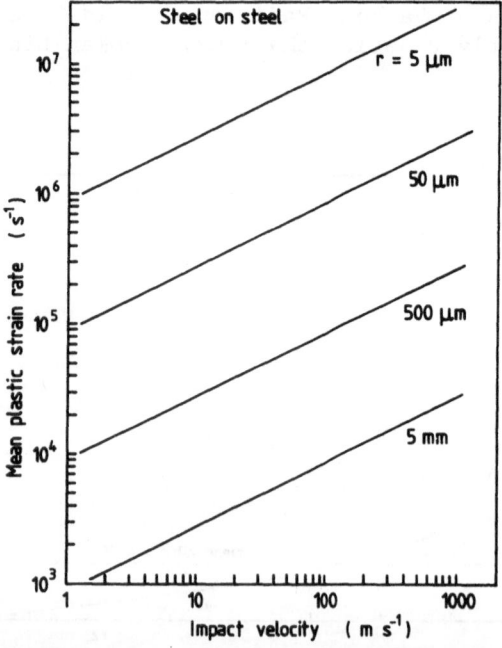

Figure 7.   Mean Plastic Strain Rate, Calculated from Equation (22),
for the Normal Impact of a Steel Sphere on a Mild Steel
Target.

macroscopic projectiles; for small spheres, such as erosive dust
particles, they are extremely high. It is noteworthy that these
very high strain rates result from quite moderate impact velocities,
because of the small linear scale of the impact geometry. Very high
impact velocities are not essential to produce deformation at very
high strain rates, provided the scale of the deformation is small.

     Few experimental measurements are available for loading times,
or total contact times, for sphere impacts in the velocity range of
interest here. However, Chaudhri et al. (33) did measure t_p for
impacts on ionic crystals at velocities up to 250 ms^{-1}. For the
impact of tungsten carbide spheres on MgO, they found quite good
agreement with values derived from equation (9), although for the
impact of glass spheres on NaCℓ and LiF, the observed times were up
to three times greater than the predicted values. The reasons for
this difference are not clear.

     Experimental measurements are also lacking, since the measure-
ment is difficult to make, for the dependence of indentation size
with time at these impact velocities. The code computations of
Evans et al. (25), however, provide an a versus t plot which agrees
within 10% with values obtained from equation (7) over the time
covered by their computations. We may conclude that the simple
constant indentation pressure model provides a fairly accurate
representation of many aspects of the process within this velocity
range.

Refinement of the Simple Model

     Apart from neglect of elastic effects, the constant indentation
pressure model fails to take account of work hardening in the target
material, or of strain-rate effects. Both may sometimes play significant rôles in impact indentation. Goodier (34) assumed that the
resisting force on the sphere was proportional to a^n (Meyer's Law),
where n is the Meyer index. For quasi-static indentation in fully
annealed metals, n has a value close to 2.5, while for fully work-
hardened metals, with a low work-hardening rate, it is close to the
value 2.0 assumed in the constant pressure model. Goodier did not,
however, follow through his analysis for a general value of n, but
put it equal to 2, effectively assuming a constant indentation
pressure. Tabor (14) showed that the mean pressure at the end of the
deformation process would be given by

$$\frac{4}{n+2} \, PV = \frac{1}{2} \, mv_o^2 \qquad\qquad (23)$$

The difference between this expression and equation (19) is in
practice not large. Mamoun (35) has analysed the problem fully,
employing Meyer's law, and gives expressions for the size of the
indentation and the plastic loading time for general n . They have
not been tested experimentally; it is not clear that they offer

significant advantages over the constant indentation pressure model.

An attempt to analyse the problem, taking account of the strain-rate dependence of flow stress, has been made by Singer and Evans (36). The final dimensions of the indentation are shown to depend on the strain-rate sensitivity of the target material. The model remains to be tested experimentally.

The oblique impact of spheres on to deforming targets has been relatively little studied in this velocity regime (37). Hutchings et al. (38) reported good agreement between a simple computational model, assuming a constant indentation pressure and a constant coefficient of friction, for the oblique impact of a rigid sphere on a deforming metal target, and experimental results obtained with 9.5 mm hard steel spheres striking mild steel targets at velocities up to 350 ms^{-1}. Indentation volumes and kinetic energy losses were quite accurately predicted. However, the method could not account for rebound velocities at high angles of incidence, since these were mainly due to elastic forces, nor, of course, could it provide the information on subsurface deformation obtainable from a code computation.

MEASUREMENT OF MECHANICAL PROPERTIES BY DYNAMIC INDENTATION

The high strain-rates associated with the impact of spheres (see Figure 7), and the simple relationship expected between indentation volume and projectile energy (equation 19), suggest that this method might be used to measure mechanical properties at high strain-rates. In the words of Lifshitz and Kolsky (39), "the elegance and simplicity of the method from an experimental point of view is, however, often outweighed by the extreme complexity of interpretation of the experimental observations." Several investigators have nevertheless attempted to use the method.

One approach is to study impacts at very low velocities, to detect the very first stage of yielding in the target and then to deduce a yield stress from Hertz's equations. This method would be expected to work best with materials exhibiting a sharp yield point. Davies (40) used this approach with success, detecting initial yielding in steels by careful optical examination of the target surface, initially plane, after impact. There are, however, severe experimental difficulties with this method, and it is not clear that the formation of the first detectable concavity in the surface does necessarily coincide with initial subsurface yielding. Lifshitz and Kolsky (39) detected the change in the coefficient of restitution at the onset of yielding, and obtained results conforming with those of Davies.

A second approach is to impact the surface at higher velocity, producing a fully plastic indentation, and then to infer the flow stress from the mean indentation pressure by a relationship of the form

$$P = cY \qquad\qquad\qquad\qquad\qquad\qquad (24)$$

where $c \simeq 3$ and is found to be approximately constant for quasi-static indentation by spheres on a wide range of metals (14). The value of P may be found either by direct measurement of the indentation force, or inferred from the indentation geometry and the energy balance. Measurement of the force was carried out by Yew and Goldsmith (41) who attached a force transducer to a large sphere, and by Goldsmith and Lyman (42) who indented the end of an instrumented Hopkinson pressure bar. In the latter case, force-time histories in both the elastic and elastic-plastic regimes were obtained; in both investigations, dynamic flow stresses higher than the static values were deduced. Other workers have derived the flow stress from measurements on the residual indentation. Tabor (14) showed that the following relationship would be expected for dynamic indentation of a non-work-hardening material:

$$PV_a = \frac{1}{2} mv_o^2 (1 - 3e^2/8) \qquad\qquad\qquad\qquad (25)$$

where e is the coefficient of restitution, and V_a is the "apparent" volume of the indentation as calculated from measurement of the chordal radius a. (V_a is not the actual final volume since elastic recovery increases the radius of curvature of the indentation, reducing its depth but having little effect on the chordal radius.) Tabor carried out a series of measurements of dynamic hardness, analysing his results by equation (25), on a range of metals at velocities up to ~ 5 ms^{-1}. Equation (25) was also used by Chaudhri (33).

If the elastic constants of the target and the sphere are known, then measurement of V_a is not necessary, and an expression can be derived for P in terms of the impact and rebound velocities and the elastic constants. This method offers clear advantages if measurements are to be made at high temperatures (43). Singer and Evans (36) also used sphere impact, at velocities up to ~ 10 ms^{-1}, to assess flow stress at high temperatures and high strain-rates, employing a strain-rate dependent model for the material behaviour. However, only preliminary results have been published.

At sufficiently high impact velocities, where $3e^2/8$ becomes much less than 1, elastic recovery may be neglected, $V \simeq V_a$, and equation (25) reduces to equation (19). The latter equation has been employed by several investigators (27,44,45) to estimate the dynamic indentation pressure under impacts at velocities between about 10 and 350 ms^{-1}. A slight modification of equation (19), to allow for work-hardening in the target material, was employed by

Mok and Duffy (46). All these workers (27,44-46) report values of dynamic indentation pressure which are higher than the quasi-static hardness, but have not made systematic comparisons with dynamic flow stress measured on identical materials by other methods.

An analogous experimental technique employs a conical, rather than a spherical, projectile. Equation (24) holds for quasi-static indentation by cones as well as by spheres. In this case, the value of c depends on the angle of the cone. The strain associated with the indentation is also a function of cone angle, being greater for sharper cones. For cones of large semi-angle β, the strain should be proportional to cot β (30), but for $\beta \gtrsim 60^0$, the relationship is more complex, as was shown experimentally for quasi-static indentation by Atkins and Tabor (47). With a conical indenter, geometrical similarity is preserved, and in this sense the method offers advantages over indentation by spheres. The equation of motion for a rigid conical striker of mass m may be written, by analogy with equation (5), as

$$m\ddot{x} = -\pi P \tan^2 \beta \cdot x^2 \qquad (26)$$

Solution for the appropriate boundary conditions gives (19) for the time of impact

$$t_p \simeq 1.4(3m/2\pi P\tan^2\beta)^{1/3} v_o^{-1/3} \qquad (27)$$

Since the mean strain is independent of impact velocity (geometrical similarity), the mean strain-rate, defined as the mean strain divided by t_p, is therefore proportional to the cube root of the impact velocity v_o.

Dynamic indentation by cones has not been widely investigated: Davis and Hunter (19) used the method to measure the dynamic hardness of a range of alloys, and found good agreement with the limited data on strain-rate sensitivity of flow stress available from other sources. Their analysis of the elastic recovery phase was subsequently reevaluated by Stilwell and Tabor (48) who showed that an equation very similar to equation (25) was valid for cone indentation:

$$PV_a = \frac{1}{2} mv_o^2 (1 - 0.363 e^2) \qquad (28)$$

Mahtab et al. (49) used dimensional analysis in an attempt to evaluate the inertial contribution to the indentation pressure, and, like Goldsmith and Yew (50), employed a Hopkinson pressure bar to measure the indentation force. Dynamic cone indentation has also been used at elevated temperatures (51).

Major uncertainties associated with indentation, either by spheres or cones, as a method of assessing mechanical properties at

high rates of strain, are in the contribution to the apparent
indentation pressure from inertial forces in the material, and from
friction at the interface. No single method, either of experimental
technique or of analysis, has so far been widely used, and it appears
that there are still difficulties in analysing the results of these
experiments. A further difficulty arises in connection with the
use of equation (24); although shown to be valid for quasi-static
indentation, it is not clear that the same value of c will apply in
the dynamic case. As will be pointed out below, the pattern of subsurface deformation is not necessarily the same in the dynamic case
as in the static, and the value of c might therefore be expected to
depend on the velocity of indentation.

IMPACT OF A DEFORMABLE PROJECTILE ON A RIGID TARGET:
TAYLOR IMPACT

Introduction

     The impact of a deformable projectile on to a plane non-
deforming target at subordnance velocities has been most widely
studied for the case of a plane-ended cylindrical projectile,
striking at normal incidence. G.I. Taylor (52) was the first to
show how, from the residual deformation of the cylinder, quantitative information could be derived about its mechanical properties
at the high rates of strain experienced in the experiment. The
use of the term "Taylor impact" to describe ballistic impact with
this geometry is widespread. Figure 8 illustrates how the projectile suffers plastic deformation on impact, "mushrooming" at the
impacted end and being reduced in length from its original overall
length, L_0, to a final length, L. Some length, X, of the projectile
remains undeformed. The simple geometry, and the fact that the
mechanical properties are derived from measurements on the recovered
projectile, with the impact velocity being the only quantity to be
measured during the experiment, make Taylor impact a potentially
attractive method for measuring high strain-rate mechanical properties. Although, clearly, the rate of strain experienced by the
projectile material depends on its location within the cylinder,
and varies continuously during impact, an estimate of the mean
strain-rate may be shown (52) to be given by:

$$\dot{\varepsilon} \simeq v_0/2(L_0 - X) \qquad\qquad (29)$$

Mean strain-rates for Taylor impact experiments, estimated in this
way, lie typically in the range $10^3 - 10^4$ s^{-1}.

     The major problem in using Taylor impact to assess mechanical
properties at these strain-rates is in selecting a suitable method
for analysing the results.

Theoretical Models for Taylor Impact

Development of the final projectile shape may be modelled by
considering the motion and interaction within the projectile of one-
dimensional elastic and plastic waves (see, e.g. refs. 1 and 53),
originating at the impact face. Most theoretical analyses employ
this approach and assume strain-rate independent material behaviour,
but vary in the other assumptions they make. Taylor (52) assumed
that the undeformed portion of the cylinder, remote from the target,
would be smoothly decelerated as a result of the rapid elastic wave
reflexions between the plastic wave-front and the free end of the

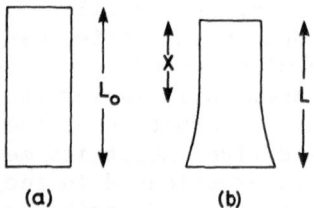

(a)   (b)

Figure 8.   Geometry of Cylindrical Projectile (a) Before and
             (b) After Normal Impact (Taylor Impact) Against
             a Plane Rigid Anvil.

projectile. He took the material to be rigid-plastic and, by
employing a momentum balance across the plastic front, derived an
expression for the yield stress of the projectile material in terms
of its density, the impact velocity, and the lengths L_o, L and X as
defined in Figure 8. Hawkyard (54) used an essentially similar
approach, but employed an energy balance, rather than a momentum
balance, across the plastic wavefront. The two theories due to
Taylor and Hawkyard have been discussed and compared by Johnson
(ref. 1, ch. 5). At low impact velocities (D $\gtrless$ 0.3) both theories
predict very nearly the same values for the reduction in length of
the cylinder on impact.

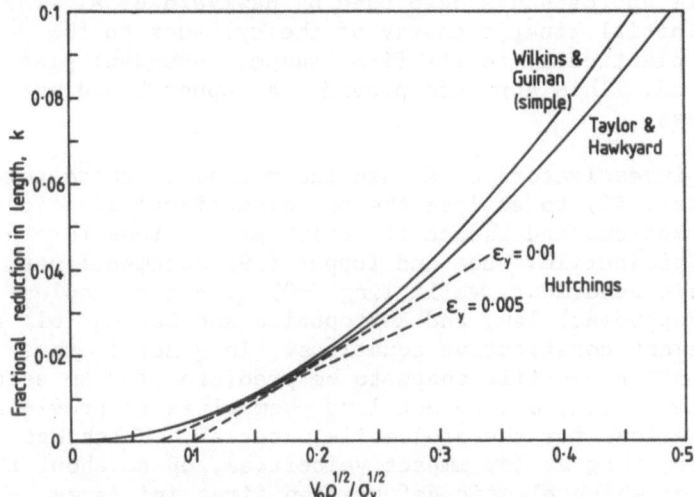

Figure 9.   Theoretical Dependence of Fractional Reduction in
            Length, k, on Dimensionless Impact Velocity in Taylor
            Impact, According to the Models of Wilkins and Guinan
            (56), Taylor (52), Hawkyard (54) and Hutchings (62).

The relationships are represented by a single solid curve in
Figure 9, which shows the theoretical dependence of the fractional
reduction in length, k, $(= (L_0 - L)/L_0)$ on the dimensionless group
$v_0\rho^{1/2}/\sigma_y^{1/2}$, which is proportional to impact velocity, v_0. σ_y is the
uniaxial yield stress of the projectile material and ρ is its
density.

    Taylor's original analysis (52) has been extended by Carley
(55, see also ref. 6, pp. 513-516) to incorporate the effects of
work-hardening in the projectile material; the theory can cope with
any assumed uniaxial stress-strain relationship, but numerical
solution of the equations is necessary. Wilkins and Guinan (56)
showed how a very simple model, in which the plastic wavefront is
assumed to be located at the target surface, yields the relationship

$$\ln(L/L_0) = \ln(1 - k) = -\rho v_0^2/2\sigma_y \qquad (30)$$

This relationship was also derived by Carley (55), and is plotted in
Figure 9 (upper solid curve). The difference between this curve and
that due to Taylor and Hawkyard (lower solid curve) is not great
over the velocity range shown here. At low velocities, both may be
approximated by

$$k \simeq \rho v_0^2/2\sigma_y \qquad (31)$$

Another simple approach has been used by Hawkyard et al. (57) who equated the initial kinetic energy of the cylinder to the work needed to deform it plastically to its final shape, redundant plastic work being neglected. This approach provides an upper bound estimate of the flow stress.

Several investigators have used the method of characteristics (see, e.g., ref. 58) to analyse the one-dimensional elastic and plastic wave motions and thence to derive predictions for the final shape of the projectile. Lee and Tupper (59) assumed linear elastic, perfect plastic behaviour, while Ting (60) used a viscoplastic (i.e. strain-rate dependent) law, and Raftopoulos and Davids (61) employed several different constitutive equations. In general, these methods permit the final projectile shape to be predicted for an assumed constitutive equation, but do not lend themselves to providing constitutive relations for the projectile material. Hutchings (62) has shown, however, that at low impact velocities, up to about three times the velocity at which plastic deformation first initiates, values of yield stress and yield strain can be readily deduced by a characteristic method. The method is possible since at these low velocities, for a linearly elastic, perfectly plastic material, the plastic wave disappears after the first elastic wave interaction. The relationship between the fractional change in length and the impact velocity predicted by this method is nearly linear, with an intercept on the velocity axis at the critical velocity below which the projectile deformation remains purely elastic. The broken lines in Figure 9 represent the predictions of this theory, for materials with yield strains of 0.01 and 0.005. This approach, while valid over a small velocity range and therefore only of limited applicability for metals (63), has proved useful for examining the dynamic yielding of polymers (64).

Theories based on computational models form a final category. That of Carley (55) has been noted above; Raftopoulos (65) and Hashmi and Thompson (66, see also ref. 67) have employed finite difference methods. These three models all treat the problem as one-dimensional. The only three-dimensional treatments have been by code computations: Wilkins and Guinan (56) and Carley (55) report results. In all these computational methods, a constitutive relation must be assumed, and then a projectile shape computed for comparison with the experimentally observed shape. A suitable constitutive relation for the projectile material is arrived at by successive refinement of the model in an iterative process. While it is not evident that this method will necessarily provide a unique constitutive model, the simplicity of the experimental method makes it suitable for checking and refining the constitutive relation to be employed in code computations (12).

Although we shall not further consider the question here, it is noteworthy that many of the theoretical treatments of the impact of a deforming projectile on a rigid target, discussed above, can be extended to cover the problem, important in terminal ballistics, in which both the projectile and the target suffer substantial deformation (see, e.g., refs. 68 and 69).

Experimental Results

Relatively few experimental results have been reported from Taylor impact experiments, in comparison with the number of theoretical treatments. Nearly all investigators have studied metals. Whiffin (70), Balendra and Travis (71), Wilkins and Guinan (56) and Azrin et al. (72) have reported measurements on a variety of alloys at room temperature, analysed by several different methods. Hutchings and O'Brien (63) used copper specimens at very low velocities (< 100 ms^{-1}). Gordon et al. (73) used composite projectiles, consisting of collinear cylinders of two different metals. High temperature measurements on metals have also been made, either by pre-heating the projectile and transferring it rapidly from the furnace to the gun barrel (57), or by reversing the impact geometry and firing a massive plane-ended projectile against a smaller stationary heated cylinder (74). Taylor impact of polymer specimens at high temperatures has been carried out by heating the projectile within the gun barrel (64).

Very few researchers have examined the deformation patterns and microstructures developed within the metallic projectiles after Taylor impact; the most detailed metallurgical examination has been that of Carrington and Gayler (75), on dural, mild steel, and a silver 7.5% copper alloy. Macroscopic flow patterns were revealed by macroetching in the steel and dural projectiles, and deformation twins were revealed in the steel by microscopic examination. Adiabatic shear bands in cylindrical specimens examined after Taylor impact have not been reported, although adiabatic shear has been observed in ogival penetrators at ordnance velocities (76) and would be expected to occur in cylindrical projectiles of suitable materials at these higher impact speeds.

Macroscopic observations of the internal deformation in metal cylinders after Taylor impact have also been reported by Hawkyard et al. (57) and Balendra and Travis (71). All these experimental observations confirm, as suggested by the computer code studies (56), that it is a considerable simplification to treat Taylor impact by a one-dimensional model.

DEFORMATION PATTERNS AND MICROSTRUCTURAL FEATURES
DUE TO BALLISTIC IMPACT

When a rigid projectile strikes a semi-infinite metal target at sub-ordnance velocity, plastic deformation in the target is localized in a zone around the indentation. Considerably more attention has been directed at zones of deformation around quasi-static indenta-

tions than at the deformation associated with indentations formed
by impact. Although, from limited experimental evidence (79), the
extent of the plastic zone appears to be about the same in both
cases, the strain distributions in the deformed regions are certainly
not identical.

Figure 10 (from ref. 26) illustrates the pattern of deformation
observed in a mild steel specimen impacted normally by a 9.5 mm
hard steel ball at ~ 270 ms^{-1} , compared with that around an indentation of the same size formed slowly by indentation in a hydraulic
press. The lines indicate the orientation of the banded ferrite-
pearlite microstructure, originally aligned parallel to the direction
of impact, and allow the pattern of subsurface flow to be deduced.
There is considerable difference between the two patterns, particularly noticeable immediately beneath the centre of the indentations,
and at the surface. In the static case, the banding remains nearly
perpendicular to the indentation surface, whereas in the dynamic
case the microstructure exhibits increasingly intense shearing in
regions towards the rim of the crater.

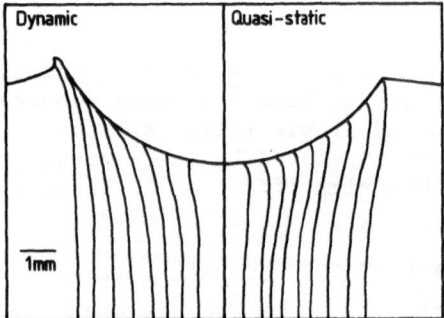

Figure 10.   Patterns of Deformation Revealed in a Banded Ferrite-
             Pearlite Steel by Etching, Resulting from the Dynamic
             and Quasi-Static Indentation of a 9.5 mm Diameter
             Steel Ball.

   The rims themselves are quite different; that formed dynamically
is raised up, and terminated by a weakly attached lip which in
certain circumstances becomes detached. Such concentrated deformation is completely absent in the quasi-static case. The difference
in the degree of "pile-up" around the two indentations, corresponding to a difference in the localization of plastic strain in the
two cases, is similar to that observed in the quasi-static indentation of work-hardened and annealed metals (14). An attempt to
account semi-quantitatively for the difference between the dynamic
and quasi-static cases, in terms of the stress-strain relationship
of the metal under the conditions of the experiment, has been made
recently by Sundararajan (80).

   Differences are also visible on the surfaces of the indentations
shown in Figure 10; that formed slowly is specularly reflective,
whereas the impression formed by impact exhibits white radial streaks
over the outer part of its surface. Similar observations of deformation patterns and surface markings have been reported in mild steel
and copper alloys by other investigators (81,82).

   Metals which are susceptible to adiabatic shear failure readily
form shear bands under projectile impact. Figure 11 (from ref. 83)
illustrates the pattern of shear bands produced by the normal impact
of a plane-ended cylindrical punch on α brass. The bands have formed

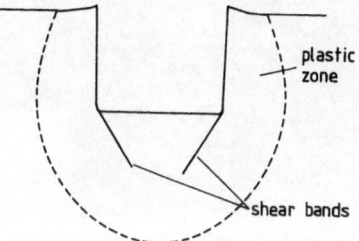

Figure 11. Pattern of deformation around a dynamic punch indentation
            in α brass. (Redrawn from ref. 83).

on a cone, of semi-angle approximately 30^0. The slip line field for plane-strain indentation by a flat rigid die includes a pair of orthogonal slip lines extending from the two corners of the die to a point on the axis of symmetry (84). These slip lines lie at an angle of 45^0 to the axis. The planes on which adiabatic shear failure is observed in Figure 11 are analogous to these slip lines, and have also been observed in other alloys (85,86). They form, however, at angles to the impact axis more acute than might be expected from the slip line analogy.

Shear bands are also observed beneath spherical indentations. In this case, a series of conical slip planes operates during the formation of the indentation, and well-developed shear bands form coaxial conical frusta. Shear bands in this geometry have been in aluminium alloys (86,87) and in pure titanium and its alloys (85,86). Patterns of shear bands of similar geometry are also commonly found in target perforation processes, which lie outside the scope of this review.

Timothy (27,86) has carried out a detailed study of adiabatic shear band formation in titanium alloys; because of their high susceptibility to adiabatic shear, bands are particularly well developed in these materials. Figure 12 shows the pattern of shear bands formed in Ti6Aℓ4V due to normal impact by a 6.35 mm steel ball

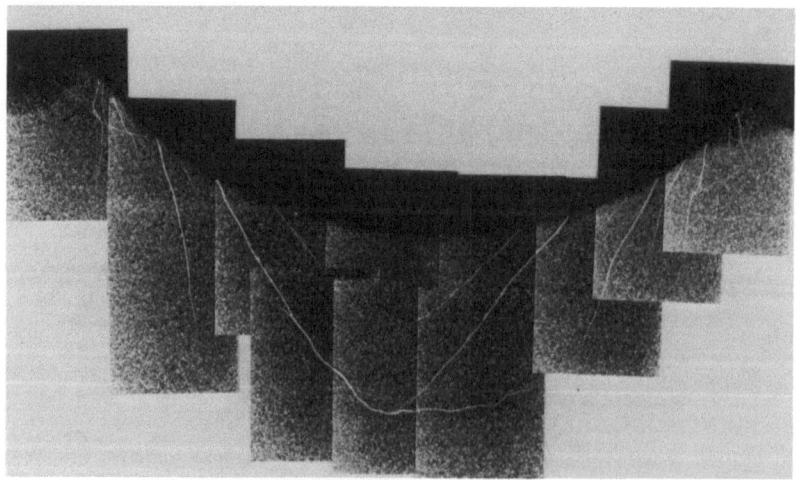

Figure 12.   Pattern of Deformation Around an Indentation Formed in Ti6Aℓ4V by the Normal Impact of a 6.35 mm Diameter Steel Ball at 318 ms^{-1}. (From ref. 86).

at 318 ms^{-1}. Most of the shear bands lie on conical surfaces, analogous to the single cone seen in Figure 11, although there is
occasional bifurcation of the bands. The crater surface, as shown
in Figure 13, exhibits a central smooth region, surrounded by a
region of rougher texture; the appearance is similar to that seen
in dynamic indentations in mild steel (26) and discussed above. The
approximately circular steps seen on the crater surface mark the
intersection of subsurface shear bands with the surface. One of
these steps in shown at higher magnification in Figure 14, which
also shows the "knobbly" nature of the roughened crater surface.
This surface texture is similar to that shown by DeMorton and
Woodward (88) to result from heating due to high speed friction. It
is thought that the interfacial displacement necessary towards the
rim of the indentation to form the spherical surface will have led
to a transition from sticking friction (no displacement) at the
centre of the crater, to sliding friction (producing a knobbly
surface) at some distance from the axis.

Not all shear bands intersecting the surface do so at a step;
Figures 15(a) and (b) show bands in Ti6Aℓ4V with and without associated steps. It is noteworthy that surface steps, where they occur,
are consistent with the formation of the step during elastic recovery, rather than during impact. For example, the directions of
the steps in Figure 13 indicate that the central conical "plug" of
material bounded by the shear bands has moved upwards relative to
the surrounding material. Under extreme conditions, where fractures
occur within the shear bands, this conical plug can become detached
from the bulk (85,87).

The impact conditions necessary for shear band formation in
titanium alloys have been investigated by Timothy (27,86). By
using spheres of different sizes and densities, the strain introduced, and the mean strain-rate, could be varied independently
from the impact velocity. In this way the validity of various
criteria for band formation could be tested. Figure 16 shows the
patterns of shear bands produced by impacts of 3.175 mm diameter
spheres of different density, but nominally constant velocity, on
Ti6Aℓ4V. Shear bands initiated (i.e. were first detectable by
optical metallography) at a particular indentation geometry, conveniently characterised by a critical value of a/r, as shown in Figure
17. Shear bands also initiated at nearly the same value of a/r
(0.6) in experiments with steel spheres in which the size was varied
between 0.79 and 6.35 mm diameter. Under the conditions of this
experiment, a critical strain criterion for shear band initiation,
as proposed by Culver (89) and Staker (90), was found to be preferable to any based on impact velocity or strain-rate since, as has
been discussed above, the strain associated with a spherical indentation will depend on a/r. Critical values of a/r for shear band
initiation have been measured experimentally for several titanium
alloys, in various heat treatment conditions and over a range of

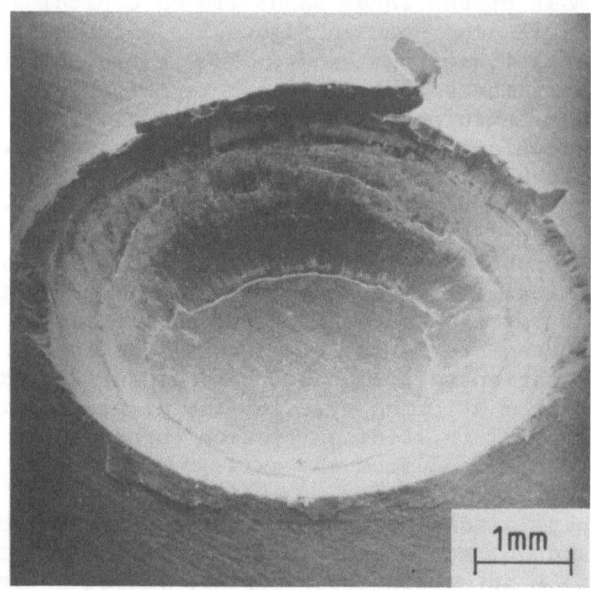

Figure 13.   **SEM Micrograph** of the Indentation Shown in Figure 12.
Note the Surface Features.

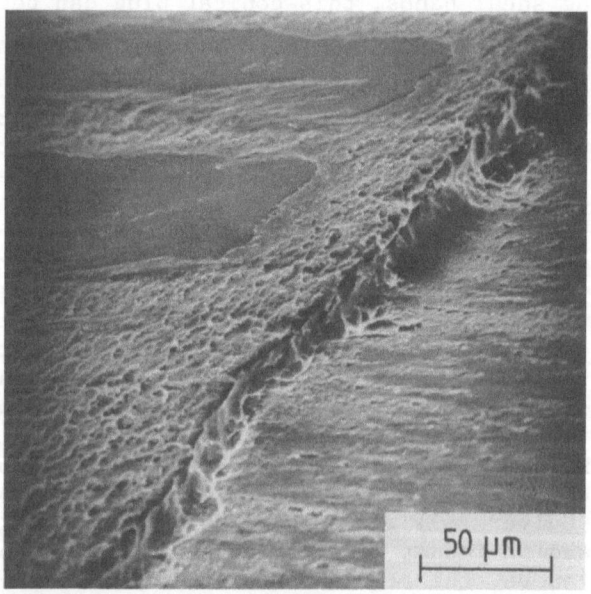

Figure 14.   **SEM Micrograph**, at Higher Magnification, of the
Surface of the Indentation Shown in Figures 12 and 13.

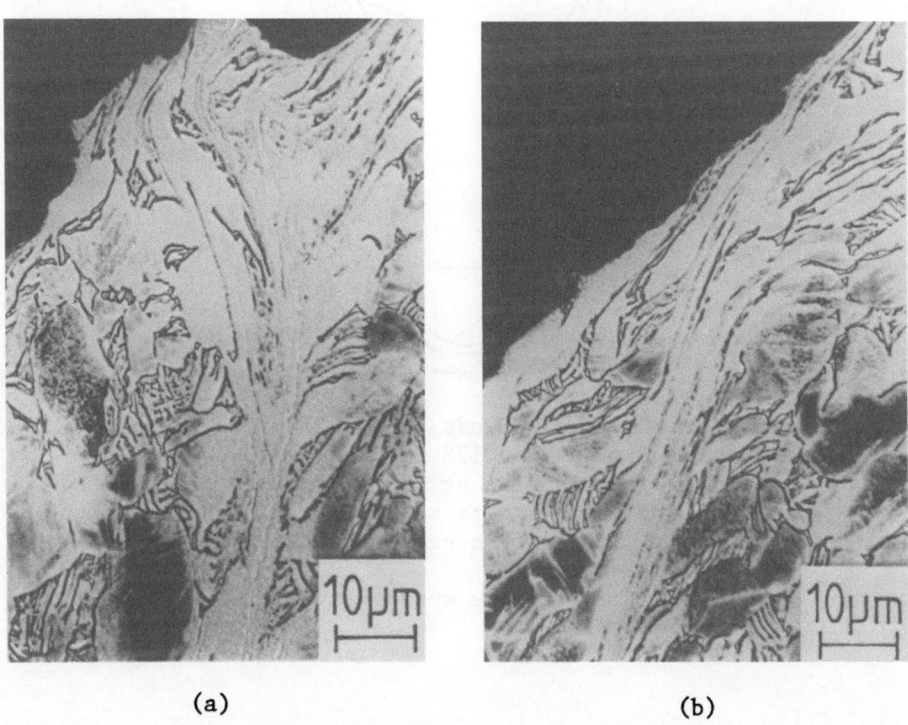

(a)   (b)

Figure 15.   Shear Bands Intersecting the Indentation Surface in
             Ti6Aℓ4V : (a) Surface Step Associated with a Band,
             (b) Intersection of a Band with a Smooth Surface.
             (From ref. 27).

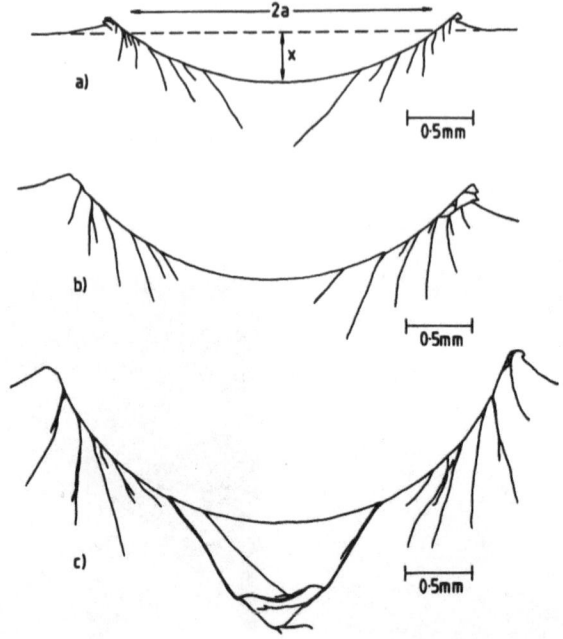

Figure 16.   Patterns of Shear Bands Beneath Impact Craters in
             Ti6Al4V formed by 3.175 mm Diameter Spheres.
             (a)   Sapphire sphere at 324 ms^{-1}.
             (b)   Hard steel sphere at 328 ms^{-1}. The lip at the left
                   hand side of the crater has been detached along
                   a shear band.
             (c)   Tungsten carbide sphere at 330 ms^{-1}.
             (From ref. 27).

temperatures (86); agreement with previously published values of
critical strain, where available, is quite good. The method appears
to provide a simple experimental method for assessing the susceptibility of alloys to adiabatic shear.

      Removal of metal from the target is important in the problem of
erosion by solid particle impingement. Shewmon and co-workers (81,
91,92) have investigated the material removal processes involved
in the impact of spheres, following earlier work by Hutchings and
Winter (93-95). Weight losses from single impacts are usually small
(typically no more than 10^{-3} times the mass of the projectile), for
spheres at normal incidence on many structural alloys at sub-ordnance

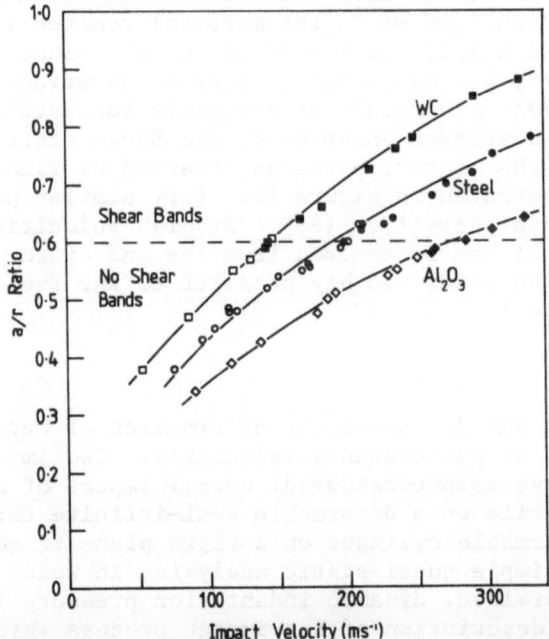

Figure 17.   Variation of a/r Ratio with Impact Velocity for 3.175 mm
             Spheres of Different Density. Open Symbols: No Shear
             Bands Detected by Optical Metallography.   Solid
             Symbols: Shear Bands Present.   (From ref. 27).

velocities. The weight loss originates in the thin, extruded lip
around the crater. In alloys particularly liable to adiabatic shear
failure, fragments of the crater rim may detach along shear bands;
an example of this occurrence is seen in Figure 16(b), where material
has been detached from the left hand side of the crater, along a
shear band nearly parallel to the free surface, similar to that
visible beneath the rim at the right. Where two impacts occur in
close proximity, it has been found for copper and aluminium alloys
(81,91,92) that a chip of metal can become detached in the region
of overlap between the two craters. Adiabatic shear then plays an
important role in the mechanism of material removal.

     In the oblique impact of a sphere, metal is pushed ahead of the
sphere forming a lip at the exit end of the crater. Under certain

conditions of velocity and impact angle this lip becomes detached; in the case of impacts on mild steel at 30^0 impact angle, at velocities between 50 and 400 ms^{-1}, the material removed in this way represented between 1/12 and 1/4 of the total volume of the crater (95). The lip may become detached along a subsurface band of intense shear (93). In metals particularly susceptible to adiabatic shear failure, subsurface shear bands are found distributed beneath the exit end of the crater; patterns observed by Timothy (86) in Ti6Aℓ4V are illustrated in Figure 18. Very similar patterns have been seen in copper–beryllium (81). At high velocities, large fragments of metal can be removed from the end of the crater, by fracture along the bands roughly parallel to the free surface, visible in Figure 18.

CONCLUSIONS

This review has discussed the deformation of metals due to ballistic impact at sub–ordnance velocities. Two impact geometries in particular have been considered: normal impact of a rigid spherical projectile on a deformable semi–infinite target, and normal impact of a deformable cylinder on a rigid plane target. For the former case, a simple quasi–static analysis, in which a constant, empirically–determined, dynamic indentation pressure is assumed to act, provides a description of the impact process which is adequate for many purposes. It cannot, however, provide information about the strain field around the indentation such as would result from

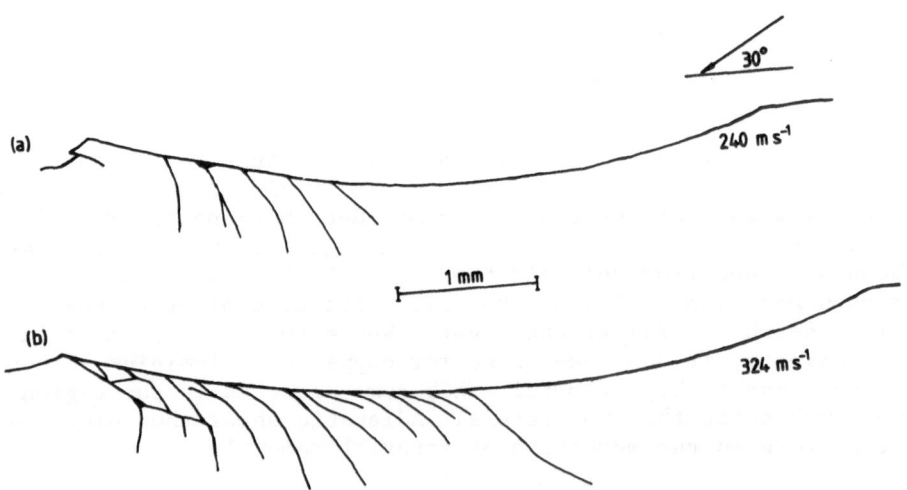

Figure 18.   Patterns of Shear Bands Formed in Ti6Aℓ4V by Oblique
             Impact (at 30^0) of 3.175 mm Hard Steel Spheres (a) at
             240 ms^{-1} and (b) at 324 ms^{-1}.   (From ref. 86).

a code computation. Dynamic indentation as a method of measuring mechanical properties at high strain-rates has been discussed, and the difficulties associated with the method have been highlighted.

Taylor impact, the normal impact of a deformable cylinder on a rigid target, has been reviewed. The technique offers attractions as a method of assessing mechanical properties at high strain-rates, but it is not simple to select a suitable method for analysing the results.

Deformation patterns and micro-structural features, notably adiabatic shear bands, associated with dynamic indentation have been discussed, and experimental results presented.

REFERENCES

1. W. Johnson, "Impact Strength of Materials," Edward Arnold, London (1972).
2. M. E. Backman and W. Goldsmith, The Mechanics of Penetration of Projectiles into Targets, Int. J. Engng. Sci. 16:1 (1978).
3. H. G. Hopkins, Dynamic Anelastic Deformation of Metals, App. Mech. Rev. 14:417 (1961).
4. W. Goldsmith, Impact: the Collision of Solids, App. Mech. Rev. 16:855 (1963).
5. G. H. Jonas and J. A. Zukas, Mechanics of Penetration: Analysis and Experiment, Int. J. Engng. Sci. 16:879 (1978).
6. E. Billington and A. Tate, "Physics of Deformation and Flow," McGraw-Hill, New York (1981).
7. J. A. Zukas, Chapter 5, in "Impact Dynamics," J. A. Zukas, T. Nicholas, H. F. Swift, L. B. Greszczuk and D. R. Curran, John Wiley, New York,(1982).
8. W. Johnson, in "The Mechanics of Solids," 303, ed. H. G. Hopkins and M. J. Sewell, Pergamon, Oxford (1982).
9. A. W. Ruff and S. M. Wiederhorn, Erosion by Solid Particle Impact, Treatise on Mat. Sci. and Tech. 16:69 (1979).
10. G. W. Meetham, ed., "Development of Gas Turbine Materials," Applied Science, London (1981).
11. National Materials Advisory Board, "Materials Response to Ultra-high Loading Rates," Rept. NMAB-356, PB80-153521, Washington D. C. (1980).
12. M. L. Wilkins, Mechanics of Penetration and Perforation, Int. J. Engng. Sci. 16:793 (1978).
13. J. A. Zukas, Chapter 10, in "Impact Dynamics," J. A. Zukas, T. Nicholas, H. F. Swift, L. B. Greszczuk and D. R. Curran, John Wiley, New York,(1982).
14. D. Tabor, "The Hardness of Metals," Clarendon Press, Oxford (1951).
15. R. M. Davies, The Determination of Static and Dynamic Yield Stress using a Steel Ball, Proc. Roy. Soc. Lond. A197:416 (1949).

16. K. L. Johnson, Reversed Plastic Flow during the Unloading of a
     Spherical Indenter, Nature 199:1282 (1963).

17. S. C. Hunter, Energy absorbed by Elastic Waves during Impact,
     J. Mech. Phys. Solids 5:162 (1957).

18. I. M. Hutchings, Energy absorbed by Elastic Waves during Plastic
     Impact, J. Phys. D.: Appl. Phys. 12:1819 (1979).

19. C. D. Davis and S. C. Hunter, Assessment of the Strain-Rate
     Sensitivity of Metals by Indentation with Conical Indenters,
     J. Mech. Phys. Solids 8:235 (1960).

20. K. L. Johnson, in "Engineering Plasticity" ed. J. Heyman and
     F. A. Leckie, Cambridge Univ. Press, Cambridge (1966).

21. P. L. Makin and I. M. Hutchings, University of Cambridge,
     unpublished work (1982).

22. D. Tabor, Mohs' Hardness Scale - A Physical Interpretation,
     Proc. Phys. Soc. B67:249 (1954).

23. A. Kelly, "Strong Solids," Oxford Univ. Press, Oxford (1966).

24. I. M. Hutchings, Some Comments on the Theoretical Treatment of
     Erosive Particle Impacts, Proc. 5th Int. Conf. on Erosion by
     Liquid and Solid Impact, Cavendish Laboratory, Cambridge (1979).

25. A. G. Evans, M. E. Gulden and M. Rosenblatt, Impact Damage in
     Brittle Materials in the Elastic-plastic response régime,
     Proc. Roy. Soc. Lond. A361:343 (1978).

26. I. M. Hutchings, The Erosion of Ductile Metals by Solid
     Particles, Ph.D. dissertation, University of Cambridge (1974).

27. S. P. Timothy and I. M. Hutchings, Microstructural Features
     associated with Ballistic Impact in Ti6Aℓ4V, Proc. 7th Int.
     Conf. on High Energy Rate Fabrication, 19, ed.
     T. Z. Blazynski, University of Leeds (1981).

28. W. Goldsmith and P. T. Lyman, The Penetration of Hard Steel
     Spheres into Plane Metal Surfaces, Trans. ASME E:Jnl of Appl.
     Mech. 27:717 (1960).

29. I. M. Hutchings, Strain Rate Effects in Microparticle Impact,
     J. Phys. D.: Appl.Phys. 10:L179 (1977).

30. K. L. Johnson, The Correlation of Indentation Experiments,
     J. Mech. Phys. Solids 18:115 (1970).

31. P. S. Follansbee, G. B. Sinclair and J. C. Williams, Modelling
     of Low Velocity, Particulate Erosion in Ductile Materials, in
     "Wear of Materials 1981," 577, A.S.M.E., New York (1981).

32. G. B. Sinclair, Private Communication to E. Yoffe (1981).

33. M. M. Chaudhri, J. K. Wells and A. Stephens, Dynamic Hardness,
     Deformation and Fracture of Simple Ionic Crystals at Very
     High Rates of Strain, Phil. Mag. 43A:643 (1981).

34. N. L. Goodier, On the Mechanics of Indentation and Cratering in
     Solid Targets of Strain-Hardening Metal by Impact of Hard and
     Soft Spheres, Poulter Res. Labs. Tech. Rept. 002-64,
     Stanford Research Inst. Menlo Park (1964).

35. M. Mamoun, Analytical Models for the Erosive-Corrosive Wear
     Process, Appendix 1 to 2nd Quarterly Rept., Jan.- Mar. 1975,
     ANL-75-XX-2, Argonne Natl. Lab., and Appendix to 3rd Quarterly
     Rept., Apr. - June 1975, Argonne Natl. Lab.

36. A. R. E. Singer and R. W. Evans, New Technique for Rapid
     Measurement of High-Temperature Flow Stress, Met. Tech.
     7:142 (1980), see also: W. Johnson and S. K. Ghosh, Met. Tech.
     8:38 (1981).
37. W. Johnson, A. K. Sengupta and S. K. Ghosh, High Velocity Oblique
     Impact and Ricochet mainly of Long Rod Projectiles: an over-
     view, Int. J. Mech. Sci. 24:425 (1982).
38. I. M. Hutchings, N. H. Macmillan and D. G. Rickerby, Further
     Studies of the oblique Impact of a Hard Sphere against a
     Ductile Solid, Int. J. Mech. Sci. 23:639 (1981).
39. J. M. Lifshitz and H. Kolsky, Some Experiments on Anelastic Re-
     bound, J. Mech. Phys. Solids 12:35 (1964).
40. see ref. 15.
41. C. Yew and W. Goldsmith, Stress Distributions in Soft Metals
     due to Static and Dynamic Loading by a Steel Sphere, Trans.
     ASME E.: J. Appl. Mech. 31:635 (1964).
42. see ref. 28.
43. F. P. Bowden and D. Tabor, "Friction and Lubrication of Solids,"
     Vol. II, 438, Oxford Univ. Press, Oxford (1964).
44. D. G. Rickerby and N. H. Macmillan, Erosion of Aluminium by
     Solid Particle Impingement at Normal Incidence, Wear 60:369
     (1980).
45. G. Sundararajan and P. G. Shewmon, The Use of Dynamic Impact
     Experiments in the Determination of the Strain Rate Sensiti-
     vity of Metals and Alloys, to be published (1982).
46. C. H. Mok and J. Duffy, The Dynamic Stress-Strain Relation of
     Metals as determined from Impact Tests with a Hard Ball, Int.
     J. Mech. Sci. 7:355 (1965).
47. A. G. Atkins and D. Tabor, Plastic Indentation in Metals with
     Cones, J. Mech. Phys. Solids 13:149 (1965).
48. N. A. Stilwell and D. Tabor, Elastic Recovery of Conical
     Indentations, Proc. Phys. Soc. 78:169 (1961).
49. F. U. Mahtab, W. Johnson and R. A. C. Slater, Dynamic Indentation
     of Copper, an Aluminium Alloy and Mild Steel with Conical
     Projectiles and Dynamic Tip Flattening of Conical Projectiles
     at Ambient Temperature, Int. J. Mech. Sci. 7:685 (1965).
50. W. Goldsmith and C. Yew, Penetration of Conical Indenters into
     Plane Metal Surfaces, Proc. 4th U. S. Nat. Cong. of Appl.
     Mechanics 177, A. S. M. E. (1962).
51. F. U. Mahtab, W. Johnson and R. A. C. Slater, Dynamic Indentation
     of Copper and an Aluminium Alloy with a Conical Projectile at
     Elevated Temperatures, Proc. Instn. Mech. Engrs. 180:285
     (1965-66).
52. G. I. Taylor, The Use of Flat-Ended Projectiles for Determining
     Dynamic Yield Stress I. : Theoretical Consideration, Proc. Roy.
     Soc. A194:289 (1948).
53. H. Kolsky, "Stress Waves in Solids, " Clarendon Press, Oxford
     (1953), repr. 1963, Dover Publications, New York.
54. J. B. Hawkyard, A Theory for the Mushrooming of Flat-ended
     Projectiles impinging on a Flat Rigid Anvil, using Energy

Considerations, Int. J. Mech. Sci. 11:313 (1969).

55. D. J. Carley, unpublished rept., A. W. R. E., Aldermaston, Berks., England (1978).

56. M. L. Wilkins and M. W. Guinan, Impact of Cylinders on a Rigid Boundary, J. Appl. Phys. 44:1200 (1973).

57. J. B. Hawkyard, D. Eaton and W. Johnson, Mean Dynamic Yield Strength of Copper and Low Carbon Steel at Elevated Temperatures from Measurements of the Mushrooming of Flat-ended Projectiles, Int. J. Mech. Sci., 10:929 (1968).

58. H. G. Hopkins, in "Engineering Plasticity," ed. J. Heyman and F. A. Leckie, 277, Cambridge Univ. Press, Cambridge (1966).

59. E. H. Lee and S. J. Tupper, Analysis of Plastic Deformation in a Steel Cylinder Striking a Rigid Target, J. Appl. Mech. 21:63 (1954).

60. T. C. T. Ting, Impact of a Non-linear Viscoplastic Rod on a Rigid Wall, Trans. ASME E.: J. Appl. Mech. 33:505 (1966).

61. D. Raftopoulos and N. Davids, Elastoplastic Impact on Rigid Targets, AIAA Journal 5:2254 (1967).

62. I. M. Hutchings, Estimation of Yield Stress in Polymers at High Strain-Rates using G. I. Taylor's Impact Technique, J. Mech. Phys. Solids 26:289 (1979).

63. I. M. Hutchings and T. J. O'Brien, Normal Impact of Metal Projectiles against a Rigid Target at Low Velocities, Int. J. Mech. Sci. 23:255 (1981).

64. S. N. Kukureka and I. M. Hutchings, Measurement of the Mechanical Properties of Polymers at High Strain-Rates by Taylor Impact, Proc. 7th Int. Conf. on High Energy Rate Fabrication, 29, ed. T. Z. Blazynski, University of Leeds (1981).

65. D. Raftopoulos, Longitudinal Impact of Two Mutually Plastically-deformable Missiles, Int. J. Solids and Structures 5:399 (1969).

66. M. S. J. Hashmi and P. J. Thompson, A Numerical Method of Analysis for the Mushrooming of Flat-ended Projectiles Impinging on a Flat Rigid Anvil, Int. J. Mech. Sci. 19:273 (1977).

67. R. L. Woodward and J. P. Lambert, A Discussion of the Calculation of Forces in the One-dimensional Finite Difference Model of Hashmi and Thompson, Int. J. Mech. Sci. 23:497 (1981).

68. R. F. Recht, Taylor Ballistic Impact Modelling applied to Deformation and Mass Loss Determinations, Int. J. Engng. Sci. 16:809 (1978).

69. R. L. Woodward, Penetration of Semi-infinite Metal Targets by Deforming Projectiles, Int. J. Mech. Sci. 24:73 (1982).

70. A. C. Whiffin, The use of Flat-ended Projectiles for determining Dynamic Yield Stress II: Tests on various Metallic Materials, Proc. Roy. Soc. A194:300 (1948).

71. B. Balendra and F. W. Travis, An Examination of the Double-frustum Phenomenon in the Mushrooming of Cylindrical Projectiles upon High-speed Impact with a Rigid Anvil, Int. J. Mech. Sci. 13:495 (1971).

72. M. Azrin, A. A. Anctil and E. B. Kula, Dynamic Mechanical Properties of Intercritically Rolled High Hardness Steel,

Mat. Sci. and Engng. 53:285 (1982).

73. P. Gordon, R. Karpp, S. C. Sanday and M. Schwartz, Influence of Dynamic Yield Point in Multimaterial Impact, J. Appl.Phys. 48:172 (1977).

74. W. H. Gust, High Impact Deformation of Metal Cylinders at Elevated Temperatures, J. Appl. Phys. 53:3566 (1982).

75. W. E. Carrington and M. L. V. Gayler, The Use of Flat-ended Projectiles for determining Dynamic Yield Stress III: Changes in Microstructure caused by Deformation under Impact at High Striking Velocities, Proc. Roy. Soc. A194:323 (1948).

76. G. J. Irwin, Metallographic Interpretation of Impacted Ogive Penetrators, Rept. DREV-R-652/72, N73-18538, Defence Research Establishment, Valcartier, Canada (1972).

77. L. E. Samuels and T. O. Mulhearn, An Experimental Investigation of the Deformed Zone Associated with Indentation Hardness Impressions, J. Mech. Phys. Solids, 5:125 (1957).

78. R. L. Woodward, Strain Fields Associated with the Indentation of Metals, J. Aust. Inst. Metals 19:128 (1974).

79. S. T. S. Al-Hassani, Mechanical Aspects of Residual Stress Development in Shot Peening, in "First International Conference on Shot Peening," 583, ed. A. Niku-Lari, Pergamon, Oxford (1982).

80. G. Sundararajan, An Analysis of the Localization of Deformation and Weight Loss during Single Particle Normal Impact, to be published (1982).

81. T. Quadir and P. Shewmon, Solid Particle Erosion Mechanisms in Copper and Two Copper Alloys, Met. Trans. 12A:1163 (1981).

82. D. R. Andrews and J. E. Field, The Erosion of Metals by the Normal Impingement of Hard Solid Spheres, J. Phys. D.: Appl Phys. 15:571 (1982).

83. J. V. Craig and T. A. C. Stock, Microstructural Damage Adjacent to Bullet Holes in 70-30 Brass, J. Aust. Inst. Metals 15:1 (1970).

84. R. Hill, "Plasticity," Clarendon Press, Oxford (1950).

85. R. E. Winter, Adiabatic Shear of Titanium and Polymethylmethacrylate, Phil. Mag. 31:765 (1975).

86. S. P. Timothy, Ph.D. Dissertation, University of Cambridge (1982).

87. T. A. C. Stock and K. R. L. Thompson, Penetration of Aluminium Alloys by Projectiles, Met. Trans. 1:219 (1970).

88. M. E. DeMorton and R. L. Woodward, The Effect of Friction on the Structure of Surfaces produced during Ballistic Tests, Wear 47:195 (1978).

89. R. S. Culver, in "Metallurgical Effects at High Strain-Rates," 519, ed. R. W. Rohde et al., Plenum, New York (1973).

90. M. R. Staker, The Relation between Adiabatic Shear Instability Strain and Material Properties, Acta Met. 29:683 (1981).

91. T. Christman and P. G. Shewmon, Erosion of a Strong Aluminium Alloy, Wear 52:57 (1979).

92. T. Christman and P. G. Shewmon, Adiabatic Shear Localization and Erosion of Strong Aluminium Alloys, Wear 54:145.

93. I. M. Hutchings, R. E. Winter and J. E. Field, Solid Particle
     Erosion of Metals: the Removal of Surface Material by
     Spherical Projectiles, Proc. Roy. Soc. A 348:379 (1976).
94. R. E. Winter and I. M. Hutchings, The Role of Adiabatic Shear
     in Solid Particle Erosion, Wear 34:141 (1975).
95. I. M. Hutchings and R. E. Winter, The Erosion of Ductile Metals
     by Spherical Particles, J. Phys. D.: Appl. Phys. 8:8 (1975).

DYNAMIC FRACTURE MECHANICS AND ITS APPLICATIONS TO MATERIAL
BEHAVIOR UNDER HIGH STRESS AND LOADING RATES

M. F. Kanninen

Stress Analysis and Fracture Section
Battelle Columbus Laboratories
Columbus, Ohio   43201

INTRODUCTION

Fracture is a key consideration in assessing material behavior under high stress and loading rates. Fracture assessments generally introduce the subject of fracture mechanics. This paper provides a review of current research in a specialization of the general subject called dynamic fracture mechanics. The objective of the paper is to help set the stage for the more detailed papers that follow in this volume, and to provide a point of departure for further work on fracture under high stress and loading rates.

Dynamic fracture mechanics encompasses two main areas of application: (1) the initiation of cracking under a rapidly applied loading, and (2) the rapid propagation of a crack in a body. While both areas have important technological applications, most current research is focused on the latter. For this reason the following concentrates on rapid crack propagation and crack arrest. It should also be recognized that two points of view on dynamic fracture mechanics are extant: continuum-based and micromechanical-based. Except for one dominant crack-like defect, the former view generally assumes the material to be continuous. The latter, in contrast, considers the failure process to develop from the initiation, growth, and coalescence of a great many material imperfections. This paper is confined to the continuum view, consistent with the author's previous work in the subject,[1,2] and with most research and application work in the field. The micromechanical view is given, for example, by Zukas, et al.[3]

CURRENT STATUS OF DYNAMIC FRACTURE MECHANICS FOR RAPID
CRACK PROPAGATION AND ARREST

The bulk of current state-of-the-art applications of dynamic
fracture mechanics to rapid crack propagation and arrest are based
upon the use of quasi-static linear elastic fracture mechanics
(LEFM). Specifically, the relation $K = K_{IC}$ is used for the initiation of unstable crack propagation and the relation $K = K_{Ia}$ for
its termination.* In both instances K is calculated as if the
crack were stationary while K_{IC} and K_{Ia} are taken as temperature-
dependent material properties. For example, the thermal shock
problem for nuclear pressure vessels, an application that is receiving a great deal of attention at present,[4] is being handled in
just this way.[5] It can be noted that this point of view does not
include (nor could it) any consideration of the unstable crack
propagation process that links the initiation and arrest points.

The quasi-static view of crack propagation requires crack
arrest to occur smoothly with an intimate connection between the
slowing down process and the static deformation state long after
arrest; i.e., crack arrest must be the reverse, in time, of crack
growth initiation. However, there is ample evidence to suggest
that crack arrest instead occurs abruptly. A point of view that
gives direct consideration to crack propagation, with crack arrest
occurring only when continued propagation becomes impossible, is
free from this erroneous assumption. Within the confines of
elastodynamic behavior, unstable crack propagation occurs in such
an approach under the condition that $K = K_{ID}$ (a), where K_{ID}, the
dynamic fracture toughness, is a temperature-dependent function of
the crack speed, $\dot{a}$. In this dynamic approach arrest occurs at the
position and time t_a for which $K < min (K_{ID})$ for all greater times
$t > t_a$, where K is the dynamically calculated value of the stress
intensity factor.

The experiments of Kalthoff, et al[6] have shown clearly that
the dynamic approach to the arrest of unstable propagation is the
more correct and that the use of K_{Ia} as a material property is
therefore not well founded. Specifically, Kalthoff, et al showed
by direct measurements (using the method of caustics for Araldite
B double cantilever beam specimens) that, while cracks propagating
at different speeds arrested at the same value of the dynamic
stress intensity factor, the K_{Ia} values were clearly dependent
upon the crack jump length. This is completely in accord with the

_____

*The use of the subscript I in designating a material fracture
property is technically correct only when plane strain conditions
are satisfied. However, the distinction is not too important for
the purposes of this paper.

theoretical analysis prediction made previously for the DCB specimen by Kanninen who showed that K_{Ia} systematically decreases with
crack jump length.[7] Nevertheless, the dynamic view is not univer-
sally accepted. There are strong advocates (e.g., Ripling, et al [8])
for the K_{Ia} as a material property approach. Others[9] appear to
be aware of the lack of fundamental validity of K_{Ia} but have nevertheless chosen to use it as an expediency on the basis that it
promises a conservative estimate of the true crack arrest property;
a property that is considerably more difficult to determine.

It should be recognized that there are many practical problems
where the difference between the predictions of the two approaches
is not great. These applications are typified by (1) the use of
K_{Ia} values measured from short crack jumps (whereupon K_{Ia} closely
approximates the minumum value of K_{ID}) and (2) component boundaries
that do not reflect stress waves back to the running crack tip. The
through-the-wall propagation of a long axial crack in a circular
cylinder (e.g., in the thermal shock problem[4],[5]) is an example
where (2) is fairly well (but, as shown by Jung and Kanninen,[10] not
exactly!) satisfied. The double cantilever beam (DCB) test specimen is one where it decidedly is not.[6],[7] The decisive observations
of run/arrest events in the DCB specimen are usually brushed aside
on the grounds that it is "too dynamic" to represent actual applications. Nevertheless, others such as Kanazawa and Machida[11] state
that dynamic considerations are indispensable for the interpretation
of run/arrest events even in wide plate tests.

It is important to understand the kinds of experimentation
that can be performed in this area. Rapid crack propagation/arrest
experiments can be classified as either direct or indirect, depending upon whether crack tip characterizing parameters are measured
during the event or are inferred from a supplementary analysis. In
the first category are the experiments on photoelastic and reflective materials where a shadow spot (caustic)[6] or a fringe pattern
[12],[13] is photographed by high speed cameras. The second category
contains experiments where only the crack growth history is
measured; e.g., by timing wires broken by the advancing crack.[14]
The resulting crack length versus time data can be used as input to,
say, a finite element computation in which details (e.g., the
dynamic stress intensity factor) that cannot be measured can then
be calculated in what has been called a "generation-phase" calculation.[1],[2]

In both types of experiments some assumption about both the
nature of the event and the constitutive behavior of the material
during rapid crack propagation is required. While this is obvious
in the indirect approach, it is equally true in the direct approach.
The size of a reflected shadow spot may indeed correspond to the
dimensions of the crack tip plastic zone, but its relation to other

features of the deformation will depend upon the material behavior at the strain rates experienced by the crack tip. It therefore appears that the question of the correct formulation for a rapidly propagating crack cannot be unequivocally answered by experimentation alone any more than it can by analysis alone.

THE POSSIBLE NONUNIQUENESS OF THE DYNAMIC FRACTURE TOUGHNESS IN UNSTABLE CRACK PROPAGATION

The application of elastodynamic fracture mechanics treatments requires that K_{ID} be a geometry- and load-independent property of the material. As already mentioned, there is ample evidence to show that K_{ID} is more nearly invariant (e.g., with respect to crack jump length) than is the quasi-static parameter K_{Ia}. Nevertheless, disquieting results have been accumulating suggesting that both load- and geometry-dependence of K_{ID} exist. A typical result showing the latter effect is given in Figure 1 which shows the dynamic fracture toughness values deduced from crack length versus time data in a series of experiments on a polymeric material. Clearly, if K_{ID} is a true material property, all of these data would lie on a single curve.

Geometry-dependence of K_{ID} has also been suggested by Kobayashi and his coworkers at the University of Washington,[12] and by Kalthoff and his associates in Germany.[15] In particular, Kalthoff's results were obtained by the method of caustics in the photoelastic material Araldite B. They show that K_{ID} values determined in the rectangular double cantilever beam (RDCB) specimens differ significantly from those in single edge notch (SEN) specimens. Note that, by the use of the combination RDCB/SEN specimen, Kalthoff was cleverly able to establish that this difference was not due to batch-to-batch material variations; a strong possibility for the large spread in the data in Figure 1.

It has been suggested in work at the University of Maryland[13] that substantial viscous damping exists during dynamic crack propagation. In order to determine if such an effect could account for the geometry-dependence observed in Figure 1, Popelar and Kanninen[16] devised a dynamic viscoelastic analysis model for the DCB specimen. The viscoelastic material properties were set by matching the post arrest stress intensity factors measured by Kalthoff.[15] The amount of damping prior to arrest could then be calculated. A comparison between their calculation and Kalthoff's measurements is shown in Figure 2.

Figure 2 shows the pronounced difference between K_{Ia} and the minimum value of K_{ID} that is found in this type of experiment. The latter value is 0.8 $MNm^{-3/2}$ (pre-arrest regime) while the former is 0.7 $MNm^{-3/2}$ (long-time ring-down value after arrest). But, of more significance for the question of the possible non-uniqueness of K_{ID},

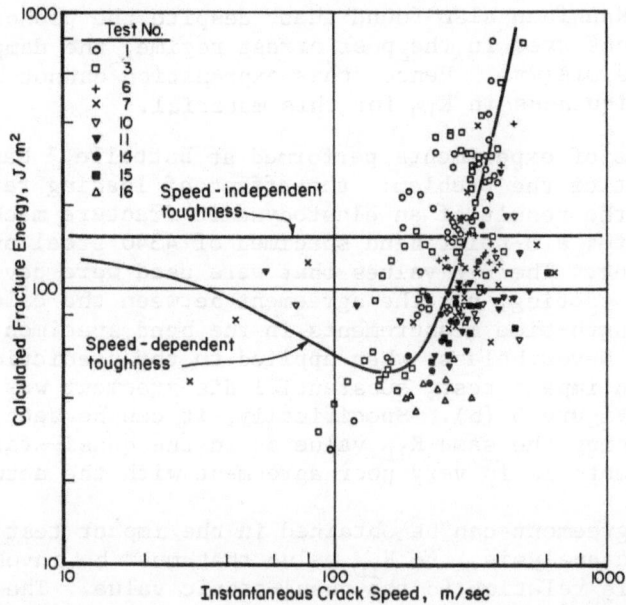

Figure 1   Dynamic Fracture Toughness Values For a Polymeric Material
            Obtained From Generation-Phase Analyses of DCB Specimens
            With Different Initial Crack Lengths and Load Levels.

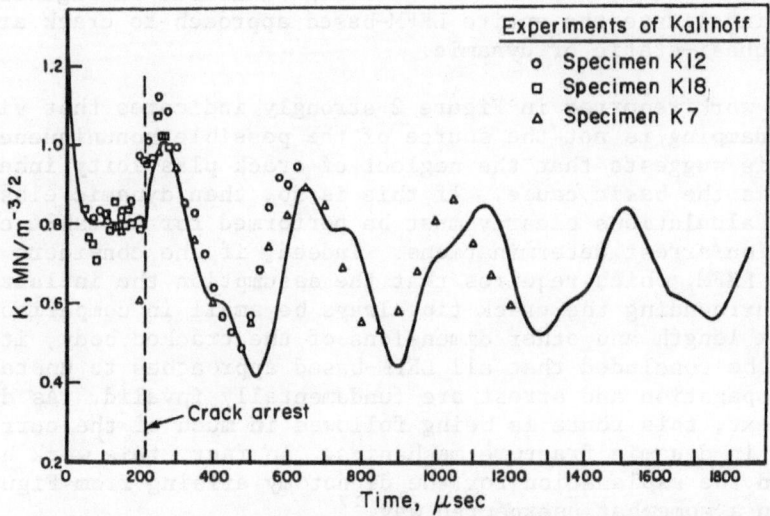

Figure 2   Comparison Between Dynamic Stress Intensity Factors
            Measured in a Polymeric DCB Test Specimen With Dynamic
            Viscoelastic Computations.

Popelar and Kanninen also found that, despite the pronounced damping that can be observed in the post arrest regime, the damping prior to arrest was minimal. Hence, this explanation cannot be the source of the nonuniqueness in K_{ID} for this material.

A series of experiments performed at Battelle[14] has examined another facet of the problem: the effect of loading rate. Figure 3 (a) shows the result of an elastodynamic fracture mechanics calculation for a 3-point bend specimen of 4340 steel under quasi-static loading. The K_{ID} values that were used were developed in DCB specimen testing, yet the agreement between the calculation and the crack length-time measurements in the bend specimen is entirely reasonable. Nevertheless, when applied to the prediction of the results of an impact test, substantial disagreement was found. This is shown in Figure 3 (b). Specifically, it can be seen that the prediction using the same K_{ID} value as in the quasi-static initiation experiments is in very poor agreement with the actual data.

While agreement can be obtained in the impact test with an elastodynamic analysis, the K_{ID} value that must be invoked bears no discernable relation to the quasi-static value. The value $K_{ID} = 170 \text{ MNm}^{-3/2}$ seems to fit the experimental data--see Figure 3 (b)-- but there is no rational reason for this choice. It was therefore concluded that a definite loading rate dependence exists that, if correct, adds considerably to the doubt that K_{ID} is a true material property.[14] Whether or not this is true, it does not provide much comfort to the proponents of the quasi-static view in that the validity of K_{Ia} rests upon its relation to K_{ID} for short crack jumps. Hence, the apparent nonuniqueness in K_{ID} exhibited in Figures 1 and 3 would invalidate the entire LEFM-based approach to crack arrest, whether quasi-static or dynamic.

The work reported in Figure 2 strongly indicates that visco-elastic damping is not the source of the possible nonuniqueness of K_{ID}. This suggests that the neglect of crack plasticity inherent in LEFM is the basic cause. If this is so, then dynamic elastic-plastic calculations clearly must be performed for unstable crack propagation/arrest determinations. Indeed, if one considers the basis of LEFM, which requires that the assumption the inelastic region surrounding the crack tip always be small in comparison to the crack length and other dimensions of the cracked body, it can possibly be concluded that all LEFM-based approaches to unstable crack propagation and arrest are fundamentally invalid. As dis- cussed next, this route is being followed in much of the current research in dynamic fracture mechanics. In fact, this work has suggested the explanation for the dichotomy arising from Figure 3, albeit in a somewhat unexpected way.[17]

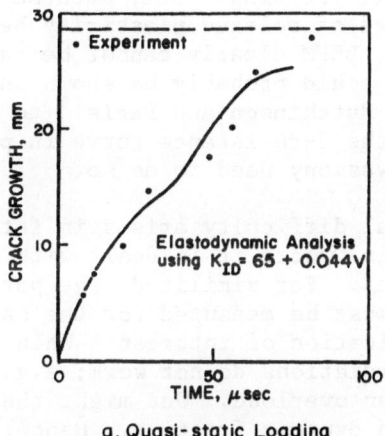

a. Quasi-static Loading

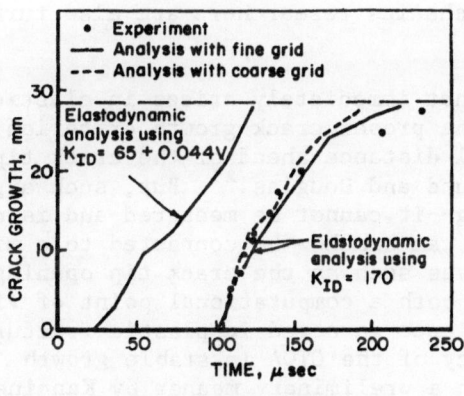

b. Impact Loading

Figure 3   Comparison Between Measured and Calculated Crack Length
Histories in 4340 Steel Bend Specimens Subjected to
Different Loading Types

DIRECTIONS OF CURRENT RESEARCH ON DYNAMIC CRACK PROPAGATION

It is generally accepted that LEFM is valid under the condition that the inelastic deformation surrounding the crack tip is
"dominated" by the elastic K field. This sets a requirement for
contained or small-scale yielding. But, because a moving crack
inevitably leaves a wake of relaxed plasticity behind, except for
very short crack jumps, LEFM clearly cannot be valid at the point
of crack arrest. This could probably be shown in an analogous
manner to that used by Hutchinson and Paris[18] to determine the
region of validity of the J-resistance curve in plastic fracture
mechanics,[19] if there was any need to do so.

The same conceptual difficulty arises in fatigue (at generally
much lower stress levels) where it is dealt with by appealing to
the idea of "similitude". For similitude the parameters governing
the crack growth rate must be measured for the same type of load
history as in the application of interest. When similitude is
violated, the fatigue relations do not work; e.g., in crack growth
retardation following an overload. One might therefore expect the
same kind of effects in dynamic fracture. Hence, just as current
advanced research efforts in fatigue are now being made via direct
consideration of plastic yielding and relaxation (e.g., Newman[20]),
dynamic fracture mechanics researchers are also turning to elastic-
plastic analyses.

A difficulty that immediately arises in elastic-plastic analyses
is in identifying the proper crack growth criterion. A critical
strain at a critical distance ahead of the crack tip is being used,
for example, by Freund and Douglas.[21] But, such a parameter is
somewhat unappealing--it cannot be measured and is not in any event
palatable unless it can somehow be connected to a micromechanical
picture. Alternatives such as the crack tip opening angle (CTOA)
are attractive from both a computational point of view, and from
the extensive experience garnered in plastic fracture mechanics
showing the constancy of the CTOA in stable growth.[19] The latter
fact was utilized in a preliminary manner by Kanninen, et al[22] to
investigate crack propagation and arrest in weld-induced residual
stress fields. Nevertheless, the use of this parameter similarly
requires a proper theoretical basis.

Nishioka and Atluri[23] are proceeding in this direction with
detailed numerical studies of various possible dynamic crack
propagation criteria to ascertain their connection to energy release
rates. Dantam and Hahn,[24] in contrast, are pursuing a more prag-
matic course in attempting to connect a ductile crack propagation
criterion to the J-resistance curve. Other work, such as that of
Achenbach, et al[25] and Freund and Douglas[21] is aimed at elucidating
the fundamental character of the crack tip singularity in elastic-
plastic dynamic crack propagation with a view towards evolving a

more basic ductile crack growth criterion. Such a result could be
used, for example, to construct a singular element for elastic-
plastic dynamic crack propagation via the finite element method.

Computations in which a crack extension criterion is specified
and the crack length history determined in a given initial value/
boundary value problem have been called "application-phase" computations.[1,2] As already mentioned, their counterparts, "generation-
phase" computations, are those in which an actual crack propagation
event is simulated in order to generate values of one or more selected crack growth criteria. Clearly, while the current lack of a
well-established elastic-plastic dynamic criterion prohibits application-phase computations, generation-phase computations on appropriate experiments are still possible.

A recent effort of this kind, albeit elastodynamic, is that of
Shmuely and Levy[26] who also suggested the possible dependence of K_{ID}
upon loading parameters and specimen geometry. The work of Shockey,
et al[27] is of more interest. They have performed experiments in
which simultaneous measurements were made of the stress intensity
factor (by the method of caustics) and of the energy dissipated at
the crack tip (by the temperature rise) and found that the two
quantities are not always equal. They have concluded that the energy
absorbed at a fast moving crack tip can be substantially less than
that available to it whereupon equating K and K_{ID} may be erroneous.
If correct, this would be an extremely important conclusion, one
that could possibly explain the seeming nonuniqueness of K_{ID} just
discussed. However, there are several aspects of this study (e.g.,
the imprecision of the connection between the measured temperature
and the energy dissipation rate, the violation of a global energy
balance) that make it suspect. In any event, a more convincing
alternative explanation exists, as follows.

Elastic-plastic dynamic generation-phase computations have
been performed for the first time only recently. The results of
Ahmad, et al[17] are of some interest in this regard because of the
insight that they bring to the dilemma possed by the elastodynamic
results discussed in the above. Specifically, Ahmad, et al developed finite element solution for precracked specimens of 4340 steel
subjected to (1) quasi-static, (2) impact, and (3) explosive loading.
The plastically deformed regions that were computed in each specimen
at the initiation of crack growth are shown in Figure 4.

It can be seen in Figure 4 that while the crack tip plastic
zones are significant in all three instances, the plastic deformations at the load point (NB; the entire plate surface is loaded
in the explosive loading experiment) differ markedly. As elastodynamic analyses cannot account for such differences, this must be
the reason for the apparent nonuniqueness of K_{ID} noted in connection with Figure 3. It can be concluded that elastic-plastic

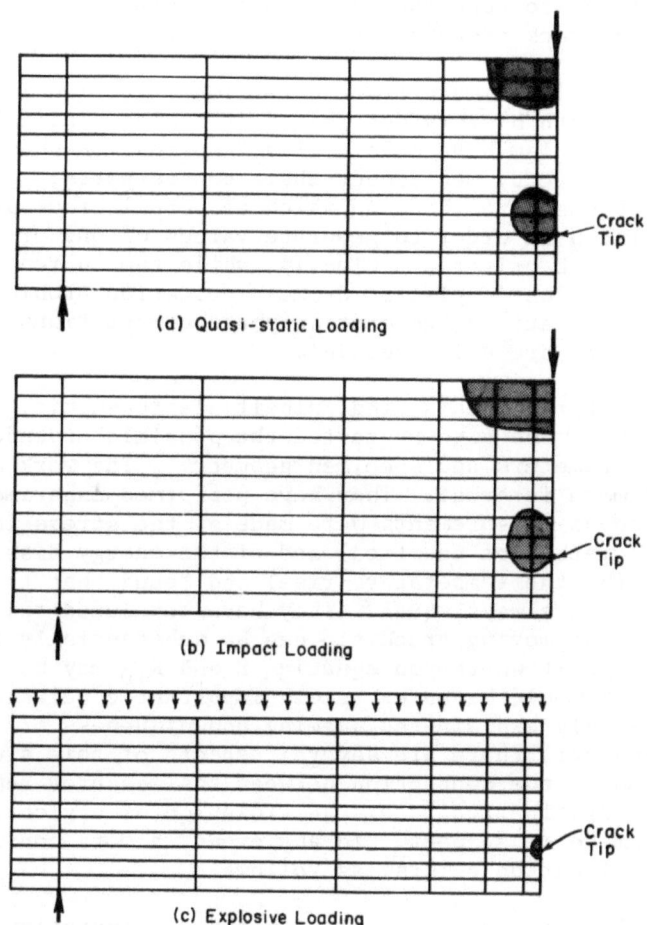

Figure 4   Computed Plastic Zones in 4340 Steel Specimens Subjected
          to Different Loading Types

analyses can be needed even when contained yielding is expected at
the crack tip (e.g., when significant plasticity is experienced
elsewhere in the body). On the other hand, as in Figure 4 (c), when
the only significant plasticity is at the crack tip, elasto-dynamic
analyses should work well. As Ahmad, et al have found, this is
indeed the case.

FUTURE WORK

One of the most important concerns clearly is the need for an appropriate inelastic criterion for rapid crack propagation and arrest. But, this is still not the only issue. The possibility exists that elastic-plastic treatments are really not appropriate at all! The reason is as follows.

Because plasticity is inherently a time-dependent process (cf, the effect of loading rate on the yield stress in tension), the plastic zone accompanying a rapidly propagating crack requires time to form. Hence, it could be much smaller than that for the same crack length and applied load under static conditions. On this basis (i.e., that the plastic zones accompanying rapid crack propagation are small), it has been suggested that very high crack arrest values are tolerable within the static LEFM approach.[28] This is an intriguing idea. But, in order to investigate it, one could not use either LEFM (for obvious reasons) or elastic-plastic analyses that do not contain the rate-dependence of yielding. A viscoplastic formulation therefore is needed.

Some work is progressing in this direction. Achenbach, for example, is making use of the Bodner-Partom[29] model for an asympotic analysis of the near tip region in the sense of the elastic-plastic treatment of Achenbach, et al.[25] Kishimoto, et al[30] have proposed the use of the J-integral. Use of all such models appears to be limited, however, by the lack of data that can be used to evaluate the constants that they contain. In addition, the Bodner-Partom and other viscoplastic models have not been adequately tested for fracture problems where the strain rates can be very much higher than in more conventional problems. So, progress towards quantitative dynamic viscoplastic assessments of rapid crack propagation and crack arrest can be expected to be slow. Nevertheless, some results can be foreseen.

The view that the viscoplastic nature of rapid crack propagation justifies the use of the quasi-static LEFM parameter for crack arrest would seem to be incorrect. The reason is that it relies upon two conflicting assumptions. On the one hand, the plastic zone of the propagating crack at the arrest point must be significantly smaller than that corresponding to the arrested crack. But, for K_{Ia} to be meaningful, there must be a direct connection between the dynamic deformation state of the crack at the instant of crack arrest and the static state that exists at some (relatively long) time later. A smooth slowing down to arrest, which is contrary to the experimental evidence in any event, would not reconcile this dichotomy. If the crack arrests with a large enough plastic zone then LEFM is invalid for arrest regardless of the plastic zone sizes that precede it.

SUMMARY AND CONCLUSIONS

Two opposing views of crack arrest exist. These are that
either: (1) crack arrest is the reverse in time of crack initiation whereupon it is characterizable by a single material property;
e.g., K_{Ia}, or (2) crack arrest is the termination of unstable
propagation whereupon it is characterized by the same speed-depen-
dent properties that govern its motion; e.g., K_{ID}. While the weight
of the evidence definitely favors the second of these two views,
there are many practical applications where the much simpler analysis procedures offered by the first will suffice. Somewhat confus-
ing the situation are doubts that have existed concerning the underlying assumptions in the second view; at least insofar as its
implementation via elastodynamic procedures is concerned. These
now appear to be resolved by elastic-plastic analyses, however.

Current research efforts are generally proceeding through the
use of elastic-plastic analyses. However, these are now being
called into question on the basis that the viscoplastic nature of
rapid crack propagation is thereby neglected. General agreement on
the proper crack growth criterion does not exist for elastic-plastic
conditions, much less for viscoplastic conditions. Moreover, the
use of viscoplastic models suffers from a lack of the proper
material constants. The resolution of these questions would therefore appear to be well in the future when analyses based upon well-
documented constitutive behavior can be coupled with precisely
conducted experiments to elucidate the load/structure/crack size –
independent properties that govern crack propagation/arrest.

ACKNOWLEDGEMENT

This paper was prepared as part of a research program supported
by the Office of Naval Research under Contract No. N00014-77-C-0576.
The author would like to express his appreciation to Dr. Yapa
Rajapakse of ONR for his continued encouragement of this work and
for making possible the fruitful interactions with other ONR dynamic
fracture mechanics contractors: Professors J. D. Achenbach, S. N.
Atluri, L. B. Freund, G. T. Hahn, and A. S. Kobayashi. The many
stimulating discussions on the subject with Mr. Milton Vagins, U. S.
Nuclear Regulatory Commission, and his suggestions on the possible
viscoplastic nature of dynamic crack propagation, are also happily
acknowledged.

REFERENCES

1. M. F. Kanninen, "A Critical Appraisal of Solution Techniques in
    Dynamic Fracture Mechanics," in: Numerical Methods in Fracture
    Mechanics, A. R. Luxmoore and D. R. J. Owen, Editors, U. Swansea
    Press, Swansea, UK (1978).
2. M. F. Kanninen, "Whither Dynamic Fracture Mechanics?," in:

Numerical Methods in Fracture Mechanics, D. R. J. Owen and A. R. Luxmoore, Editors, Pineridge Press, Swansea, UK (1980).

3. J. A. Zukas, T. Nicholas, H. F. Swift, L. B. Greszczuk, and D. R. Curran, Impact Dynamics, Wiley, New York (1982).

4. M. L. Wald, "Steel Turned Brittle by Radiation Called a Peril at 13 Nuclear Plants", New York Times, September 27, 1981; see also D. L. Basdekos, "The Risk of a Meltdown", New York Times, March 29, 1982.

5. R. D. Cheverton, S. E. Bolt, D. A. Canonico, P. P. Holtz, S. K. Iskander, R. K. Nanstad, and W. J. Steizman, "Fracture Mechanics Data Deduced from Thermal Shock and Related Experiments with LWR Pressure Vessel Material", in: Aspects of Fracture Mechanics in Pressure Vessels and Piping, S. G. Sampath and S. S. Palusamy, Editors, ASME PVP-Vol 58, (1982).

6. J. F. Kalthoff, J. Beinart, and S. Winkler, "Measurements of Dynamic Stress Intensity Factors for Fast Running and Arresting Cracks in Double-Cantilever Beam Specimens", in: Fast Fracture and Crack Arrest, G. T. Hahn and M. F. Kanninen, Editors, ASTM STP 627, Philadelphia (1977).

7. M. F. Kanninen, "An Analysis of Dynamic Crack Propagation and Arrest for a Material Having a Crack Speed Dependent Fracture Toughness", in: Prospects of Fracture Mechanics, G. C. Sih, H. C. Van Elst, and D. Broek, Editors, Noordhoff, Leyden (1974).

8. E. J. Ripling, J. H. Mulherin, and P. B. Crosley, "Crack Arrest Toughness of Two High Strength Steels (AISI 4140 and AISI 4340):, Met. Trans., 13A:657 (1982).

9. A. R. Rosenfield, P. N. Mincer, C. W. Marschall, and A. J. Markworth, "Recent Advances in Crack-Arrest Technology", unpublished manuscript (1982).

10. J. Jung and M. F. Kanninen, "An Analysis of Dynamic Crack Propagation and Arrest in a Nuclear Pressure Vessel under Thermal Shock Conditions", in: Aspects of Fracture Mechanics in Pressure Vessels and Piping, S. G. Sampath and S. S. Palusamy, Editors, ASME PVP-Vol 58 (1982).

11. T. Kanazawa and S. Machido, "Fracture Dynamics Analysis on Fast Fracture and Crack Arrest Experiments, Fracture Tolerance Evaluation", T. Kanazawa, A. S. Kobayashi, and K. Iido, Editors, Toyoprint Co., Ltd., Japan (1982).

12. A. S. Kobayashi and S. Mall, "Dynamic Fracture Toughness of Homalite-100", Exp. Mech., 18:11 (1978).

13. A. Shukla, W. L. Fourney, and J. W. Dally, "Mechanisms of Energy Loss During a Fracture Process", Proceedings of the Society for Experimental Stress Analysis (1981).

14. M. F. Kanninen, P. C. Gehlen, C. R. Barnes, R. G. Hoagland, G. T. Hahn, and C. H. Popelar, "Dynamic Crack Propagation Under Impact Loading", in: Nonlinear and Dynamic Fracture Mechanics, N. Perrone and S. N. Atluri, Editors, ASME AMD-Vol 35 (1979).

15. J. F. Kalthoff, et al, Fraunhofer-Institut fur Werkstoffmechanik, Progress Report No. 9 to the Electric Power Research Institute on RP886-4 (1980).

16. C. H. Popelar and M. F. Kanninen, "On the Energy Loss During Dynamic Crack Propagation", unpublished manuscript (1982).

17. J. Ahmad, C. R. Barnes, and M. F. Kanninen, "Analysis of Crack Initiation and Growth in a Ductile Steel Under Dynamic Loading", manuscript in preparation (1982).

18. J. W. Hutchinson and P. C. Paris, "The Theory of Stability Analysis of J-controlled Crack Growth", in: Elastic-Plastic Fracture, J. D. Landers, J. A. Begley, and G. A. Clarke, Editors, ASTM STP 668 (1979).

19. M. F. Kanninen, C. H. Popelar, and D. Broek, "A Critical Survey on the Application of Plastic Fracture Mechanics to Nuclear Pressure Vessels and Piping", Nuclear Eng. and Des., 67:27 (1981).

20. J. C. Newman, Jr., "Fatigue-Crack Propagation and Closure Under Variable Amplitude Loading", in: Fracture Tolerance Evaluation, T. Kanazawa, A. S. Kobayashi and K. Iida, Editors, Toyoprint Co. Ltd., Japan (1982).

21. L. B. Freund and A. S. Douglas, "The Influence of Inertia on Elastic-Plastic Antiplane-Shear Crack Growth", J. Mech. Phys. Solids, 30:59 (1982).

22. M. F. Kanninen, F. W. Brust, J. Ahmad, and I. S. Abou-Sayed, "The Numerical Simulation of Crack Growth in Weld-Induced Residual Stress Fields", in: Residual Stress and Stress Relaxation, Proceedings of the 28th Army Sagamore Research Conference (1982).

23. T. Nishioka and S. N. Atluri, "Path-Independent Integrals, Energy Release Rates, and General Solutions of Near-Tip Fields in Mixed-Mode Dynamic Fracture Mechanics", unpublished manuscript (1982).

24. V. Dantam and G. T. Hahn, "Definition of Crack Arrest Performance of Tough Alloys", in: Fracture Tolerance Evaluation, T. Kanazawa, A. S. Kobayashi, and K. Iida, Editors, Toyoprint Co. Ltd., Japan (1982).

25. J. D. Achenbach, M. F. Kanninen, and C. H. Popelar, "Crack-Tip Fields for Fast Fracture of an Elastic-Plastic Material", J. Mech. Phys. Solids, 29:211 (1981).

26. M. Shmuely and C. Levy, "The Dependence of the Dynamic Material Toughness on the Velocity and on Some Loading Parameters", Int. Journ. of Fracture, 19:221 (1982).

27. D. A. Shockey, J. F. Kalthoff, W. Klemm, and S. Winkler, "Simultaneous Measurements of Stress Intensity and Toughness for Fast Running Cracks in Steel", Exp. Mech., in press (1982).

28. M. Vagins, private communication (1982).

29. S. D. Bodner and Y. Partom, "Constitutive Equations for Elastic-Viscoplastic Strain-Hardening Materials", J. Applied Mech., 42:385 (1975).

30. K. Kishimoto, S. Aoki, and M. Sakata, "Use of J-Integral in Dynamic Analysis of Cracked Linear Viscoelastic Solids by Finite-Element Method", J. Applied Mech., 49:75 (1982).

ENERGY RELEASE RATES AND PATH INDEPENDENT

INTEGRALS IN DYNAMIC FRACTURE

Satya N. Atluri

Center for the Advancement of Computational Mechanics
School of Civil Engineering
Georgia Institute of Technology
Atlanta, Georgia 30332

ABSTRACT

The topics of energy-release rate to a crack-tip propagating under mixed-mode non-steady conditions with an arbitrary velocity, and path-independent integrals which may be used to calculate such energy-release rates, are critically examined, and certain new results obtained recently are summarized.

INTRODUCTION

It is now well-known that the so-called J integral[1,2]: (i) is in fact the component along the crack-axis of a path-independent vector integral, (ii) is applicable only in elasto-statics, and (iii) in only the elasto-static case has the meaning of the rate of energy release per unit quasistatic crack extension.

Recently the author has derived[3] some general conservation laws for both finite-elastic solids, as well as for those characterized by rate-type inelastic constitutive laws, wherein body forces, inertia, and arbitrary crack-face conditions are accounted for. On the basis of these conservation laws, the author had also investigated[3] path-independent integrals in the case of dynamic crack propagation in elastic as well as inelastic solids.

The contour integral[3], eventhough path-independent, was shown not to be equivalent, in general, to the rate of energy release for a propagating crack under unsteady conditions with a non-constant velocity of propagation. The integration path for the contour integral previously defined[3] is fixed in space. On the other hand,

211

Kishimoto et al[4] introduced an integral, $\hat{J}_k$, which was argued incorrectly, to have the meaning of an energy release rate. Here we also discuss the result of Bui[5] for a contour integral involving a path which rigidly translates along with the crack-tip at the same velocity, and contrast it with the present result.

Finally we discuss a new path-independent integral recently introduced in[6], involving a space fixed contour, which has the meaning of energy release rate under general mixed-mode conditions for a crack propagating with non-constant velocity under unsteady conditions. This integral is shown to reduce to that of Sih[7], which is valid for steady-state crack propagation at constant velocity. Further, eventhough the subject of path-independent integrals is not discussed in 8, it is shown that the energy release rate expression given by Freund[8] is in agreement with the present result.

A discussion of the utility of the path-independent integrals in computing the time-dependent mixed-mode stress-intensity factors for propagating cracks is also presented.

Path-Independent Integrals in Dynamic Fracture Mechanics

Eventhough the direct tensor formalism employed in 3 makes the application to three-dimensional problems rather transparent, here we consider only two-dimensional linear elastodynamic crack propagation. We consider a fixed cartesian coordinate system x_i (i=1,2,3) such that x_1 is along the crack surface, x_2 normal to the crack face, and x_3 along the crack front. We consider the general unsteady propagation of the crack at non-constant velocity c(t). At any time t, the field equation and boundary conditions are:

$$\sigma_{ij,j} + \rho f_i = \rho \ddot{u}_i \; ; \quad \sigma_{ij} = \sigma_{ji} \tag{1}$$

$$\varepsilon_{ij} = \tfrac{1}{2} [u_{i,j} + u_{j,i}] \tag{2}$$

$$\sigma_{ij} = \partial W / \partial \varepsilon_{ij} \tag{3}$$

$$\sigma_{ij} n_j = \bar{t}_i \quad \text{at} \quad S_t \tag{4}$$

$$u_i = \bar{u}_i \quad \text{at} \quad S_u \tag{5}$$

where σ_{ij} is the stress tensor, ε_{ij} the strain tensor, ( ),$_j$ denotes partial differentiation with respect to x_j, W is the strain energy density per unit volume (which for a linear elastic homogeneous material is a single-valued function of ε_{ij} alone), $\bar{t}_i$ are prescribed tractions at S_t, and $\bar{u}_i$ are prescribed displacements at S_u.

Consider a closed volume V which is free from singularities or
other defects. Provided the field conditions (1-5) above are valid,
the following conservation law holds for this volume V[3]:

$$0 = \int_V [\frac{\partial W}{\partial x_k} - \frac{\partial}{\partial x_i}(\sigma_{ij}u_{j,k}) - \rho(f_j - \ddot{u}_j)u_{j,k}] \, dV$$

$$+ \int_{S_t} [n_i\sigma_{ik} - \bar{t}_k] ds + \int_{S_u} n_i\sigma_{ij}(u_{j,k} - \bar{u}_{j,k}) ds \qquad (6)$$

Eq. (6) can be verified from Eqs. (1-5) and the additional identity:

$$\frac{\partial W}{\partial x_k} = \sigma_{ij}\epsilon_{ij,k} = \sigma_{ij} \cdot \tfrac{1}{2}[u_{i,jk} + u_{j,ik}]$$

$$= \sigma_{ij}u_{j,ik} \qquad (7)$$

Consider the case when V includes the crack-tip as shown in Fig. 1.

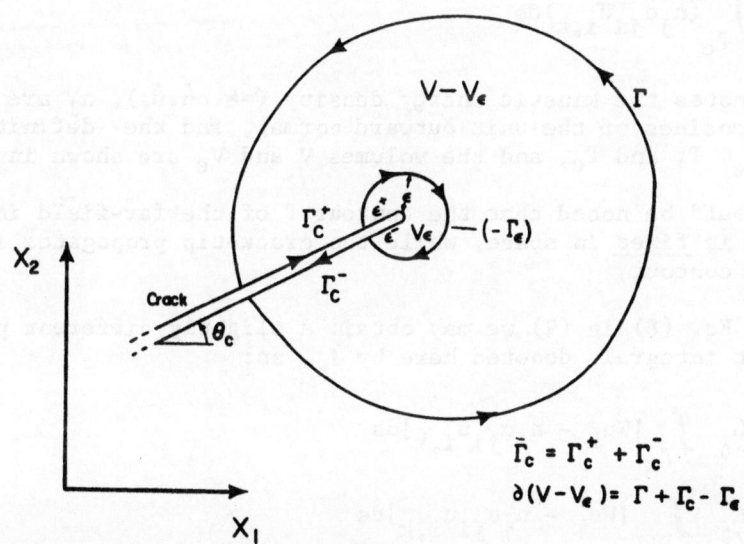

$$\bar{\Gamma}_c = \Gamma_c^+ + \Gamma_c^-$$
$$\partial(V-V_\epsilon) = \Gamma + \bar{\Gamma}_c - \Gamma_\epsilon$$

Fig. 1

Near the crack-tip, i.e. the point of singularity, W is singular and is of order r^{-1} where r is the radial distance from the crack-tip. Thus $\partial W/\partial x_k$ is of the order r^{-2}. Likewise, near the propagating crack-tip, σ_{ij} is $0(r^{-1/2})$ and $u_{i,k}$ is $0(r^{-1/2})$, while $\ddot{u}_i$ is of $0(r^{-3/2})$[6]. Thus the integrands in the volume integral in Eq. (6) are non-integrable. Thus the divergence theorem cannot be applied to the entire volume V, but only to the domain $(V-V_\varepsilon)$ in the limit $\varepsilon \to 0$ as indicated in Fig. 1.

Before applying the divergence theorem to the conservation law of the type of Eq. (6) for a volume $V-V_\varepsilon$, we note the identity:

$$\int_{V-V_\varepsilon} \rho\ddot{u}_i u_{i,k} dV = \int_{V-V_\varepsilon} \{ \frac{d}{dt}(\rho\dot{u}_i u_{i,k}) dV - \frac{\partial}{\partial x_k}(\tfrac{1}{2}\rho\dot{u}_i\dot{u}_i) \} dV \tag{8}$$

Using (8) in (6) and applying the divergence theorem to the domain $(V-V_\varepsilon)$ we obtain by definition the k-th component of a path-independent vector, denoted here by J_k, as:

$$J_k \equiv \mathop{L}_{\varepsilon \to 0} \int_{\Gamma_\varepsilon} [n_k(W-T) - n_j\sigma_{ji}u_{i,k}]ds$$

$$= \mathop{L}_{\varepsilon \to 0} \int_{\Gamma+\Gamma_c} [n_k(W-T) - n_j\sigma_{ji}u_{i,k}]ds$$

$$+ \mathop{L}_{\varepsilon \to 0} \int_{V-V_\varepsilon} \frac{d}{dt}(\rho\dot{u}_i u_{i,k}) dV - \int_{\Gamma_c} (\bar{t}_i u_{i,k})ds$$

$$- \int_{\Gamma_c} (n_j\sigma_{ji}\bar{u}_{i,k})ds \tag{9}$$

where T denotes the kinetic energy density $(=\tfrac{1}{2}\rho\dot{u}_i\dot{u}_i)$, n_j are the direction cosines of the unit outward normal, and the definition of paths Γ_ε, Γ, and Γ_c, and the volumes V and V_ε are shown in Fig. 1.

It should be noted that the contour Γ of the far-field integral in Eq. (9) is _fixed_ in space, while the crack-tip propagates into this fixed contour.

Using Eq. (8) in (9) we may obtain a slightly different path-independent integral, denoted here by $\hat{J}_k$, as:

$$\hat{J}_k \equiv \mathop{L}_{\varepsilon \to 0} \int_{\Gamma_\varepsilon} [Wn_k - n_j\sigma_{ji}u_{i,k}]ds$$

$$= \mathop{L}_{\varepsilon \to 0} \int_{\Gamma+\Gamma_c} [Wn_k - n_j\sigma_{ji}u_{i,k}]ds$$

$$+ \mathop{L}_{\varepsilon \to 0} \int_{V-V_\varepsilon} \rho\ddot{u}_i u_{i,k} dV - \int_{\Gamma_c} [\bar{t}_i u_{i,k} + n_j\sigma_{ji}\bar{u}_{i,k}]ds \tag{10}$$

Eq. (10) was given in a slightly less general form in 4, eventhough
the definition of $\hat{J}_k$ as the limit of the integral over Γ_ϵ does not
appear to have been stressed in Ref. 4.   Once again, Γ in Eq. 10 is
a space-fixed contour.

On the other hand, considering a far-field contour Γ to be a
rigid path surrounding the crack-tip and in translation at the same
velocity c_1 (along the x_1-axis) as the crack-tip, a path-independent
integral, denoted here by $J_1{}^*$, was derived in Ref. 5:

$$J_1{}^* = \int_\Gamma [Wn_1 - n_k\sigma_{kj}u_{j,1} - \tfrac{1}{2}\rho\dot{u}_i\dot{u}_in_1 - \rho\dot{u}_ju_{j,1}c_1n_1]ds$$

$$+ \frac{D}{Dt} \int_V \rho\dot{u}_ju_{j,1}dV \tag{11}$$

where V is the area enclosed by Γ.   Since $\dot{u}_ju_{j,1}$ is $0(r^{-1})$ near the
crack-tip, one may show[9] that the time derivative of the integral
over the moving volume V in Eq. 11 becomes:

$$\frac{D}{Dt} \int_V \rho\dot{u}_ju_{j,1}dV = \underset{\epsilon\to0}{L} \int_{V-V_\epsilon} \frac{d}{dt}(\rho\dot{u}_ju_{j,1})dV + \int_\Gamma\rho\dot{u}_ju_{j,1}c_1n_1ds$$

$$- \underset{\epsilon\to0}{L} \int_{\Gamma_\epsilon}\rho\dot{u}_ju_{j,1}c_1n_1ds \tag{12}$$

The last integral on the right hand side of Eq. (12) can be evaluated
from the asymptotic singular field near the propagating crack-tip:

$$\dot{u}_i \simeq -c_1u_{i,1} \quad ; \quad -\rho\dot{u}_ju_{j,1}c_1n_1 \simeq n_1\rho\dot{u}_j\dot{u}_j = 2Tn_1 \tag{13}$$

Using (12) and (13) in (11) and comparing with (9) [upon noting that
(11) is written for the case when the crack faces are free from any
prescribed conditions] we obtain:

$$J_1{}^* = \int_\Gamma [Wn_1 - \sigma_{jk}n_ku_{j,1} - Tn_1]ds + \underset{\epsilon\to0}{L} \int_{V-V_\epsilon} \frac{d}{dt}(\rho\dot{u}_ju_{j,1})dV$$

$$+ \underset{\epsilon\to0}{L} \int_{\Gamma_\epsilon} 2Tn_1ds$$

$$= J_1 + \underset{\epsilon\to0}{L} \int_{\Gamma_\epsilon} 2Tn_1ds \tag{14}$$

We emphasize again that in the direct evaluation of $J_1{}^*$ from Eq. (11)
for a propagating crack, one must choose a rigid Γ that moves along
with the crack-tip.

It has been shown[3], from first principles, that the rate of energy release to a propagating crack-tip can be expressed by:

$$G = (c_k/c) \ G_k$$

$$G_k = \lim_{\varepsilon \to 0} \int_{\Gamma_\varepsilon} [(W + T)n_k - t_i u_{i,k}]ds \tag{15}$$

where c_k denotes the component of crack velocity in the x_k direction, and $T = 1/2 \ (\rho \dot{u}_i \dot{u}_i)$ where $\dot{u}_i$ is the absolute velocity of a material particle.

Now comparing Eqs. (9,10,14 and 15) one finds that:

$$J_k = G_k - \lim_{\varepsilon \to 0} 2 \int_{\Gamma_\varepsilon} (\tfrac{1}{2}\rho \dot{u}_i \dot{u}_i)n_k ds \tag{16}$$

$$\hat{J}_k = G_k - \lim_{\varepsilon \to 0} \int_{\Gamma_\varepsilon} (\tfrac{1}{2}\rho \dot{u}_i \dot{u}_i)n_k ds \tag{17}$$

$$J_1{}^* = G_1 \tag{18}$$

For stationary cracks in dynamic elastic fields the kinetic energy density, $T(= \tfrac{1}{2}\rho \dot{u}_i \dot{u}_i)$, is nonsingular, since the material velocity $\dot{u}_i$ is of the order $(r^{+1/2})$. Thus the last terms in Eqs. (16,17) vanish as Γ_ε shrinks to zero. Thus for __stationary__ cracks in elastodynamic fields, J_k, $\hat{J}_k$, as well as $J_1{}^*$ have the meaning of appropriate energy release rates.

On the other hand, for dynamically propagating cracks, the kinetic energy density is singular near the crack-tip since $\dot{u}_i$ is of the order $(r^{-1/2})$ near the crack-tip. Thus the last integrals in Eqs. (16,17) lead to __finite__ values for propagating cracks. Thus neither J_k nor $\hat{J}_k$ of Eqs. (16) and (17) respectively has the meaning of an energy release rate. However J_k does have[3] the meaning of the rate of change of the Lagrangean of the dynamic system due to unit crack growth.

An examination of Eqs. (9,10, and 16) reveals the possibility of deriving a path-independent vector integral, using space-fixed far-field contours, which has the meaning of an energy release rate in the general case of non-constant velocity, non-steady crack propagation. To this end, we first observe that:

$$\int_{V-V_\varepsilon} \frac{\partial}{\partial x_k} (\tfrac{1}{2}\rho \dot{u}_i \dot{u}_i) dV \equiv \int_{V-V_\varepsilon} \rho \dot{u}_i \dot{u}_{i,k} \ dV$$

$$= \int_{\Gamma+\Gamma_c} (\tfrac{1}{2}\rho \dot{u}_i \dot{u}_i) n_k ds - \int_{\Gamma_\varepsilon} (\tfrac{1}{2}\rho \dot{u}_i \dot{u}_i) n_k ds \tag{19}$$

Likewise for any two volumes V_2 and V_1 enclosed by far-field contours Γ_2 and Γ_1 respectively,

$$\int_{V_2-V_1} \frac{\partial}{\partial x_k} (\tfrac{1}{2}\rho \dot{u}_i \dot{u}_i) dV = \int_{\Gamma_2+\Gamma_{c2}} (\tfrac{1}{2}\rho \dot{u}_i \dot{u}_i) n_k ds - \int_{\Gamma_1+\Gamma_{c1}} (\tfrac{1}{2}\rho \dot{u}_i \dot{u}_i) n_k ds \tag{20}$$

Considering, for simplicity, the entire crack faces to be traction free, one may from a close examination of Eqs. (10) and (19), define a path-independent integral J_k' [6] such that:

$$
\begin{aligned}
J_k' \equiv G_k &= \underset{\varepsilon \to 0}{L} \int_{\Gamma_\varepsilon} [(W + T)n_k - n_j \sigma_{ji} u_{i,k}] ds \\
&\equiv \underset{\varepsilon \to 0}{L} \Bigg\{ \int_{\Gamma+\Gamma_c} [(W + T)n_k - n_j \sigma_{ji} u_{i,k}] ds \\
&\quad + \int_{V-V_\varepsilon} [\rho \ddot{u}_i u_{i,k} - \rho \dot{u}_i \dot{u}_{i,k}] dV \Bigg\}
\end{aligned}
\tag{21}
$$

Path-independency of the extreme right-hand side of Eq. (29) is clear when one verifies from the elastodynamic equilibrium, Eq. (1), and Eq. (19) that

$$
\begin{aligned}
&\int_{\Gamma_2+\Gamma_{c2}} [(W + T)n_k - n_j \sigma_{ji} u_{i,k}] ds + \int_{V_2-V_1} [\rho \ddot{u}_i u_{i,k} - \rho \dot{u}_i \dot{u}_{i,k}] dV \\
&- \int_{\Gamma_1+\Gamma_{c1}} [(W + T)n_k - n_j \sigma_{ji} u_{i,k}] ds = 0
\end{aligned}
\tag{22}
$$

At this point it is worth comparing the presently derived expressions for G_k (Eq. (15)) and J_k' (Eq. (21)), with those derived in Refs. 7 and 8, for a mode I problem under steady-state constant-velocity crack-propagation (with velocity c_1). For a mode I crack-propagation at constant velocity c_1 along the x_1 axis, one may introduce a moving coordinate system, $X_1 = x_1 - c_1 t$, $X_2 = x_2$ etc. Thus the displacements can be expressed as: $u_i (X_1, X_2, t)$. The absolute velocity and acceleration respectively can be written, in general, as:

$$\dot{u}_i = \frac{\partial u_i}{\partial t} + \frac{\partial u_i}{\partial X_1} \frac{\partial X_1}{\partial t} = \frac{\partial u_i}{\partial t} - c_1 \frac{\partial u_i}{\partial x_1} \tag{23}$$

and $\quad \ddot{u}_i = \frac{\partial^2 u_i}{\partial t^2} + c_1^2 \frac{\partial^2 u_i}{\partial x_1^2} - 2c_1 \frac{\partial^2 u_i}{\partial x_1 \partial t} \tag{24}$

Under steady-state conditions (as seen by an observer moving with the crack-tip) at constant crack-velocity c_1 the above can be approximated everywhere in the domain as:

$$\dot{u}_i \simeq -c_1 \frac{\partial u_i}{\partial x_1} \quad ; \qquad \ddot{u}_i = c_1^2 \frac{\partial^2 u_i}{\partial x_1^2} \qquad\qquad (25a,b)$$

Further, the asymptotic (singular) velocity field near the crack-tip can be approximated even in the non-steady case as:

$$\dot{u}_i = -c_1 \frac{\partial u_i}{\partial x_1} \qquad\qquad (26)$$

The energy-release-rate expression given by Sih[7] for mode I crack propagation, using the present notation, is:

$$G_1 \equiv \mathop{L}_{\varepsilon \to 0} \int_{\Gamma_\varepsilon} (Wn_1 + \tfrac{1}{2}\rho c_1^2 \frac{\partial u_i}{\partial x_1} \frac{\partial u_i}{\partial x_1} n_1 - n_j \sigma_{ji} \dot{u}_{i,1}) ds \qquad (27a)$$

$$= \int_{\Gamma} (Wn_1 + \tfrac{1}{2}\rho c_1^2 \frac{\partial u_i}{\partial x_1} \frac{\partial u_i}{\partial x_1} n_1 - n_j \sigma_{ji} \dot{u}_{i,1}) ds \qquad (27b)$$

Using Eqs. (25a) and (26) in Eq. (15) it is seen that Eq. (20a) is a special case of the present general expression, Eq. (15) for steady-state crack-propagation at constant velocity c_1. Further, the path-independency, viz, the equality of the integrals in over Γ_ε and Γ as in Eqs. (27a,b) respectively is predicated on the satisfaction of the simplified equation of motion

$$\sigma_{ij,j} = \rho c_1^2 \frac{\partial^2 u_i}{\partial x_1^2} \quad \text{in } V \qquad\qquad (28)$$

as also pointed out by Sih[7]. It is seen that Eq. (28) pertains to the steady-state case with constant velocity crack propagation.

On the other hand, the energy-release rate expression given by Freund[8] is:

$$G = \mathop{L}_{\varepsilon \to 0} \int_{\Gamma_\varepsilon} (\sigma_{ij} n_j \dot{u}_i + \tfrac{1}{2}\sigma_{ij} u_{i,j} c_1 n_1 + \tfrac{1}{2}\rho \dot{u}_i \dot{u}_i c_1 n_1) ds \qquad (29)$$

Upon substituting the asymptotic value for $\dot{u}_i$ from Eq. (26), one sees that Eq. (29) reduces to:

$$G = c_1 \mathop{L}_{\varepsilon \to 0} \int_{\Gamma_\varepsilon} [(W + T)n_1 - n_j \sigma_{ji} u_{i,1}] ds \qquad\qquad (30)$$

which agrees with the corresponding mode I result in Eq. (15).

Finally we compare the present path-independent integral J_k' to the results of Sih[7] and Freund[8]. It should be remarked that J_k' evaluated from the far-field contour as in Eq. (21) is valid for non-steady propagation at non-constant velocity. If one invokes the steady-state assumption and constant-velocity propagation (with velocity c_1) one may get everywhere in the domain,

$$T = \tfrac{1}{2}\rho \dot{u}_i \dot{u}_i \approx \tfrac{1}{2}\rho c_1^2 \frac{\partial u_i}{\partial x_1} \frac{\partial u_i}{\partial x_1}$$

$$\rho \ddot{u}_i u_{i,k} \approx \rho c_1^2 \frac{\partial^2 u_i}{\partial x_1^2} \frac{\partial u_i}{\partial x_k}$$

$$\rho \dot{u}_i \dot{u}_{i,k} \approx \rho c_1^2 \frac{\partial u_i}{\partial x_1} \frac{\partial^2 u_i}{\partial x_1 \partial x_1} \tag{31}$$

Thus, for steady-state crack propagation at constant velocity c_1 along the x_1 axis, the far-field contour integral in Eq. (21) can be simplified as:

$$(J_1')_{\text{steady-state}} = (G_1)_{\text{steady-state}}$$

$$= \int_\Gamma [(W + \tfrac{1}{2}\rho c_1^2 \frac{\partial u_i}{\partial x_1} \frac{\partial u_i}{\partial x_1}) n_1 - n_j \sigma_{ji} u_{i,1}] ds$$

$$+ \int_{V-V_\varepsilon} (\rho c_1^2 \frac{\partial^2 u_i}{\partial x_1^2} \frac{\partial u_i}{\partial x_1} - \rho c_1^2 \frac{\partial u_i}{\partial x_1} \frac{\partial^2 u_i}{\partial x_1^2}) ds$$

$$\equiv \int_\Gamma [(W + \tfrac{1}{2}\rho c_1^2 \frac{\partial u_i}{\partial x_1} \frac{\partial u_i}{\partial x_1}) n_1 - n_j \sigma_{ji} u_{i,1}] ds \tag{32}$$

which agrees with the corresponding result of Sih[7].

From a knowledge of the asymptotic strain and stress fields near an (arbitrarily) propagating crack-tip, the energy release rate G_k can be directly evaluated from Eq. (15) and expressed in terms of the mixed mode stress intensity factors $K_I(t)$, $K_{II}(t)$, and $K_{III}(t)$. From Eq. (18) it is seen that $J_1^* = G_1$; while from Eq. (21) it is seen that $J_k' = G_k$. Eventhough as noted in Eqs. (16 and 17) both J_k and $\hat{J}_k$ are not equal to G_k, Eqs. (16 and 17) can be used to relate J_k and $\hat{J}_k$ to the mixed-mode stress-intensity factors.

The near-tip fields in general mixed-mode crack propagation have recently been succintly presented in Ref. 6. Using these solutions, one may directly evaluate G_k from Eq. (15) as[6]:

$$G_1 = \frac{1}{2\mu} \left\{ K_I^2 A_I(c) + K_{II}^2 A_{II}(c) + K_{III}^2 A_{III}(c) \right\} \tag{33}$$

$$G_2 = -\frac{1}{\mu} K_I K_{II} A_{IV}(c) \tag{34}$$

where μ is the shear modulus and

$$A_I(c) = [\beta_1(1 - \beta_2^2)]/D(c)$$

$$A_{II}(c) = [\beta_2(1 - \beta_2^2)]/D(c)$$

$$A_{III}(c) = 1/\beta_2$$

$$A_{IV}(c) = \frac{(\beta_1-\beta_2)(1-\beta_2^2)}{[D(c)]^2}\left[\frac{\{4\beta_1\beta_2 + (1+\beta_2^2)^2\}\,(2+\beta_1+\beta_2)}{2[(1+\beta_1)(1+\beta_2)]^{\frac{1}{2}}} - 2(1+\beta_2^2)\right]$$

$$\beta_1^2 = 1 - (c^2/c_d^2) \ ; \qquad \beta_2^2 = 1 - (c^2/c_S^2)$$

$$c_d^2 = \frac{\kappa + 1}{\kappa - 1}\frac{\mu}{\rho} \ ; \qquad c_S^2 = \frac{\mu}{\rho}$$

$$\kappa = \begin{cases} (3-\nu)/(1+\nu) & : \quad \text{plane stress} \\ (3-4\nu) & : \quad \text{plane strain} \end{cases}$$

$$D(c) = 4\beta_1\beta_2 - (1 + \beta_2^2)$$

From Eqs. (18) and (21) it is seen that

$$J_1' = G_1 \ ; \qquad J_2' = G_2 \ ; \qquad J_1^* = G_1$$

On the other hand, using Eq. (16) one may derive[6] that:

$$J_1 = \frac{1}{2\mu}\{K_I^2 F_I(c) + K_{II}^2 F_{II}(c) + K_{III}^2 F_{III}(c)\} \tag{35}$$

$$J_2 = \frac{-K_I K_{II}}{\mu} F_{IV}(c) \tag{36}$$

where,

$$F_I(c) = \frac{\beta_1(1-\beta_2^2)}{[D(c)]^2}\left[4\beta_1 - \frac{1}{\beta_1}(1+\beta_2^2) - \frac{4(\beta_1-\beta_2)(1+\beta_2^2)}{[(1+\beta_1)(1+\beta_2)]^{\frac{1}{2}}}\right]$$

$$F_{II}(c) = \frac{\beta_2(1-\beta_2^2)}{[D(c)]^2}\left[4\beta_2 - \frac{1}{\beta_2}(1+\beta_2^2) - \frac{4(\beta_2-\beta_1)(1+\beta_2^2)}{[(1+\beta_1)(1+\beta_2)]^{\frac{1}{2}}}\right]$$

$$F_{III}(c) = \frac{1}{\beta_2^2}$$

and $F_{IV}(c) = \dfrac{[4\beta_1\beta_2 + (1+\beta_2^2)^2]}{2[D(c)]^2} \; \dfrac{(1-\beta_2^2)(\beta_1^2 - \beta_2^2)}{[(1+\beta_1)(1+\beta_2)]^{\frac{1}{2}}}$

Thus the path-independent integrals J_k' or J_k (both of which involve space-fixed contours into which cracks propagate) are convenient and highly useful in computing dynamic stress-intensity factors for arbitrarily propagating cracks using numerical methods such as the finite element method. As usual, since the integrals involve contours in the far-field, detailed modeling of the crack-tip region is not necessary. Such applications have already been presented[10].

ACKNOWLEDGEMENTS

The results presented herein were obtained during the course of investigations supported by the Office of Naval Research under contract N00014-78-C-0636, to Georgia Tech. The author thanks Dr. Y. Rajapakse for his interest in and encouragement of this work. The assistance of Ms. B. Bolinger in the preparation of this manuscript is sincerely appreciated.

REFERENCES

1. J. D. Eshelby, The Continuum Theory of Lattice Defects, in: "Solid State Physics," Vol. III, Academic Press, pp. 79-144, (1956).
2. J. R. Rice, A Path-Independent Integral and the Approximate Analysis of Strain Concentration by Notches and Cracks, J. Appl. Mech., Vol. 35, pp. 376-386, (1968).
3. S. N. Atluri, Path-Independent Integrals in Finite Elasticity and Inelasticity, with Body Forces, Inertia, and Arbitrary Crack-Face Conditions, Engg. Frac. Mech., Vol. 16, No. 3, pp. 341-364, (1982).
4. K. Kishimoto, S. Aoki, and M. Sakata, On the Path-Independent Integral Ĵ, Engg. Frac. Mech., Vol. 13, pp. 841-850, (1980).
5. H. D. Bui, Stress and Crack-Displacement Intensity Factors in Elastodynamics, Fracture, Vol. 3, ICF4, pp. 91-95, Waterloo, (1979).
6. T. Nishioka, and S. N. Atluri, Path-Independent Integrals, Energy Release Rates, and General Solutions of Near-Tip Fields in Mixed-Mode Dynamic Fracture Mechanics, Report No. GIT-CACM-SNA-82-14, March '82, Georgia Tech, also Engg. Frac. Mech., (In Press).
7. G. C. Sih, Dynamic Aspects of Crack Propagation, in: "Inelastic Behavior of Solids," (Eds. M.F. Kanninen, et al) McGraw-Hill, pp. 607-639, (1970).
8. L. B. Freund, Energy Flux into the Tip of an Extending Crack in an Elastic Solid, J. Elas., Vol. 2, No. 4, pp. 341-349, (1972).
9. H. D. Bui, Private Communication to the Author, 21 April, (1982).

10. T. Nishioka, and S. N. Atluri, A Numerical Study of the Use of
     Path Independent Integrals in Elasto-Dynamic Crack Propagation,
     Report No. GIT-CACM-SNA-82-15, March 1982, Georgia Tech, also
     Engg. Frac. Mech., (In Press).

HIGH RATE DEFORMATION IN THE

FIELD OF A CRACK

R. Hoff, C.A. Rubin and G.T. Hahn

Vanderbilt University
Department of Mechanical and Materials Engineering
Nashville, TN   37235

INTRODUCTION

As part of a study of the crack arrest capabilities of tough steels[1], efforts are underway to simulate rapid crack extension and arrest in elastic-plastic finite element models. As a first step, stationary cracks in compact tension specimens have been modelled and the effects of loading rate, strain rate sensitivity and inertia on J_I have been examined. The aim of this work is to examine those features of the plastic zone influential in determining the toughness, namely, the size of the process zone, and the crack tip opening displacement.

Plasticity associated with a stationary crack has been characterized as occurring in two separate zones, as shown in Fig. 1. The larger region, called the plastic zone, features small plastic strains in the range $0 < \varepsilon_p < 0.1$. The size of the plastic zone (at $\theta = 0°$) is given by a characteristic dimension, r_o. Levy, et al.[2] have determined r_o for a nonhardening material with a semi-infinite crack in an infinite plate. In cases where the plastic zone is small compared to the specimen dimensions

$$r_o \approx 0.036 \ EJ_I/\sigma_o^2(1- \nu^2). \tag{1}$$

Even closer to the crack tip is the heavily-strained process zone, where plastic strains range roughly from 0.1 to 1.0. This intensely non-linear zone can be given a characteristic dimension, w, which can be related to the crack tip opening displacement, δ, by

$$w = \beta\delta . \tag{2}$$

Rice[3] suggests 1.9 as a possible value of β, and Paris[4] states it is of the order of 2. Since the process zone has no "obvious" boundary, the value of β depends on how the process zone is defined; for instance, a critical strain value may be chosen to define the process zone boundary. For convenience, $\beta = 1$ has been used in this analysis.

The crack tip opening displacement, δ, can be expressed as a function of the J-integral[5] by:

$$\delta = d_n J/\sigma_o,\tag{3}$$

for a nonlinear elastic, power law hardening material. The constant, d_n, is a function of material properties, α, σ_o, E and n (Ramberg-Osgood[6]), where

$$\varepsilon/\varepsilon_o = \alpha(\sigma/\sigma_o)^n.\tag{4}$$

For small values of strain, ε can be interpreted as the total strain; for large values of strain, ε can be interpreted as the plastic strain. Using values for a steel such as A533B, a value of d_n can be determined[5] as $d_n = 0.52$. Equation (3) can be substituted into (2) and then divided by (1) giving the relative size of the process and the plastic zones:

$$w/r_o = 13.14\ \sigma_o/E.\tag{5}$$

If E = 197 GPa and σ_o = 415 MPa as in the case of A533B steel at 93°C, then

$$w/r_o \approx 0.03.\tag{6}$$

The plastic strain rates are related to the strain gradients and the extent of the plastic and process zones. If it is assumed that the equivalent plastic strain is 0.0 at $r = r_o$ (the plastic zone boundary), and varies linearly up to a value of 0.005 at $r = w$ (the process zone boundary), then the plastic strain can be expressed as:

$$\varepsilon_p \approx 0.005(r_o - r)/(r_o - w).\tag{7}$$

After making appropriate substitution from (6) and (1), and differentiating with respect to time:

$$\dot{\varepsilon}_p \approx 0.005(r/r_o)(\dot{J}_I/J_I).\tag{8}$$

An "average" value of plastic strain rate could be calculated at the middle of the plastic zone, i.e. where $r = 0.5 r_o$.

This gives

$$\dot{\varepsilon}_p \approx 0.0025(\dot{J}_I/J_I) = 0.005(\dot{K}_I/K_I) \tag{9}$$

for the plastic zone, and is equivalent to the ε_p equation of Wilson[14].

Another expression for the strain rate in the plastic zone can be derived from the J-field solutions of Hutchison[7], Rice and Rosengren[8]. The equivalent plastic strain is

$$\varepsilon_p = (\sigma_o/E)[(EJ)/(\alpha\sigma_o^2 I_n r)]^{n/n+1} \tilde{\sigma}_e^{\,n}(n,\theta). \tag{10}$$

(Note that equation (10) assumes elastic strains to be negligible.) The constants I_n and $\tilde{\sigma}_e$ are given in Refs. [8,9], and the remaining values are defined by the Ramberg-Osgood model of eq. (4). Differentiating (10) with respect to time, and substituting the appropriate values for A533B ($\alpha = 1.12$, $n = 9.71$)[10] at $\theta = 0$, yields

$$\dot{\varepsilon}_p = 3.24 \times 10^{-11} \, r^{-0.9066} J^{-0.09337} \dot{J}. \tag{11}$$

Again, an "average" value of plastic strain can be calculated in the middle of plastic zone ($r = 0.5r_o$) giving

$$\dot{\varepsilon}_p \approx 0.00024(\dot{J}/J) = 0.00048 \, (\dot{K}_I/K_I). \tag{12}$$

The plastic strain rate predicted by (9) is 10 times larger than that predicted by (12). This is due to (i) the nature of the linear approximation used in (9), and (ii) the fact that the J-field is not a good approximation where plastic strains are small (such as in most of the plastic zone). It is not clear which model is better or more reliable.

A similar linear approximation could be used to determine the order of magnitude of plastic strains in the process zone. Here the assumptions, that $\varepsilon_p = 0.005$ at $r = w$ and $\varepsilon_p = 0.25$ at $r = 0$, are suggested by finite element results of McMeeking[11] for a material where $n = 10$. Typical strain rates in the center of the process zone for $\theta = 0$ are

$$\dot{\varepsilon}_p \approx 0.35(\dot{J}_I/J_I) = 0.7(\dot{K}_I/K_I). \tag{13}$$

The plastic strain rates can directly be calculated for $\theta = 0$ using the J-field solution of (11). The result given below is not reliable since $r = w/2$ [18].

$$\dot{\varepsilon}_p \approx 0.007(\dot{J}_I/J_I) = 0.014(\dot{K}_I/K_I). \tag{14}$$

In view of equations (9) and (13), plastic strain rate can be

determined at the onset of crack extension of ductile materials such
as A533B. Table 1 gives typical values of plastic strain rates for
"slow" and "fast" loading rates. Later, these shall be compared
with finite element computations.

Ductile crack extension proceeds by void nucleation and growth
in the process zone. Toughness is controlled by void spacing which
determines w_c, the critical size of the process zone. From equations (2) and (3) one can deduce J_{Ic} in terms of the critical process
zone size

$$J_{Ic} = w_c \sigma_o / \beta d_n \approx 2\sigma_o w_c. \tag{15}$$

Changes in the yield stress, for constant values of w_c, will alter
J_{Ic} according to

$$[(J_{Icb})/(J_{Ica})] = [K^2_{Icb}/K^2_{Ica}] = \sigma_{ob}/\sigma_{oa}. \tag{16}$$

The subscripts a and b in (16) refer to two different loading rates.

Experimental data is available from a number of sources[12-16]
which relate the variation in K_{Ic} to the loading rate $\dot{K}_I$. Some of
these data are plotted in Figure 2. Using the strain rate expression
(9), and the yield stress data given in Refs.[12-16], comparisons
are made between the changes in yield stress and the changes in
K_I, as a function of K_I-rate. These comparisons are given in
Table 2. Equation (16) suggests that the last two columns in
Table 2 should have equal values. In fact, good agreement is
achieved only for AISI 1018. The lack of agreement may have 2
sources. The first one is connected with the relative contribution
of the flow stress in the plastic and process zones. The yield
stress ratios in Table 2 are based on the strain rates in the
plastic zone. Since the strain rates in the process zone are 10^2
to 10^3 times the values for the plastic zone, the yield stress
ratios appropriate for the process zone would be larger for materials whose rate sensitivity increases with strain rate (see Figure
4b). Consequently, a significant contribution by the resistance
to flow in the process zone could account for the greater rate
sensitivity of the K_{Ic}-values of A533B and AISI 1020. The second
source is the rate sensitivity of w_c, which is neglected by Equation (16). The increases in the flow stresses and normal stresses
associated with the higher strain rates can facilitate void nucleation and reduce the void spacing and w_c. A large reduction in
w_c in the face of the increase in σ_o could account for the
relatively low K_{Ic}-ratios displayed by AISI 1018 and the PA-6
aluminum alloy.

The existing analyses do not clearly distinguish the relative
contributions of the changes in the plastic and process zones. The
fact that the rate of strain hardening (the value of n) influences

both δ and w[5,11], when the large strains and hardening are largely confined to the process zone, is evidence that the rate sensitivity of flow in the process zone cannot be ignored. The finite element calculations described in the following sections were undertaken to illuminate this issue.

FINITE ELEMENT MODELLING

A stationary crack in a compact tension specimen has been modelled using finite elements. Figure 3 shows a plot of the mesh. To limit the complexity of the model, the elements nearest the crack tip were sized to 1 mm. The mesh was composed of 8-noded isoparametric elements. Numerical studies by deLorenzi, Shih[19], Hoff and Byrne[20] have shown that these elements are suitable for use in the fully plastic range.

The model shown in Fig. 3 has 3500 degrees-of-freedom, and consequently the analysis times for the elastic-plastic problem are very long. As a preliminary step, the same mesh configuration was used in an analysis employing 4-noded isoparametric elements, which resulted in execution times of about one-tenth of the 8-noded analysis. The 4-noded isoparametric elements tend to be too stiff in the fully plastic range, as discussed in Ref. [19], and this will be borne out in the results.

The mesh was constructed so that the stationary crack analysis could later be extended into the advancing crack regime. The method of crack extension, as implemented by Kanninen et al.[17], involves the release of constraints at the crack tip. This technique gives no "obvious" crack tip opening displacement for the stationary crack problem. The authors have selected the separation of the crack faces 1 mm away from the crack tip to be the CTOD, as shown in Figure 6. A more refined approach could be effected by imposing displacements on the boundary of a region very close to the crack tip, as has been done by Sorensen[22].

An important feature of the analysis is the incorporation of the strain rate dependence of the flow stress. Considerable experimental data is available for A533B steel[23] as a result of testing for nuclear applications. The uniaxial stress-strain curve is given in Figure 4a[17], and corresponds to a nominal strain rate of 10^{-3}/s. The relationship between the instantaneous flow stress, $\bar{\sigma}$, and the plastic strain rate, $\dot{\varepsilon}_p$, is given by the Malvern[24] equation

$$\bar{\sigma}/\sigma_0 = [\dot{\varepsilon}_p/D]^{1/p} + 1. \tag{17}$$

The parameters D and p must be determined experimentally, and σ_0 is the flow stress at zero strain rate. Using the data in Ref. [23] at 93C, the rate dependence relationship is plotted in Fig. 4b.

RESULTS

The mesh configuration and material model, described in the previous section, were incorporated in 6 analyses; 3 different loading rates were employed with 4-noded and 8-noded elements. Table 3 summarizes the analyses which have been conducted.

Analysis 1 is static using σ-ϵ relations for A533B for an infinitesimally slow rate of loading. A displacement is applied to the load point in increments of 0.2 mm until a maximum displacement of 2.4 mm is reached. The maximum equivalent stress intensity factor, K_J, is calculated from

$$K_J = [JE/(1-\nu^2)]^{1/2}. \tag{18}$$

Analysis 2 is a dynamic analysis. A constant velocity of 0.008 m/s is applied to the load point, until a displacement of 2.4 mm is reached. This results in an almost constant K-rate, but not a constant J-rate.

Analysis 3 is also a dynamic analysis. A constant velocity of 0.8 m/s (100 times faster than Analysis 2) is applied to the load point. Previous studies using 8.0 m/s as a loading rate had been attempted, but the inertia of the model caused unwanted vibrations which made the interpretation of results very difficult.

Results for the static analysis are plotted in Figure 5. Fig. 5a shows a dimensionless J-integral as a function of a dimensionless load line displacement. A comparison with experimental results by Andrews and Shih[25] reveal that the numerically determined J-integral values are low. Figure 5b shows a dimensionless crack tip opening displacement as a function of a different dimensionless load line displacement. Results from the present study are again smaller than the experimental results of [25]. However, the finite element results of this study are in good agreement with the finite element results of Shih[18]. Equation (3) suggests that δ varies linearly with the J-integral. Fig. 7 plots this relationship and shows remarkable agreement with the static finite element results. The degree of agreement could be fortuitous since the definition of δ is somewhat arbitrary, but is does suggest that the definition of δ in Fig. 6 is reasonable.

The plastic strains in the crack tip vicinity are of interest since they essentially determine the plastic strain rates in a dynamic analysis. The magnitude of the plastic strain, as expressed by tensor product

$$\epsilon_p = [(3/2)\epsilon^P_{ij}:\epsilon^P_{ij}]^{1/2}, \tag{19}$$

is shown in Figure 8, as well as a comparison with the J-field solution of equation (10) at an angle of 45°. Also shown in Figure 8 are results from the slip line solution of Rice and Johnson[26], which were later duplicated by McMeeking[11] using finite elements, for a blunt notch. A comparison shows that the present finite element results underpredict the plastic strains within a region of about 3δ from the crack tip. This is not surprising since the mesh is not sufficiently fine to capture this large strain gradient. It follows that the present finite element analysis only shows the effects of plastic strain rate in the plastic zone, and it is not refined enough to show the effects of plastic strain rate in the process zone.

The effect of loading rate is apparent in Fig. 9. If a critical value of crack tip opening displacement is chosen as a failure criterion, then higher values of J can be tolerated at higher loading rates. In fact, for the range of strain rates occurring in the plastic zone, as given in equation (9), the "average" increases in yield stress would be 11% and 21%, as given by the Malvern equation (17). In view of equation (16) one would expect equal increases in J-integral over this range. Fig. 9 shows that the J-integral values increase by 14% and 26% for a CTOD of 0.28 mm. Although the agreement between the two predictions of yield stress increase looks good, it is possible that, even at these relatively low loading rates the inertia of the specimen may have some effect on the CTOD. Therefore, it may be desirable, in future analysis, to attempt to separate the effects of inertia and strain rate sensitivity on the crack tip opening displacement.

The equivalent plastic strain rate is given in Fig.10. Results are plotted for $\theta = 45°$, since at this angle, the plastic strains are larger and comparisons are more obvious. Equation (10) can be differentiated with respect to time (at $\theta = 45°$) in the same sense that equation (11) was derived. Values of the equivalent strain rate from the J-field, in Fig. 10, are in reasonable agreement with dynamic finite element results. Differences increase as the plastic strains become small, since the J-field is only valid where the elastic strains can be neglected. Here again, the J-field solution is only valid over a very small region.

CONCLUSIONS

1) The loading rate sensitivity of the fracture toughness for the ductile, fibrous mode, as expressed by J_{Ic}, should correspond with the rate sensitivity of the flow stress in the plastic and process zones, when the critical crack tip opening displacement is fixed.

2) The 1 mm, 8-noded finite element mesh used to model the region near the crack tip describes the plastic zone, but is not

sufficiently refined for an adequate representation of the process zone of A533B steel at the J_I = 250 KJ/m^2 level.

3) Numerical results obtained with the finite element model confirm that in the absence of a process zone, the increase of J_I with loading rate at constant CTOD corresponds closely with the increase in the flow stress in the plastic zone.

4) Experimental measurements reported in the literature, contain examples of J_{Ic}-values with a greater rate sensitivity, and values which are less rate sensitive than the flow stress in the plastic zone. These cases are associated with either the 10^2- to 10^3-fold higher strain rates generated in the process zone, or the possible negative rate depending on the void spacing.

5) Values of the equivalent (total) strain rate in the plastic zone derived from the J-field agree with equivalent plastic strain rate values from the finite element model in the range $3\delta < r < 0.5 \ r_o$.

ACKNOWLEDGEMENTS

This work has been supported by the Office of Naval Research, Structural Mechanics Division, under Contract No. N00014-80C-0521. The authors would like to thank Dr. Y. Rajapakse of ONR for his support and encouragement.

All finite element computations were conducted with the general purpose nonlinear code ABAQUS. The authors wish to thank Hibbitt, Karlsson and Sorensen, Inc. for use of the program. Discussions, and J-field results of L. B. Freund are gratefully acknowledged. The authors also wish to thank Ms. Wieger for her work on the manuscript.

TABLE 1:   Estimates of Plastic Strain Rates

Loading Rate	$\dot{J}_I$ [kJ·m^{-2}·s^{-1}]	$\dot{K}_I$ [MPa·m$^{1/2}$·s^{-1}]	$\dot{\varepsilon}_p$ [s^{-1}] Plastic Zone	$\dot{\varepsilon}_p$ [s^{-1}] Process Zone
Slow	2	1	3×10^{-5}	4×10^{-3}
Fast	2×10^5	10^5	3	400

Tabulated values are for a ductile steel such as A533B.
J_{Ic} = 180kJ/m^2; K_{Ic} = 200 MPa·m$^{1/2}$.

TABLE 2:   Comparison of the Strain Rate Dependence of Yield
Strength and the K_I-Rate Dependence of K_{Ic} for Fibrous
Mode Crack Extension in Several Alloys

Alloy	Ref.	K_I-Range $[MPa \cdot m^{1/2} \cdot s^{-1}]$	$\dot\varepsilon_p$-Range $[s^{-1}]$	$[\sigma_{ob}/\sigma_{oa}]^{1/2}$	K_{Icb}/K_{Ica}
A533B(@177C)	12,13	$2-2 \times 10^5$	$5 \times 10^{-5}-5$	1.09	1.60
AISI 1020(HR)	14	$1-2 \times 10^6$	$6 \times 10^{-5}-80$	1.36	1.65
AISI 1018(CR)	15	$1-2 \times 10^6$	$4 \times 10^{-5}-80$	1.14	1.08
PA6 Aluminum	16	$0.5-2 \times 10^6$	$6 \times 10^{-5}-350$	1.14	0.66

Subscript a refers to the lowest strain rate and b refers to the
highest rate.

TABLE 3:   Summary of Finite Element Analyses

Analysis Number	Analysis Type	$\dot{v}_L$ [m/s]	v_{Lmax} [mm]	J_{Imax} [kJ/m^2]	K_{Jmax} [MPa$\cdot$m$^{1/2}$]	$\dot{K}_J$ [MPa$\cdot$m$^{1/2}$/s]
1	Static	0.0	2.4	245	230	0
2	Dynamic	0.008	2.4	254	235	689
3	Dynamic	0.8	2.4	260	238	71000

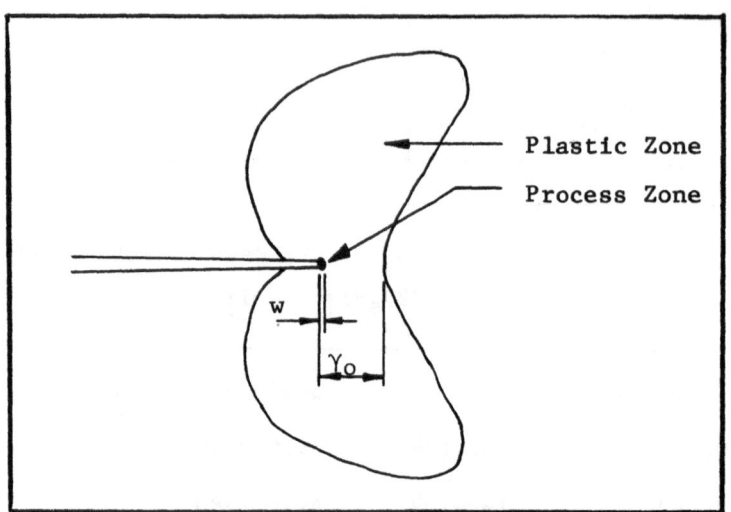

Figure 1:   Plastic and process zones in the vicinity
of a crack.

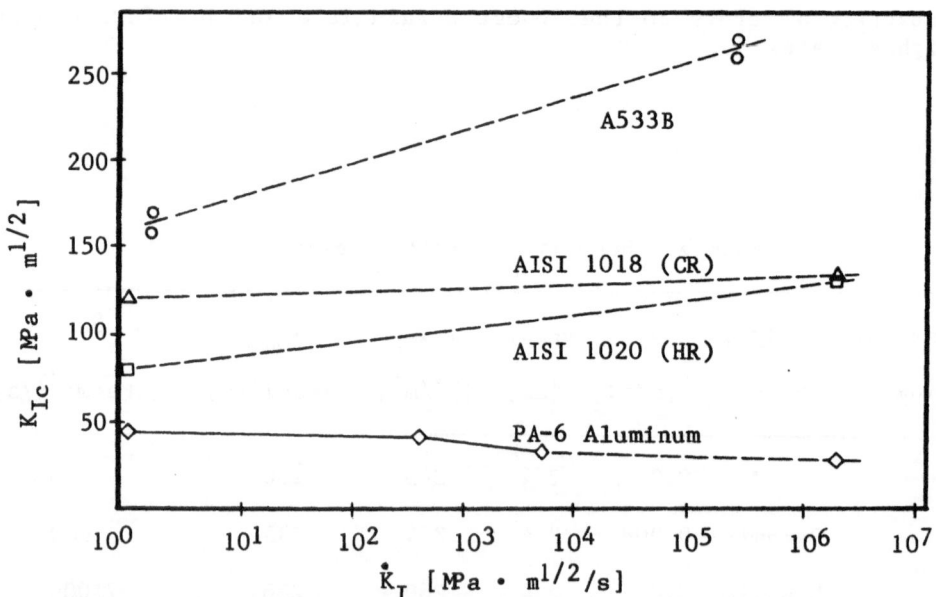

Figure 2:   Variation of K_{Ic} with K_I-rate for different
materials for ductile, fibrous crack
extension.

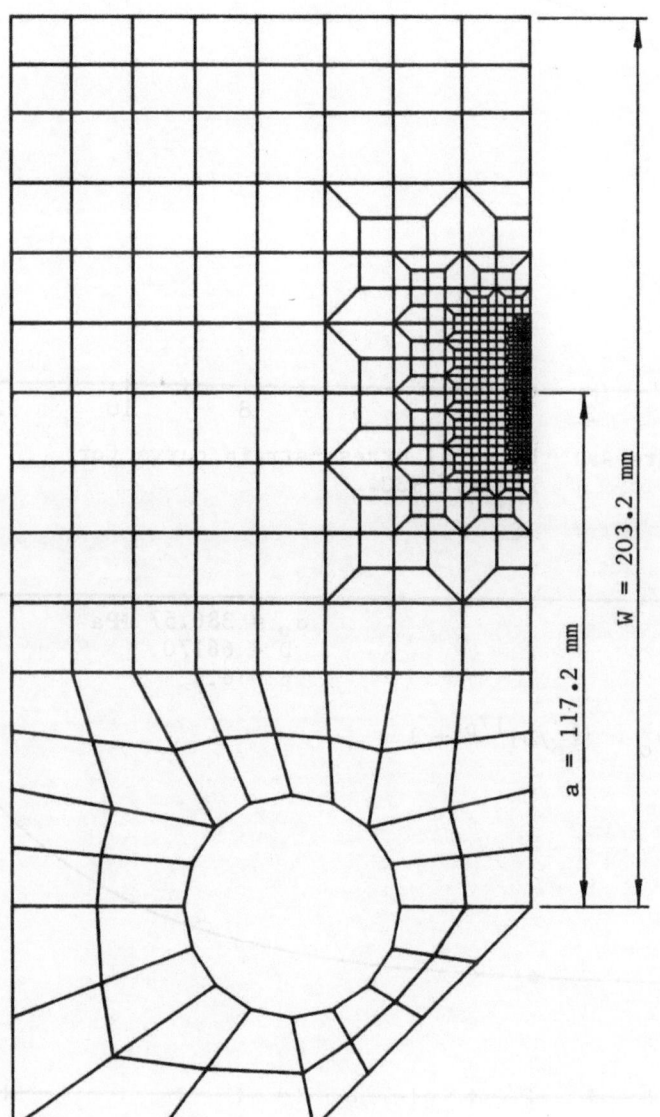

Figure 3:   Finite element model of compact tension specimen.

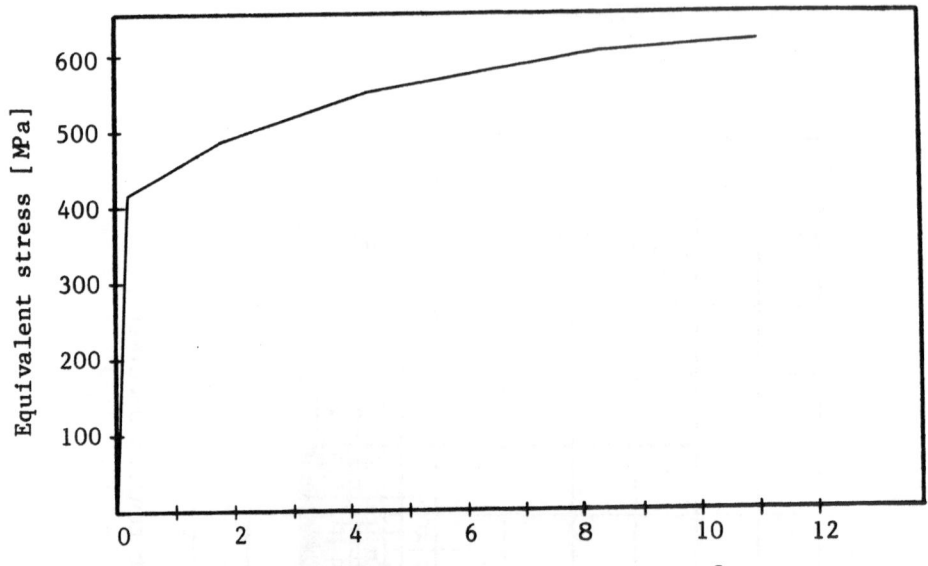

Figure 4a: Uniaxial stress-strain curve for
            A533B @ 93C.

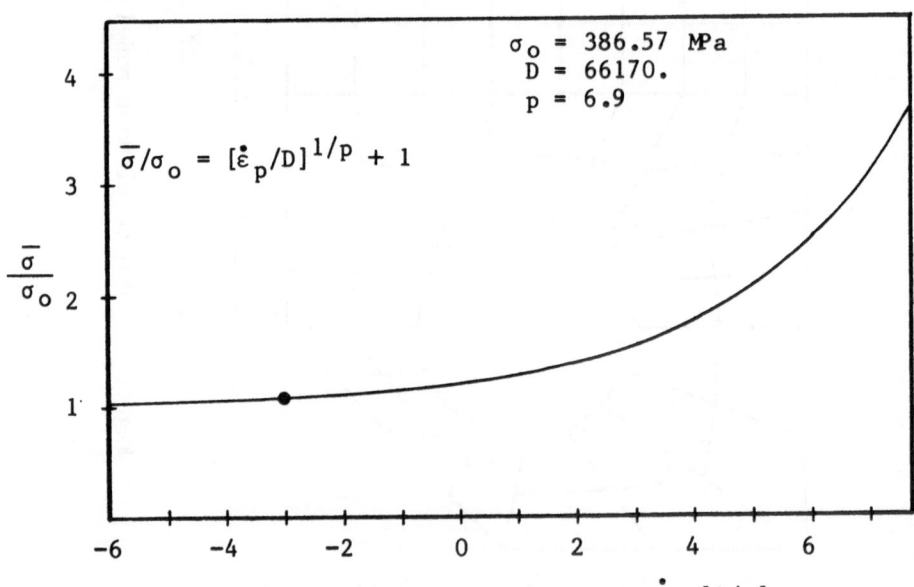

Figure 4b: Variation of flow stress with plastic
            strain rate, for A533B @ 93C.

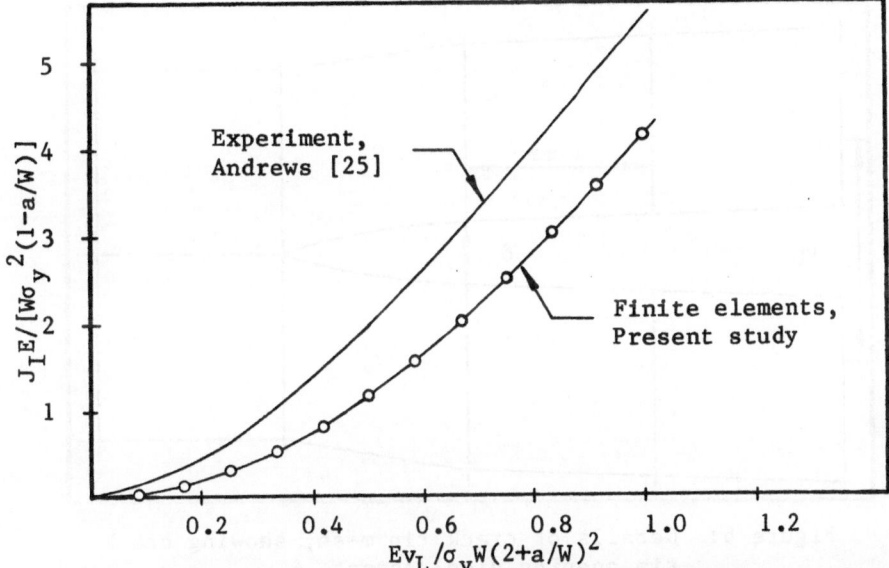

Figure 5a: Variation of the J-integral with
load line displacement for the
static analysis.

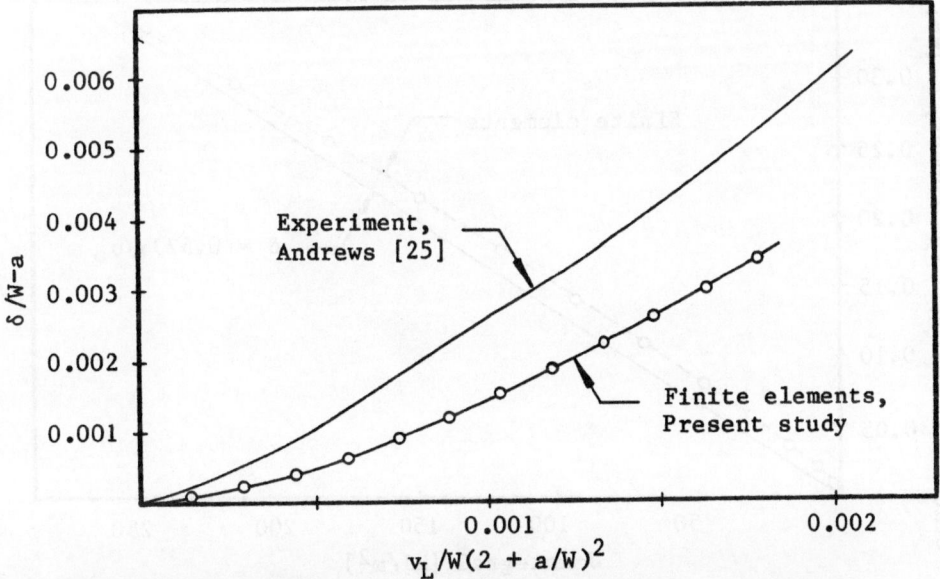

Figure 5b: Variation of crack tip opening displacement
with load line displacement for the static
analysis.

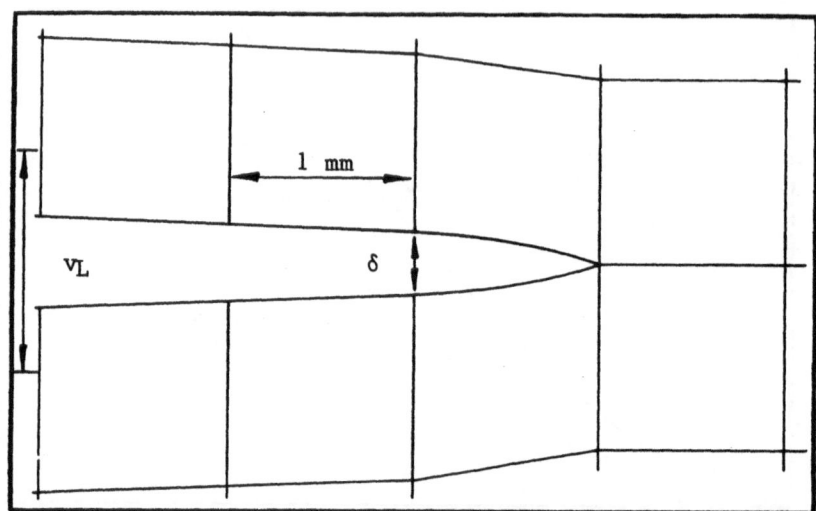

Figure 6:   Details of crack tip mesh, showing crack
            tip opening displacement, δ.

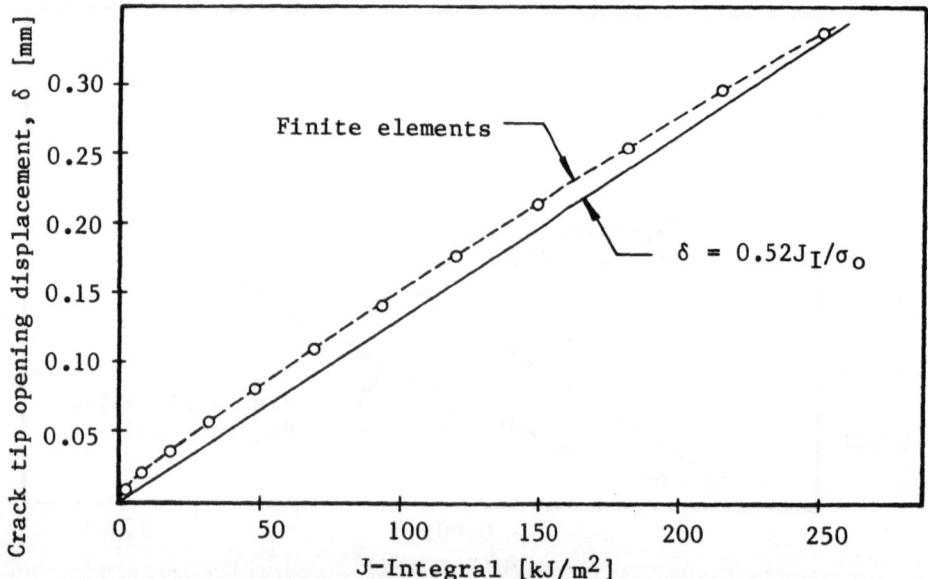

Figure 7:   Relationship between crack tip opening
            displacement and J-integral for a static
            finite element analysis for A533B @ 93C.

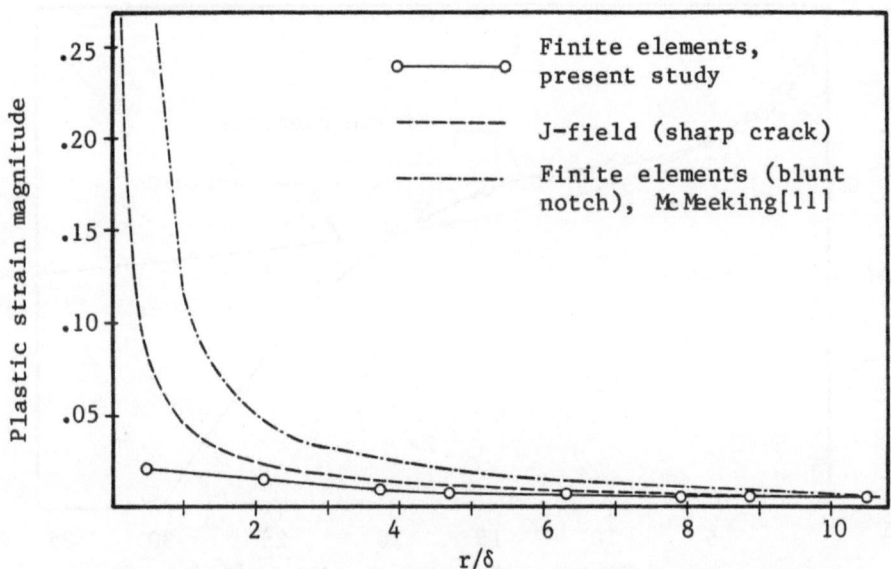

Figure 8:   Variation of plastic strains with distance
            from the crack tip @ θ = 45°.

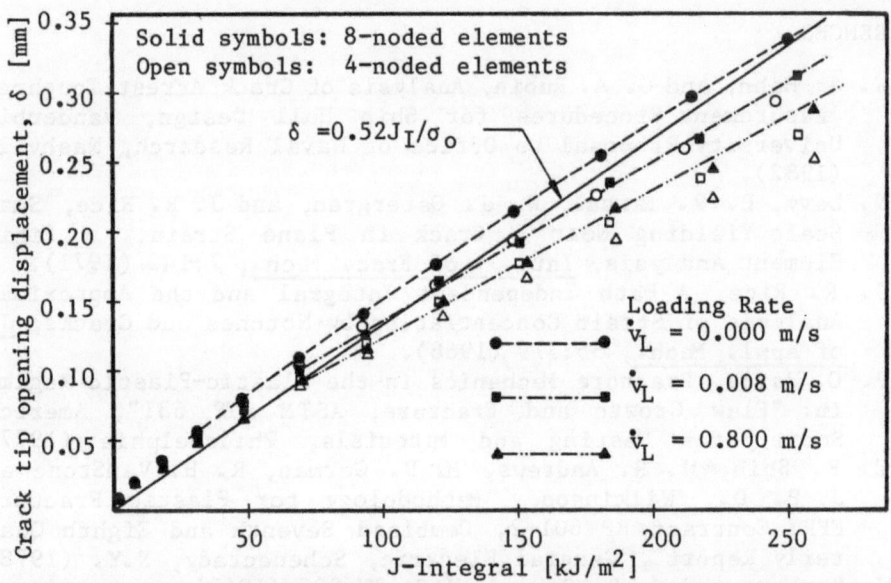

Figure 9:   Relationship between crack tip opening
            displacement and J-integral for statically
            and dynamically loaded analyses.

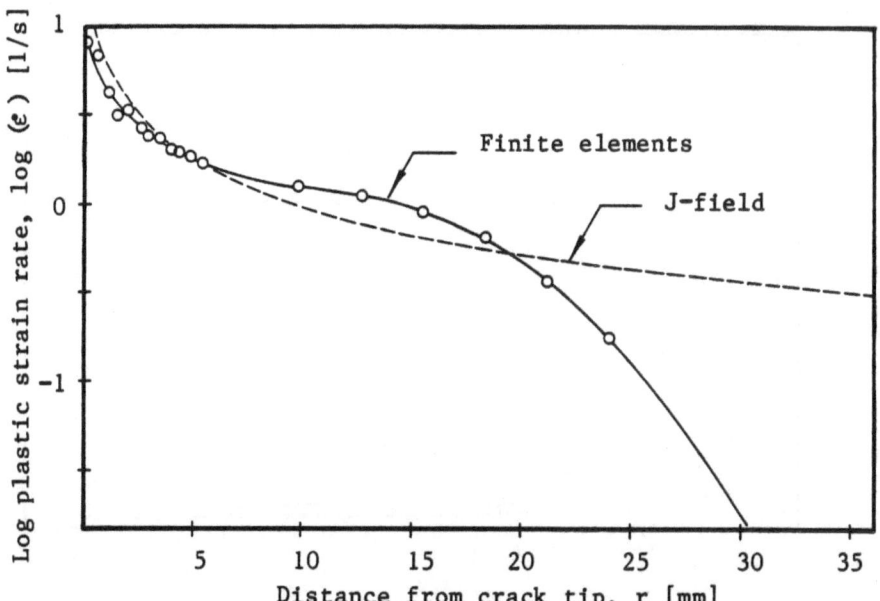

Figure 10:   Relationship between plastic strain rate
             and distance from crack tip for θ = 45°,
             $\dot{v}_L$ = 0.8 m/s and J_I = 245 kJ/m^2.

REFERENCES

1. G. T. Hahn, and C. A. Rubin, Analysis of Crack Arrest Toughness
    Measurement Procedures for Ship Hull Design, Vanderbilt
    University Proposal to Office of Naval Research, Nashville
    (1982).
2. N. Levy, P. V. Marcal, W. J. Ostergren, and J. R. Rice, Small
    Scale Yielding Near a Crack in Plane Strain: A Finite
    Element Analysis, Int. J. of Frac. Mech., 7:143 (1971).
3. J. R. Rice, A Path Independent Integral and the Approximate
    Analysis of Strain Concentration by Notches and Cracks, J.
    of Appl. Mech., 35:379 (1968).
4. P. C. Paris, Fracture Mechanics in the Elastic-Plastic Regimes
    in: "Flaw Growth and Fracture, ASTM STP 631", American
    Society for Testing and Materials, Philadelphia (1977).
5. C. F. Shih, W. R. Andrews, M. D. German, R. H. VanStone and
    J. P. D.   Wilkinson, "Methodology for Plastic Fracture,
    EPRI Contract RP 601-2, Combined Seventh and Eighth Quar-
    terly Report", General Electric, Schenectady, N.Y. (1978).
6. W. Ramberg and W. R. Osgood, NACA TN 902 (1943).
7. J. W. Hutchinson, Singular Behaviour at the End of a Tensile
    Crack in a Hardening Material, J. Mech. Phys. Solids, 16:13
    (1968).

8. J. R. Rice and G. F. Rosengren, Plane Strain Deformation Near a
    Crack Tip in a Power-Law Hardening Material, J. Mech.
    Phys. Solids, 16:1 (1968).

9. C. F. Shih, "Elastic Plastic Analysis of Combined Mode Crack
    Problems", Ph.D. Thesis, Harvard University (1973).

10. V. Kumar, M. D. German, and C. F. Shih, "An Engineering Approach
    for Elastic-Plastic Fracture Analysis, EPRI NP-1931",
    Electric Power Research Institute, Palo Alto, Ca. (1981).

11. R. M. McMeeking, "Finite Deformation Analysis of Crack Tip
    Openings in Elastic-Plastic Materials and Implications for
    Fracture Initiation", Brown University Report COO-3084/44,
    Providence, R.I. (1976).

12. W. L. Server, Static and Dynamic Fibrous Initiation Toughness
    Results for Nine Pressure Vessel Materials, in: "Elastic-
    Plastic Fracture, ASTM STP 668", American Society for
    Testing and Materials, Philadelphia (1979).

13. W. L. Server, W. Oldfield, and R. A. Wullaert, "Experimental
    and Statistical Requirements for Developing a Well-Defined
    K_{IR} Curve", EPRI NP-372, Electric Power Research Institute
    (1977).

14. M. L. Wilson, R. H. Hawley, and J. Duffy, The Effect of Loading
    Rate and Temperature on Fracture Initiation in 1020 Hot-
    Rolled Steel, Eng. Frac. Mech., 13:371 (1980).

15. L. S. Costin, The Effect of Loading Rate and Temperature on the
    Initiation of Fracture in a Mild Rate-Sensitive Steel, J.
    Eng. Mat. Tech., 101: 258 (1979).

16. J. Klepaczko, Application of the Split Hopkinson Pressure Bar
    to Fracture Dynamics, Inst. Phys. Conf., 47:201 (1979).

17. M. F. Kanninen, E. F. Rybicki, R. B. Stonesifer, D. Broek, A. R.
    Rosenfield, C. W. Marschall, and G. T. Hahn, Elastic-Plastic
    Fracture Mechanics for Two-Dimensional Stable Crack Growth
    and Instability Problems, in: "Elastic-Plastic Fracture,
    ASTM STP 668", American Society for Testing And Materials,
    Philadelphia (1979).

18. C. F. Shih, H. G. deLorenzi, and W. R. Andrews, Studies on Crack
    Initiation and Stable Crack Growth, in: "Elastic-Plastic
    Fracture, ASTM STP 668", American Society for Testing and
    Materials, Philadelphia, (1979).

19. H. G. deLorenzi, and C. F. Shih, Int. J. of Frac. Mech., 13:507
    (1977).

20. R. Hoff and T. P. Byrne, "Residual Stress Analysis - A Compari-
    son of Finite Element Results with Closed-Form Solutions",
    Ontario Hydro Research Div. Report 81-232-K, Toronto, Ont.
    (1981).

21. R. S. Barsoum, Int. J. Num. Meth. Eng., 11:85 (1977).

22. E. P. Sorensen, A Numerical Investigation of Plane Strain Sta-
    ble Crack Growth Under Small-Scale Yielding Conditions, in:
    "Elastic-Plastic Fracture, ASTM STP 668", American Society
    for Testing and Materials, Philadelphia, (1979).

23. R. O. Ritchie, W. L. Server, and R. A. Wullaert, Critical Frac-
     ture Stress and Fracture Strain Models for the Prediction
     of Lower and Upper Shelf Toughness in Nuclear Pressure
     Vessel Steels, Met. Trans. A, 10A:1557 (1979).
24. L. E. Malvern, Experimental Studies of Strain Rate Effects and
     Plastic Annealed Aluminum, in: "Proc. ASME Coll. on Be-
     havior of Materials under Dynamic Loading", :81 (1965).
25. W. R. Andrews and C. F. Shih, Thickness and Side Groove Effects
     on J- and δ-Resistance Curves for A533-B Steel at 93°C,
     in: "Elastic-Plastic Fracture, ASTM STP 668", American
     Society for Testing and Materials, Philadelphia, (1979).
26. J. R. Rice and M. A. Johnson, in: "Inelastic Behavior of So-
     lids", Ed. by M. F. Kanninen, W. F. Adler, A. R. Rosenfield,
     and R. I. Jaffee, McGraw-Hill, New York, :641 (1970).

DYNAMIC CRACK CURVING AND CRACK BRANCHING

M. Ramulu and A. S. Kobayashi

University of Washington
Department of Mechanical Engineering
Seattle, WA 98195

INTRODUCTION

The dynamic crack curving criterion proposed by Ramulu and Kobayashi[1] is based on a micro-mechanic model of continuous micro-flaw growth and coalescence in the vicinity of the moving crack tip. It is a dynamic extension of the crack curving criterion proposed by Streit and Finnie.[2] When an off-axis micro-flaw connects with the crack tip, the crack is momentarily kinked or bifurcated. Figure 1 shows several attempts for a rapidly propagating crack to branch in a Homalite-100 plate. Also shown is a successful crack branching under presumably a favorable state of stress. This crack curving criterion together with a crack branching stress in-

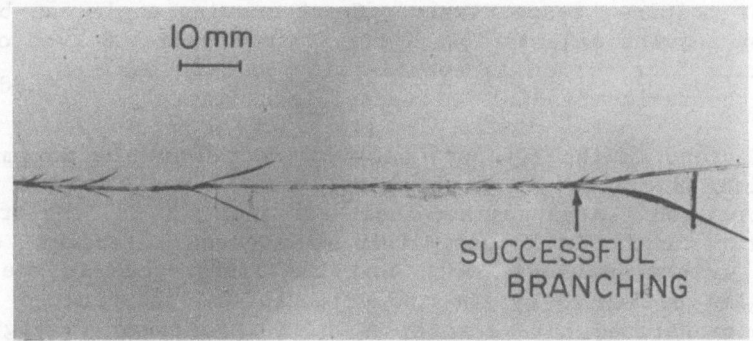

10 mm

SUCCESSFUL
BRANCHING

Figure 1.   Incipient Crack Branching in Homalite-100,
            SEN Specimen.

tensity factor were used to predict the crack branching angles in fracturing Homalite-100 plates.[3] The crack branching criterion was also used to predict dynamic crack branching in a pressurized steel pipe and in fracture specimens where the crack branched at the onset of crack propagation.[4]

The purpose of this paper is to describe; (i) the new dynamic crack curving criterion and (ii) the new crack branching criterion which utilizes the former.

DYNAMIC CRACK CURVING

Historically, dynamic crack curving of a straight crack was studied by Yoffe[5] and Cragg[6] who used the dynamic singular state of stress to predict curving at a crack velocity, c, which is 38 percent of the dilatational wave velocity, c_1. Experimental results, however, show that both crack curving and crack branching generally occur at lower crack velocities of $c/c_1 =0.2-0.28$.[1,3,7,8]
Growth and coalescence of off-axis micro-cracks, on the other hand, require the introduction of a finite radial distance, r_o, from the crack tip where such micro-fracture occurs and which in turn requires the introduction of the second order term, commonly referred to as the remote stress component of σ_{ox}, in the crack tip stress field. The fracture criterion which utilizes this stress field could be based on the maximum circumferential stress, $\sigma_{\theta\theta}$, or minimum strain energy density, S, criterion. Both criteria together with the necessary field equations are discussed in some detail in previous publications.[1,3,4]

Figure 2 shows the nondimensionalized angular variations in $\sigma_{\theta\theta}$ and S in polar coordinates for various crack velocities. Plots of these non-dimensionalized quantities of $\sigma'_{\theta\theta} = \sqrt{r}\sigma_{\theta\theta}/K_I$ and $S' = 4\mu S/K_I^2$ show the strong influence of the second order term, σ_{ox}, on the angular distribution of $\sigma'_{\theta\theta}$ and S' where r, K_I and μ are the crack tip radial distance, mode I dynamic stress intensity factor and shear modulus, respectively. Crack curving angle can be predicted by angular orientation of the minimum radial S' vector when the minimum S criterion is considered. For the maximum $\sigma_{\theta\theta}$ criterion, the maximum radial $\sigma'_{\theta\theta}$ vector is sought.

Variations in the fracture angle predicted by the maximum $\sigma_{\theta\theta}$ and minimum S criteria are shown in Figure 3 for the pure mode I crack extension. At a crack velocity of $c/c_1 = 0.38$, the predicted fracture angle is approximately 63 degrees regardless of the fracture criterion used. Note that the differences in the fracture angles predicted by the two criteria are accentuated by the non-dimensionalized parameter of A which is defined in Figure 2. At A = 0.16 and $c/c_1 = 0.0001$, the fracture angle, θ_c, predicted by minimum S for a Poisson's ratio of $\nu = 0.34$, is 90 degrees, whereas the maximum $\sigma_{\theta\theta}$ yields an angle of 21.5 degrees. Similar

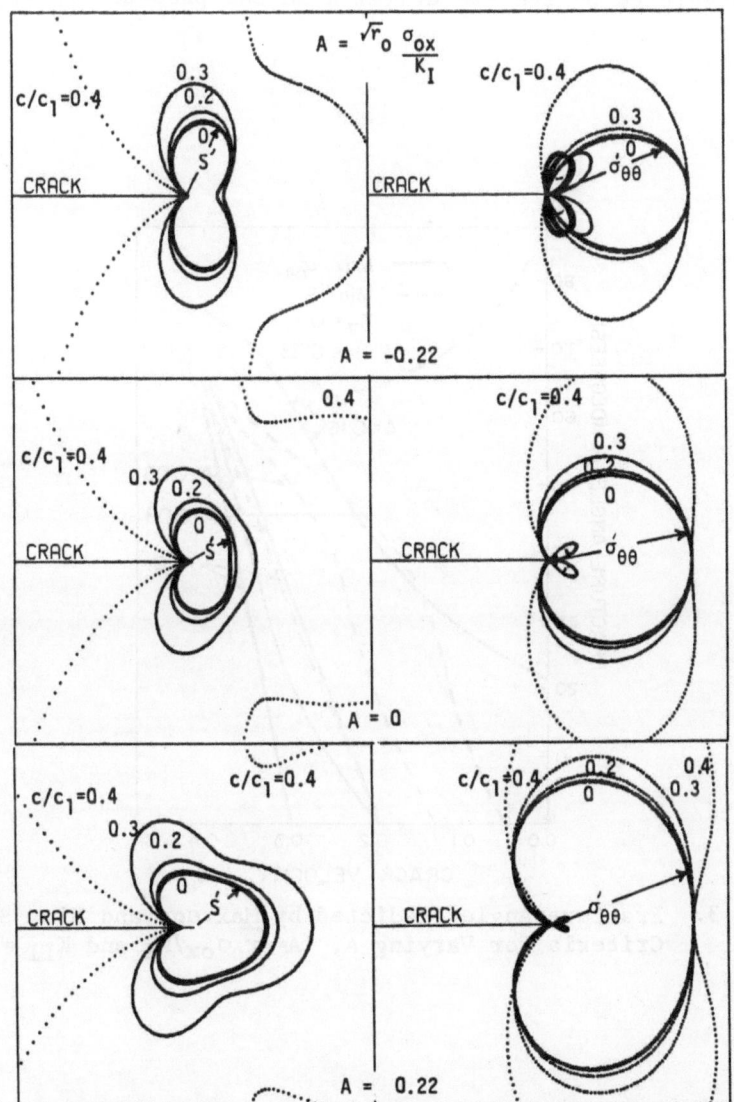

Figure 2.   Polar Plots of Angular Variations of Non-dimensionalized
S' and $\sigma_{\theta\theta}$ for Varying Crack Velocities, c/c_1.

$$\sigma'_{\theta\theta} = \sqrt{r}\ \sigma_{\theta\theta}/K_I,\ S' = 4\mu S/K_I^2\ \text{and}\ r = r_0 = 1.3\ \text{mm}$$

static results in the presence of biaxial load of $A = 0.15 (k - 1)$ and $k = \sigma_{xx}/\sigma_{yy} = 2.06$ were reported by Eftis et al.[9,10] These results are in contrast with the close agreements in fracture angles predicted by the two criteria in the presence of K_{II}.[1,3,11]

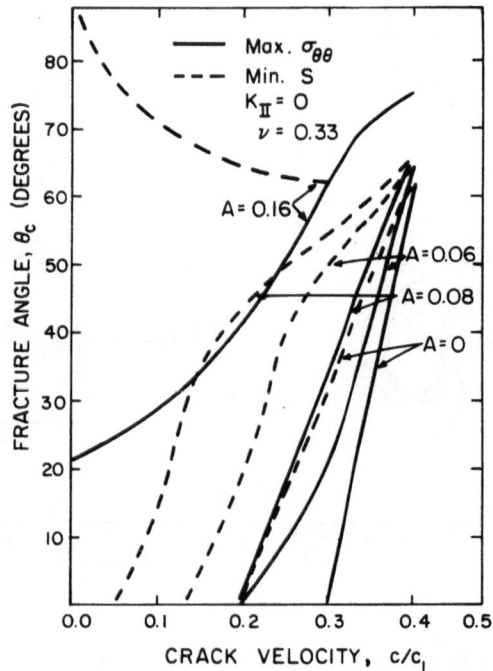

Figure 3. Fracture Angle Predicted by Max.$\sigma_{\theta\theta}$ and Min. S Criteria for Varying A, $(A=\sqrt{r_0}\sigma_{ox}/K_I$ and $K_{II} = 0)$.

Detailed discussions on crack curving under mixed mode fracture can also be found in References 1, 3 and 11.

Extensive dynamic photoelastic analysis of dynamic crack curving and crack branching in Homalite-100[1] and polycarbonate[11] frac-

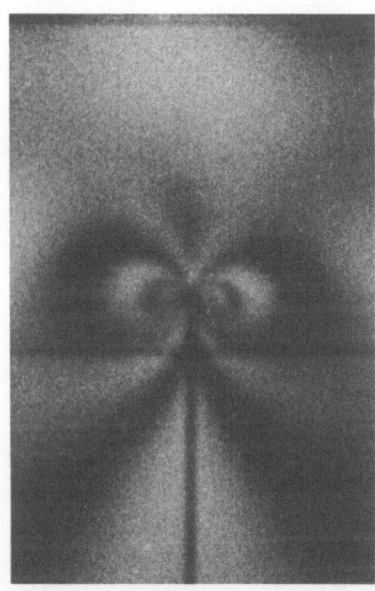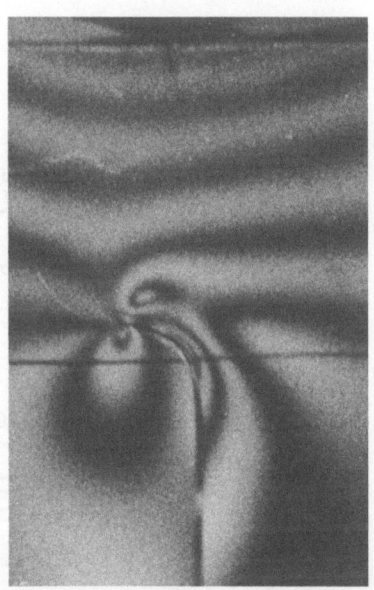

70 μ seconds   185 μ seconds

Figure 4. Typical Dynamic Isochromatics of a Curved Crack,
           Homalite-100,Impacted Notch Bend Specimen. No.
           1-C042574

ture specimens showed that the characteristic radial distance of
r_o is a material property, r_c. These results also showed that the
crack curved at the predicted fracture angle when r_o associated
with the instantaneous dynamic crack tip stress was less than the
critical r_c. For Homalite-100 and polycarbonate sheets considered
in this investigation, r_c = 1.3 and 0.5 mm, respectively.

Figure 4 shows two typical dynamic isochromatics associated
with a propagating crack in an impacted notch bend specimen.[12]
Figure 5 shows the variations in dynamic stress intensity factors,

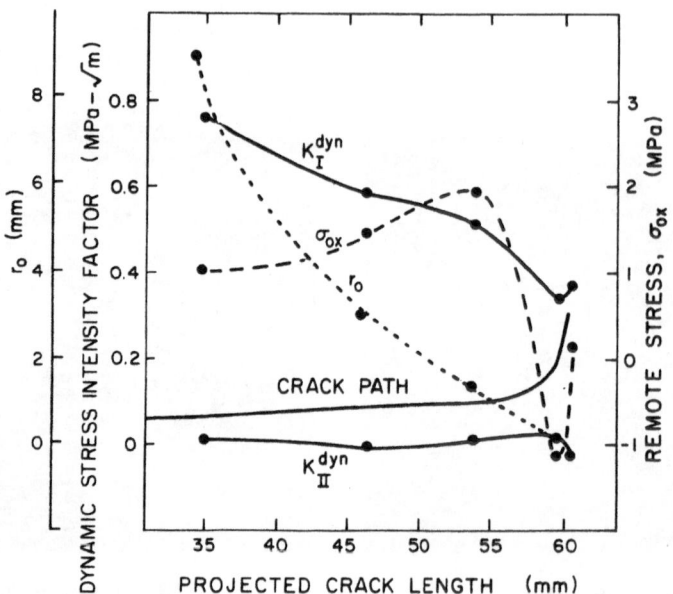

Figure 5.   Dynamic Stress Intensity Factors and r_o of a Curved
             Crack in a Homalite-100 Impacted Notch Bend Specimen.
             Specimen No. 1-C042574.

and r_o with crack extension. Curving of the crack path, which is
also shown in Figure 5, is related to r_o and commences when r_o =
1.3 mm with vanishing K_{II} and is in agreement with the previous
result.

DYNAMIC CRACK BRANCHING

     The dynamic crack branching criterion based on the above, rely
heavily on the experimental conclusion that dynamic crack branch
ing occurs when the dynamic stress intensity factor reaches a critical branching stress intensity factor, K_{Ib}.[3,13-16] This dynamic

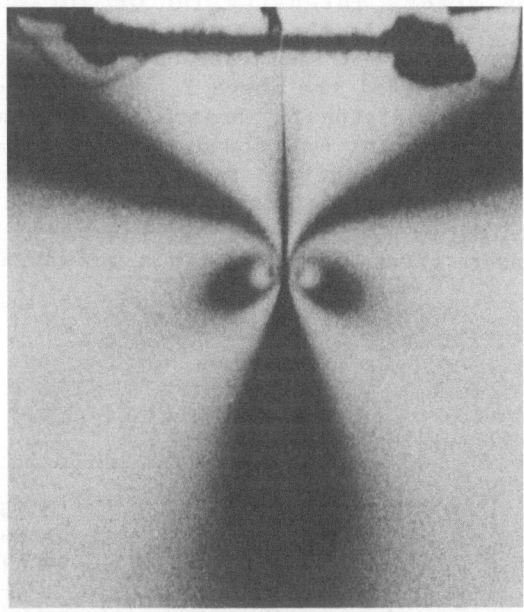

12 μ seconds

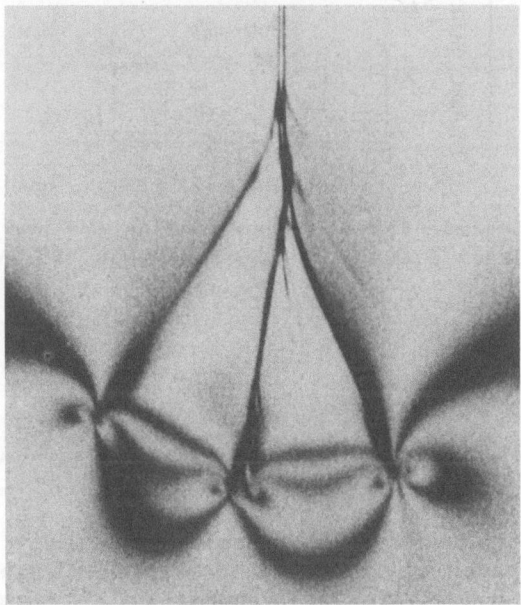

53 μ seconds

Figure 6.   Typical Crack Branching Dynamic Isochromatic Patterns
of a Homalite-100 Single-Edge Notched Specimen. Specimen No. B11.

crack branching stress intensity factor for Homalite-100 is
K_{Ib}/K_{IC} = 4.85 where K_{IC} is the fracture toughness of 0.415 MPa$\sqrt{m}$.
Reference 3 also reports on the presence of dynamic stress inten-
sity factor which exceeded the above K_{Ib} thus indicating that K_{Ib}
is only a necessary condition for branching. The sufficiency condition for crack branching was found to be $r_o \leq r_c$ where r_o is
also used to compute the fracture angle.   As mentioned previously, these necessary and sufficient conditions for crack branching,
which were verified by photoelastic experiments, were also used
to predict dynamic crack branching in a pressurized steel pipe

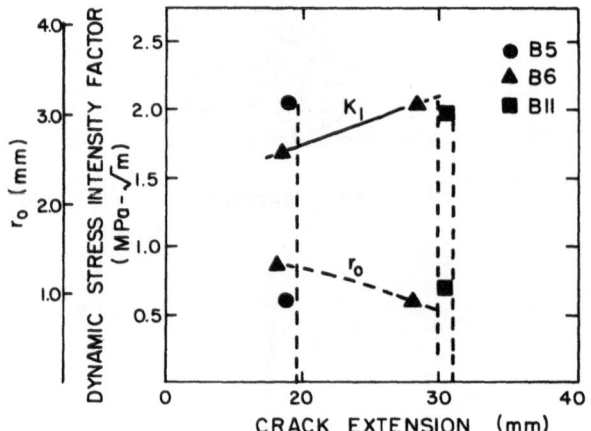

Figure 7.   Dynamic Stress Intensity Factors and r_o Prior to
            Crack Branching.   Specimen Nos. B5, B6 and B11.

and static crack branching in a wedge loaded double cantilever
beam specimen.

     Figure 6 shows two typical dynamic isochromatic patterns asso-
ciated with a branching crack in a Homalite-100 single edged notch
specimen loaded under fixed grip condition.   Figure 7 shows the
dynamic stress intensity factors and r_o prior to crack branching.
The extrapolated K_{Ib} = 2.1 MPa$\sqrt{m}$ which is in agreement with previous results.   The predicted crack branching angle of 30 degrees
compares favorably with the measured fracture angle.

CONCLUSIONS

   The dynamic crack curving and crack branching criteria have been reviewed and additional experimental evidences verifying these criteria have been presented. These criteria are:

   1.   Dynamic crack curving occurs when r_o is less than r_c which is a material property.

   2.   The necessary and sufficient condition for crack branch ing is $K^{dyn} > K_{Ib}$ and $r_o \leq r_c$.

ACKNOWLEDGEMENT

   The work reported here was obtained under ONR Contract No. 0014-76-C-0000 NR 064-478. The authors wish to acknowledge the support and encouragement of Dr. Y. Rajapakse, ONR, during the course of this investigation.

REFERENCES

1. M. Ramulu and A. S. Kobayashi, Dynamic Crack Curving - A Photoelastic Evaluation, to be published in Experimental Mechanics.

2. R. Streit and I. Finnie, An Experimental Investigation of Crackpath Stability, Exp. Mech., 20:17, (1980).

3. M. Ramulu, A. S. Kobayashi and B. S.-J. Kang, Dynamic Crack Branching - A Photoelastic Evaluation, presented at 15th National Symposium on Fracture Mechanics, University of Maryland, July 7 - 9, (1982).

4. M. Ramulu, A. S. Kobayashi and B. S.-J. Kang, Dynamic Crack Curving and Branching in Line-Pipe, to be presented at 1982 ASME WAM, PVP Division, Phoenix, AZ, November 14-19, (1982).

5. E. H. Yoffe, The Moving Griffith Crack, Phil. Mg., 42:739, (1951).

6. J. W. Craggs, On the Propagation of a Crack in an Elastic Brittle Material, J. Mech. and Phy. Solids, 8:66, (1960).

7. G. R. Irwin, J. W. Dally, T. Kobayashi, W. L. Fourney, M. J. Etheridge and H. P. Rossmanith, On the Determination of the a-K Relationships for Birefrigent Polymers, Exp. Mech., 19:121 (1979).

8. J. Congleton, Practical Application of Crack Branching Mea
surements, <u>Dynamic Crack Propagation</u>, ed. by G. C. Sih, Noordhoff Publ., pp. 427-438 (1973).

9. J. Eftis, N. Subramanian and H. Liebowitz, Crack Border Stress
and Displacement Equations Revisited, <u>Eng. Frac. Mech.</u>, 9:189
(1977).

10. J. Eftis, N. Subramanian and H. Liebowitz, Biaxial Load
Effects on the Crack Border Elastic Strain Energy and Strain
Energy Release Rate, <u>Eng. Frac. Mech.</u>, 9:753 (1977).

11. Y. J. Sun, M. Ramulu, A. S. Kobayashi and B. S.-J. Kang, Further Studies on Dynamic Crack Curving, <u>Developments in Theoretical and Applied Mechanics</u>, Vol. XI, ed. by T. J. Chang and
G. R. Karr, University of Alabama in Huntsville, pp. 203-218,
(1982).

12. A. S. Kobayashi and C. F. Chan, A Dynamic Photoelastic Analysis of Dynamic-Tear-Test Specimen, <u>Exp. Mech.</u>, 18:176 (1976).

13. A. S. Kobayashi and M. Ramulu, Dynamic Stress Intensity Factors for Unsymmetric Dynamic Isochromatics, <u>Exp. Mech.</u>, 21:41
(1981).

14. T. Kobayashi and J. W. Dally, The Relation Between Crack Velocity and Stress Intensity Factor in Birefringent Polymers,
<u>Fast Fracture and Crack Arrest</u>, ed. by G. T. Hahn and M. F.
Kanninen, ASTM STP 627, 257 (1977).

15. J. W. Dally, Dynamic Photoelastic Studies of Fracture, <u>Exp.
Mech.</u>, 19:349 (1979).

16. A. S. Kobayashi, B. G. Wade, W. B. Bradley and S. T. Chiu,
1974, Crack Branching in Fracturing Homalite-100 Plates, <u>Eng.
Fract. Mech.</u>, 6:81 (1974).

CRACK PROPAGATION AND ARREST UNDER DYNAMIC LOADING CONDITIONS IN

AN ELASTIC-PLASTIC STRIP

H.K. Chung and J. D. Achenbach

Department of Civil Engineering
Northwestern University
Evanston, IL.   60201

ABSTRACT

Arrest of a rapidly propagating Mode-I crack has been investigated for dynamic loading conditions and a crack which starts
to propagate in an elastic environment but then enters a region
of elastic-plastic material behavior. Fields of stress and
deformation for the transient crack propagation and arrest problem
have been obtained numerically by a finite difference procedure.
The extensional strain just ahead of the arrested crack tip has
been computed for application of a critical strain criterion of
crack arrest.

INTRODUCTION

In earlier papers the influence of dynamic effects on arrest
of fast fracture has been investigated for an essentially brittle
fracture process. Here we consider the case that the loading is
applied rapidly, and the crack tip starts to propagate under
brittle conditions, but then enters a transition zone to a region
of elastic-plastic material behavior.

The fields of stress and deformation near the tip of a
stationary crack in elastic-plastic materials have been studied in
some detail for quasi-static conditions [1]. The analysis of
fields near a growing crack is more difficult, and only a few
continuum plasticity solutions are available [2]. The main
difficulty stems from the history-dependent nature of the deformation and the feature that the material experiences a non-propor-
tional straining history when it is loaded and unloaded as the
crack passes by.

251

By means of asymptotic methods it has been possible to investigate dynamic fields of stress and deformation in the immediate vicinity of a propagating crack tip. Investigations of the dynamic near-tip fields in an elastic perfectly-plastic material have been presented by Slepyan [3] and Achenbach and Dunayevsky [4], for both anti-plane and in-plane deformations.

Asymptotic methods have also been found very useful for the analysis of the dynamic near-tip fields in the presence of strain hardening. As shown by Achenbach, Burgers and Dunayevsky [5], for strain hardening the governing equations are elliptic when the crack-tip speed is less than a certain critical value. The usual separation-of-variables asymptotic analysis can then be carried out, which yields singularities of the general type r^p ($-1 < p < 0$) for the stresses and the strains. As the crack-tip speed increases (or alternatively as the strain-hardening curve becomes flatter) the nature of the governing equations becomes, however, hyperbolic, and the near-tip fields change character. Indeed in the limit of elastic perfectly-plastic behavior the stresses become bounded. For linear strain hardening specific results have been obtained by Achenbach and Kanninen [6] and Achenbach, Kanninen and Popelar [7]. These results are very similar to the ones obtained by Amazigo and Hutchinson [8] for the corresponding quasi-static problem.

The solutions that were obtained in Refs.[3] and [4] show anomalies in the transition from the dynamic to the quasi-static solution. As the crack-tip speed decreases the expressions for the stresses reduce to the ones for the corresponding quasi-static solution, as might be expected on the basis of intuitive reasoning. This is however not true for the strains, which become unbounded in the limit of vanishing crack-tip speed. In Ref.[9] it was shown that the transition from dynamic to quasi-static conditions with decreasing crack-tip speed is effected because the dynamic solution is asymptotically valid in a small edge zone, which shrinks on the crack tip in the limit of vanishing crack-tip speed. Reference [9] also contains an exact solution for the Mode-III dynamic field in the plane of the crack. The same type of solution has also been obtained by Freund and Douglas [10].

Arrest of crack propagation has been investigated by Aboudi and Achenbach [11]-[12], who used a set of elastic-viscoplastic constitutive equations proposed by Bodner and Partom [13].

In the present paper the constitutive relations are based on J_2-flow theory, with the von Mises yield criterion and a bilinear relation between effective stresses and effective strains.

GOVERNING EQUATIONS

     The geometry of a propagating crack in the center-plane of a
strip is shown in Fig. 1. Let $V(t)$ denote the time-dependent
velocity of the crack tip, and let a system of coordinates
(x_1, x_2, x_3) move with the crack tip. The system of stationary
coordinates (x, y, z) is related to the (x_1, x_2, x_3) system by

$$x_1 = x - \int_0^t V(t)dt \; ; \; x_2 = y \; ; \; x_3 = z \; . \qquad (2.1)$$

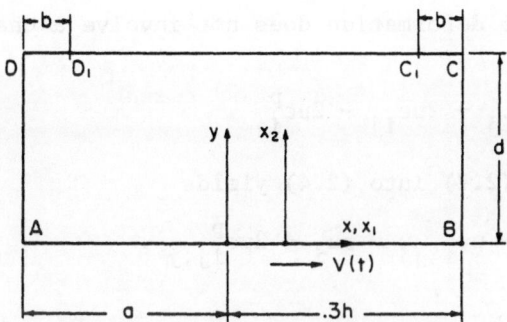

Fig. 1:   Upper half of the strip, with initial
           location of the crack tip.

In the moving coordinate system (x_1, x_2, x_3), derivatives with
respect to time become

$$(\dot{\,}) = \frac{\partial}{\partial t} - V(t)\frac{\partial}{\partial x_1} \; , \qquad (2.2)$$

$$(\ddot{\,}) = \frac{\partial^2}{\partial t^2} - 2V(t)\frac{\partial^2}{\partial x_1 \partial t} + V^2(t)\frac{\partial^2}{\partial x_1^2} - \frac{dV}{dt}\frac{\partial}{\partial x_1} \; . \qquad (2.3)$$

The equation of dynamic equilibrium then is

$$\sigma_{ij,j} = \rho\ddot{u}_i ,$$ (2.4)

where $i,j = 1,2,3$, ρ is the mass density, and $(\ddot{\ })$ is the time derivative defined by (2.3)

The total strain ε_{ij}, the elastic components of strain ε_{ij}^{el} and the plastic components of strain ε_{ij}^{P} are related by

$$\varepsilon_{ij} = \varepsilon_{ij}^{el} + \varepsilon_{ij}^{P}, \text{ where } \varepsilon_{ij} = \frac{1}{2}(u_{i,j} + u_{j,i}) .$$ (2.5)

The components of the stress tensor are related to the elastic strains by

$$\sigma_{ij} = \lambda\varepsilon_{kk}^{el}\delta_{ij} + 2\mu\varepsilon_{ij}^{el} .$$ (2.6)

Since the plastic deformation does not involve a change of volume, we have

$$\sigma_{ij} = \lambda\varepsilon_{kk}\delta_{ij} + 2\mu\varepsilon_{ij} - 2\mu\varepsilon_{ij}^{P} .$$ (2.7)

Substitution of (2.7) into (2.4) yields

$$(\lambda+\mu)u_{j,ji} + \mu u_{i,jj} = \rho\ddot{u}_i + 2\mu\varepsilon_{ij,j}^{P} .$$ (2.8)

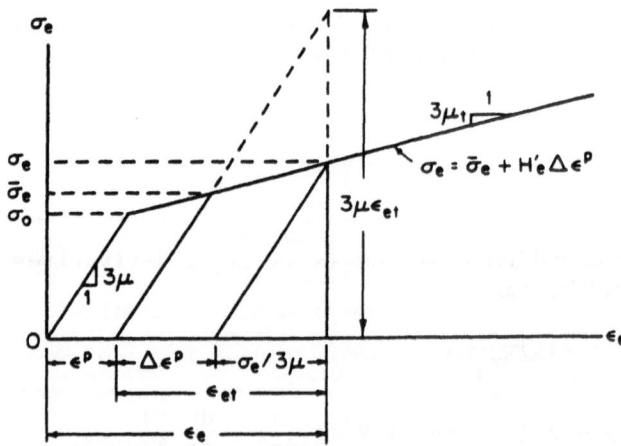

Fig. 2: Relation between effective stress and effective strain; $H'_e = 3\mu\mu_t/(\mu-\mu_t)$.

The constitutive equations for incremental plasticity used here are based on J_2-flow theory and a bilinear relation between effective stress, σ_e, and effective strain, ε_e as shown in Fig.2. We have

$$\sigma_e = \left(\frac{3}{2} s_{ij}s_{ij}\right)^{\frac{1}{2}}, \quad s_{ij} = \sigma_{ij} - \frac{1}{3}\sigma_{kk}\sigma_{ij}. \qquad (2.9a,b)$$

For elastic loading or unloading, Hooke's law yields

$$s_{ij} = 2\mu e_{ij}^{el}, \quad e_{ij} = \varepsilon_{ij} - \frac{1}{3}\varepsilon_{kk}\delta_{ij} \qquad (2.10a,b)$$

It then easily follows from (2.9) and (2.10) that

$$\sigma_e = 3\mu\varepsilon_e^{el}, \quad \varepsilon_e^{el} = \left(\frac{2}{3} e_{ij}^{el}e_{ij}^{el}\right)^{\frac{1}{2}}. \qquad (2.11a,b)$$

The plastic strain increments are related to the components of the current stress deviator by

$$d\varepsilon_{ij}^P = s_{ij}d\lambda. \qquad (2.12)$$

By using the definition of σ_e, Eq.(2.9a), we then easily find that

$$d\lambda = \frac{3}{2}\frac{d\varepsilon^P}{\sigma_e}, \quad \text{where } d\varepsilon^P = \left(\frac{2}{3} d\varepsilon_{ij}^P d\varepsilon_{ij}^P\right)^{\frac{1}{2}}. \qquad (2.13a,b)$$

3. FORMULATION OF THE PROBLEM

The geometry of the strip is shown in Fig. 1 where $a = 0.2h$, $b = .03$ and $d = 0.24h$. At time $t = 0$, the segment C_1D_1 of the upper face of the strip and the corresponding segment on the lower face are subjected to the stretching displacements of the form

$$u_2^i(t) = 0.012[1-\cos(10\pi\bar{t})] \, \sigma/\mu, \quad 0 \leq \bar{t} \leq 0.1 ,$$
$$= 0.024 \, \sigma/\mu, \qquad\qquad\qquad \bar{t} \geq 0.1 , \qquad (3.1)$$
$$u_1 \equiv 0 ,$$

where $\bar{t} = c_T t/h$, h is the scaling length and σ is the scaling stress. The transverse faces of the strip (AD and BC) remain free of stresses. The boundary conditions (3.1) generate wave motions in the strip.

The strip consists of elastic and elastic plastic parts. The initial locations of the boundaries are as follows

$$x/h \leq .001 \quad \text{elastic material}$$

$$.001 \leq x/h \leq .006 \quad \text{plastic material (transition zone) (3.2)}$$

$$x/h \geq .006 \quad \text{elastic, perfectly-plastic material .}$$

In the transition zone the plastic tangent modulus for shear varies from $\mu_t = \mu$ at $x_E = 0.001h$ to $\mu_t = 0$ at $x_P = 0.006h$ according to

$$\mu_t = \mu(1 - \sin\frac{\pi}{2}\frac{\ell}{\ell_o}) , \tag{3.3}$$

where $\ell_o = x_P - x_E = 0.005h$, and $0 \leq \ell \leq \ell_o$. The initial yield stress is set $\sigma_o = 0.88\sigma$.

In the numerical procedure the stress is continuous at the point O, while at the first node to the left of O the stresses σ_2 and σ_{21} are equal to zero. Hence the actual crack tip is in-between O and that first node. The numerical method yields bounded stresses at O. It is assumed that the crack tip begins to propagate at the instant that the equivalent stress at the point O reaches its maximum. The crack-tip speed increases according to

$$V = V_E \sin^2 (\frac{\pi}{2} \frac{t}{t_m}) \tag{3.4}$$

where $V_E = 0.1 c_T$, $t_m = 0.01 h/c_T$ and $0 \leq t \leq t_m$. As soon as O has propagated over a distance of 0.001h, it enters the transition zone and the tip speed decreases. Crack propagation stops when point O reaches the elastic-perfectly plastic portion of the material. The tip velocity in the transition zone is expressed by

$$V = V_E[1-\sin^2 (\frac{\pi}{2} \frac{t}{t_m})] , \tag{3.5}$$

where $V_E = 0.1 c_T$, $t_m = 0.05 h/c_T$, and $0 < t \leq t_m$.

4.   NUMERICAL METHOD

In the moving coordinate system, the displacement equations of motion for deformations in plane strain take the following form

$$\left(\frac{c_L^2}{c_T^2} - \frac{V^2}{c_T^2}\right)u_{1,11} + u_{1,22} + \left(\frac{c_L^2}{c_T^2} + 1\right)u_{2,22} + 2\frac{V(t)}{c_T^2}\frac{\partial}{\partial t}u_{1,1} + \frac{1}{c_T^2}\frac{dV}{dt}u_{1,1}$$

$$- \frac{1}{c_T^2}\frac{\partial^2}{\partial t^2}u_1 = 2(\varepsilon_{11,1}^P + \varepsilon_{12,2}^P) \qquad (4.1)$$

$$\left(1 - \frac{V^2}{c_T^2}\right)u_{2,11} + \frac{c_L^2}{c_T^2}u_{2,22} + \left(\frac{c_L^2}{c_T^2} + 1\right)u_{1,12} + 2\frac{V(t)}{c_T^2}\frac{\partial}{\partial t}u_{2,1} + \frac{1}{c_T^2}\frac{dV}{dt}u_{2,1}$$

$$- \frac{1}{c_T^2}\frac{\partial^2}{\partial t^2}u_2 = 2(\varepsilon_{12,1}^P + \varepsilon_{22,2}^P) \qquad (4.2)$$

where $c_L^2 = (\lambda + 2\mu)/\rho$ and $c_T^2 = \mu/\rho$.

By virtue of symmetry, it suffices to consider only the upper half of the domain. In the plane of the crack ($x_2 = 0$) the boundary conditions are:

$$\sigma_{21}(x_1,0) = 0 \qquad (4.3)$$

$$u_2(x_1,0) = 0 \qquad \text{for } x_1 \geq 0 \qquad (4.4)$$

$$\sigma_2(x_1,0) = 0 \qquad \text{for } x_1 < 0 \qquad (4.5)$$

In the numerical procedure, the derivatives in the displacement equation of motion (2.8) are replaced by their central difference approximations to yield an explicit three-level scheme of the unknown displacement components at time t + Δt. The resulting system of equations is tridiagonal. If the field variables are known at time t and t-Δt throughout the region, then the displacement components at time t+Δt can be obtained by employing a direct inversion algorithm [14], and no iteration is needed. At the outset of the computations it is assumed that the plastic strain is zero everywhere. A plastic strain is, however, immediately obtained in the vicinity of the moving crack tip. Once plastic strains have been computed, they are substituted in the equation of motion to compute the displacement components and hence the total strains ε_{ij} for the next time step.

In the moving coordinate system the flow rule (2.12) yields

$$\frac{\partial \varepsilon_{ij}^P}{\partial t} - V(t)\frac{\partial \varepsilon_{ij}^P}{\partial x_1} = \left[\frac{\partial \lambda}{\partial t} - V(t)\frac{\partial \lambda}{\partial x_1}\right]s_{ij} . \qquad (4.6)$$

Equation (4.6) can be integrated along a characteristic curve
defined by

$$\frac{dx_1}{dt} = - V(t) \tag{4.7}$$

If points A, B and E in Fig. 3 are fixed grid points, and $\overline{EF}$ is
the characteristic curve defined by (4.7), then we obtain the
location of point F as $x_F = x_o + V(t_o)\Delta t$ in the tx_1-plane. The
increments of plastic strain along the characteristic $\overline{EF}$ follow
from (4.6) as

$$\Delta\varepsilon^p_{ij} = \Delta\lambda s_{ij} \ , \ j = 1,2 \tag{4.8}$$

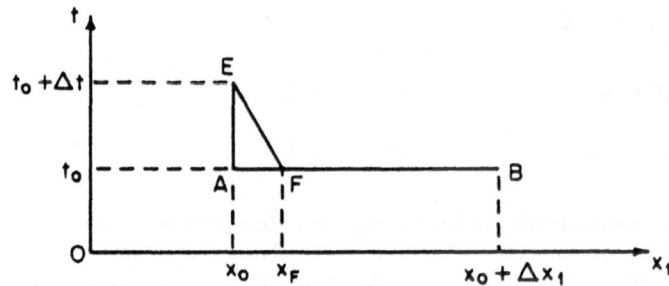

Fig. 3:   Characteristic curve in the tx_1-plane.

with given plastic strains $(\varepsilon^p_{ij})_F$ at point F, the plastic strains
at point $E(x_o,y,t_o + \Delta t)$ become

$$(\varepsilon^p_{ij})_E = (\varepsilon^p_{ij})_F + \Delta\varepsilon^p_{ij} \tag{4.9}$$

The plastic strains $(\varepsilon^p_{ij})_F$ are obtained from a linear inter-
polation between the corresponding values at points A and B. The
characteristic curve defined by (4.7) indicates the trace of the

material point in the plane. Thus, in Fig. 3 the point F is the previous position of the material point E. The plastic strain at F is the preceding plastic strain for point E.

When the total strain ε_{ij} at the current state of deformation at time $t = t_o + \Delta t$ (see Fig. 3) has been computed from eq.(2.8), by using ε_{ij}^P from the preceding state at time $t = t_o$, the current deviatoric strain can be written as

$$e_{ij} = e_{ij}^{el} + \varepsilon_{ij}^P + \Delta\varepsilon_{oj}^P . \tag{4.10}$$

Here ε_{ij}^P is the accumulated plastic strain at F up to the preceding state and $\Delta\varepsilon_{ij}^P$ is a discrete increment of plastic strain defined by (4.8), but still to be computed. Following Ref.[15] we define the equivalent or effective modified total strain as

$$\varepsilon_{et} = [\frac{2}{3}(e_{ij}-\varepsilon_{ij}^P)(e_{ij}-\varepsilon_{ij}^P)]^{\frac{1}{2}} . \tag{4.11}$$

By the use of (4.10), (2.10a), and (4.8) we subsequently obtain

$$\Delta\lambda = \frac{3}{2}\frac{\varepsilon_{et}}{\sigma_e} - \frac{1}{2\mu} . \tag{4.12}$$

Combining (4.8) with (2.13b) yields

$$\Delta\varepsilon^P = \varepsilon_{et} - \frac{1}{3\mu}\sigma_e. \tag{4.13}$$

The bilinear relation between σ_e and ε_e shown in Fig. 2 now suggests the following definition of ε_e:

$$\varepsilon_e = \varepsilon^P + \Delta\varepsilon^P + \frac{1}{3\mu}\sigma_e , \qquad \varepsilon^P = \Sigma \Delta\varepsilon^P . \tag{4.14a,b}$$

Here ε^P, which is the equivalent plastic strain up to the preceding state at $t = t_o$, is a known quantity. Substitution of (4.13) in (4.14a) yields

$$\varepsilon_e = \varepsilon^P + \varepsilon_{et}. \tag{4.15}$$

The equivalent stress can subsequently be obtained from the stress-strain curve, by using ε_e from eq.(4.15). Finally, the current deviatoric stress components are obtained from (2.10a), (4.8), (4.10) and (4.12) as

$$s_{ij} = \frac{2}{3} \frac{\sigma_e}{\varepsilon_{et}} (e_{ij} - \varepsilon_{ij}^P) , \tag{4.16}$$

while the plastic strain increment then follows from (4.8) and (4.16) as

$$\Delta\varepsilon_{ij}^P = \left(1 - \frac{1}{3\mu} \frac{\sigma_e}{\varepsilon_{et}}\right) e_{ij} - \left(\varepsilon_{ij}^P\right). \tag{4.17}$$

Equation (4.17) is then substituted into eq.(4.9) to obtain the plastic strain for the current state at E.

For elastic unloading or elastic reloading, $\Delta\lambda = \Delta\varepsilon^P = \Delta\varepsilon_{ij}^P \equiv 0$. From eqs.(4.11) and (4.13) we obtain that the point $e_{ij} = \varepsilon_{ij}^P$ in strain space corresponds to the origin in stress space. Thus, for $\varepsilon_{ij}^P > 0$ we have $\sigma_e \leq 0$ when $e_{ij} \leq \varepsilon_{ij}^P$, while for $\varepsilon_{ij}^P < 0$ we have $\sigma_e \leq 0$ when $e_{ij} \geq \varepsilon_{ij}^P$.

A kinematical hardening model has been used to account for reverse plastic loading, as shown in Fig. 2. For reverse plastic loading ε_{et}, $\Delta\varepsilon^P$ and σ_e are negative in eqs.(4.13)-(4.17). The computational procedure for reverse plastic loading at the current state defined by $t = t_o + \Delta t$ is then the same as described above.

5.   RESULTS

The time increment was chosen as $\Delta t = .0025 h/c_T$, while results were computed for three choices of Δx_1 and Δx_2, namely, $\Delta x_1 = \Delta x_2 = 0.02h$, 0.01h and 0.008h, respectively. It was found that at $(x_1, x_2) = (.04h, 0)$ the total strains ε_{11} and ε_{22}, show very small differences for the various values of the spatial increments. The stress near the crack tip shows a smaller difference between the results for $\Delta x_1 = \Delta x_2 = .01h$ and .008h than for $\Delta x_1 = \Delta x_2 = .02h$ and .01h. The results for $\Delta x_1 = \Delta x_2 = .008h$ are discussed below.

The application of the boundary conditions (3.1) gives rise to stress waves which propagate towards the crack in the center-plane of the strip. The wave motion reaches the crack tip at $t = .139h/c_T$. The stress components σ_{11} and σ_{22} have been com-

puted at the point 0. Initially, this point is stationary, i.e.,
$V(t) \equiv 0$. Figure 4 shows that the maximum stresses are reached
at $t = .205h/c_T$ which is the arrival time of the maximum value of

the stress pulse. At that time the point 0 starts to propagate
according to Eq.(3.4) until the position $x = .001h$ is reached at
time $t = .223h/c_T$. The crack-tip speed subsequently decreases

according to Eq.(3.5) until the position $x = 0.006h$ is reached at

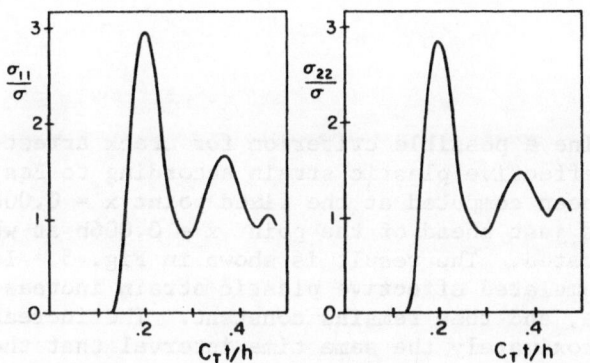

Fig. 4:   Stresses at moving point $x_2 = 0$, $x_1 = 0$.

time $t = .325h/c_T$, when the crack-tip is arrested. Thus in the
time interval $.205h/c_T < t < .325h/c_T$ the stresses shown in Fig. 4
have been computed at the moving point 0. It is noted that the
stresses show an oscillatory behavior. There is a correlation
between the times corresponding to peaks and valleys and the
arrival times of signals that are reflected and diffracted within
the cracked strip.

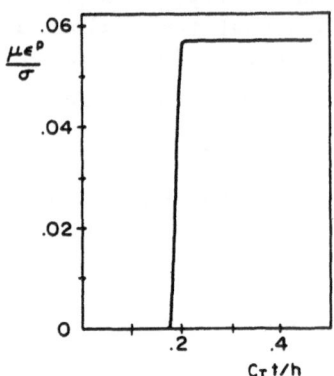

Fig. 5:   Effective plastic strain at the fixed point y=0,
           x = .008h

        To examine a possible criterion for crack arrest the
accumulated effective plastic strain according to Eqs.(4.13) and
(4.14b) has been computed at the fixed point x = 0.008h, y = 0.
This point is just ahead of the point x = 0.006h at which the
crack is arrested.   The result is shown in Fig. 5.   It is noted
that the accumulated effective plastic strain increases to a
maximum value, and then remains constant.   The increase takes
place in approximately the same time interval that the stresses at
0 increase.   Once the crack starts to propagate the stress level
decreases and no additional plastic strain is obtained.   A
criterion which can conveniently be used in conjunction with the
present results is based on a proposal of McClintock and Irwin
[16], also briefly discussed by Rice et al [17].   It postulates
that crack growth will proceed such that a critical level of the
accumulated effective plastic strain is exceeded or maintained at
a characteristic distance Δ ahead of the crack tip in the plane
of the crack.   Clearly, if the maximum value shown in Fig. 5 is
less than the critical value, then the crack will have arrested
as assumed in the analysis of this paper.

ACKNOWLEDGEMENT

        This work was carried out in the course of research sponsored
by the Air Force Office of Scientific Research under Grant
AFOSR78-3589 to Northwestern University.

REFERENCES

1.   J. R. Rice, Elastic-plastic fracture mechanics, in "The
        Mechanics of Fracture," F. Erdogan, ed., AMD Vol. 19,
        The American Society of Mechanical Engineers, New York,
        (1975).
2.   J. R. Rice, Elastic-plastic crack growth, in "Mechanics of
        Solids," H.G. Hopkins and M.J. Sewell, eds., Pergamon
        Press, Oxford and New York (1982).
3.   L. I. Slepyan, Crack dynamics in an elastic-plastic body,
        Izv. Akad. Nauk SSSR, Mekhanika, Tverdogo Tela,
        11:144(1956).
4.   J. D. Achenbach and V. Dunayevsky, Fields near a rapidly
        propagating crack tip in an elastic perfectly-plastic
        material, J. Mech. Phys. Solids, 29: 283 (1981).
5.   J. D. Achenbach, P. Burgers and V. Dunayevsky, Near-tip
        plastic deformations in dynamic fracture problems, in
        "Nonlinear and Dynamic Fracture," N. Perrone and
        S.N. Atluri, eds., AMD Vol. 35, ASME, New York (1979).
6.   J. D. Achenbach and M. F. Kanninen, Crack-tip plasticity in
        dynamic fracture mechanics, in "Fracture Mechanics",
        N. Perrone, H. Liebowitz, D. Mulville and W. Pilkey,
        eds., The Univ. of West Virginia Press, Charlottesville
        (1978).
7.   J. D. Achenbach, M. F. Kanninen and C. H. Popelar, Near-tip
        fields for fast fracture in an elastic-plastic material,
        J. Mech. Phys. Solids, 29:211 (1981).
8.   J. C. Amazigo and J. W. Hutchinson, Crack tip fields in
        steady crack growth with linear strain hardening,
        J. Mech. Phys. Solids, 25:81 (1981).
9.   J. D. Achenbach and V. Dunayevsky, Crack-tip plasticity for
        rapid crack propagation, in "Advances in Fracture
        Research," D. Francois, et al., eds., Pergamon Press,
        Oxford and New York (1980).
10.   L. B. Freund and A. S. Douglas, The influence of inertia on
        elastic-plastic antiplane-shear crack growth,
        J. Mech. Phys. Solids, 30:59 (1982).
11.   J. Aboudi and J. D. Achenbach, Arrest of Mode-III fast
        fracture by a transition from elastic to viscoplastic
        material properties," J. of Appl. Mech. 48:509 (1981).
12.   J. Aboudi and J. D. Achenbach, Arrest of fast Mode-I fracture
        in an elastic-viscoplastic transition zone, Engineering
        Fracture Mechanics, in press.
13.   S. R. Bodner and Y. Partom, Constitutive equations for
        elastic-viscoplastic strain-hardening material,
        J. Appl. Mech, 42:385(1975).
14.   E. Isaacson and H. B. Keller, "Analysis of Numerical Methods,"
        John Wiley & Sons, New York (1966).

15. A. Mendelson, "Plasticity: Theory and Application," The
        MacMillan Co., New York (1968).
16. F. A. McClintock and G. R. Irwin, Plasticity aspects of
        fracture mechanics, in "Fracture Toughness Testing and
        its Applications," ASTM STP 381, American Society for
        Testing and Materials, Philadelphia (1965).
17. J. R. Rice, W. J. Drugan and T-L Shaw, Elastic-plastic
        analysis of growing cracks, in "Fracture Mechanics:
        Twelfth Conference," ASTM STP 700, American Society for
        Testing and Materials, Philadelphia (1980).

OPTICAL DETERMINATION OF THE INTENSITY OF CRACK TIP

DEFORMATION IN A POWER-LAW HARDENING MATERIAL

L. B. Freund, A. J. Rosakis and C. C. Ma

Division of Engineering
Brown University
Providence, RI 02912

ABSTRACT

The value of the J-integral is adopted as a plastic strain
intensity factor which characterizes the crack tip deformation field
in a power-law hardening material. For a planar fracture specimen
containing an edge crack and subjected to loading which results
in mode I deformations, the lateral contraction of the specimen in
the crack tip region is calculated in terms of J from the plane
stress HRR asymptotic field of nonlinear fracture mechanics. The
caustic curve which would be generated by reflection of incident
parallel light from points of the deformed specimen surface lying
well within the crack tip plastic zone is thus determined for
different values of the hardening exponent n. The value of J is
found to be proportional to the maximum transverse diameter of the
caustic curve to the power $(3n + 2)/n$.

INTRODUCTION

An optical method, known as the shadow spot method or the
method of caustics, has been established as a standard experimental
technique for measuring the elastic stress intensity factor in
planar fracture specimens of materials for which the concepts of
elastic fracture mechanics apply [1]. When a large plate containing
an edge crack is loaded so that the response is nominally elastic
and the crack opening occurs in mode I, the stress and deformation
fields very near the crack tip assume the familiar universal spatial
distribution. Only the magnitude of the near tip field varies with
load and geometry, and this magnitude is customarily the mode I
elastic stress intensity factor. Within the framework of plane
stress analysis, the deformed shape of the specimen surface near

265

the crack tip is thus also known up to a scalar amplitude which is
equivalent to the stress intensity factor. The success of the
method of caustics is based on the fact that, with a suitable
optical arrangement, the light pattern obtained by reflecting
parallel incident light from the specimen surface near the crack
tip provides a direct measure of the stress intensity factor.

It is clear that applicability of the method does not hinge
on the material in the crack tip region responding in a linear
elastic manner. Instead, the key feature is that the deformed shape
of the specimen surface (which is the reflecting surface) in the
crack tip region is known up to a scalar amplitude. Asymptotic
elastic-plastic analyses of near-crack-tip fields in power law
hardening materials suggest that this situation may prevail for
these cases as well. Within the framework of plane stress analysis,
with small strains and proportional stress histories for stationary
cracks, the value of Rice's J-integral has been proposed as a
plastic intensity factor [2]. The viewpoint is adopted here that
J provides a suitable scalar amplitude for the deformed shape of
the surface of an elastic-plastic fracture specimen, and a means
of measuring this amplitude is proposed.

CAUSTICS BY REFLECTION

The x_1, x_2-plane coincides with the undeformed polished surface of a planar fracture specimen, and the deformed shaped of the
specimen surface is given by $x_3 = f(x_1, x_2)$; see Fig. 1. A family of
parallel light rays is incident on the deformed reflecting surface
and, if certain geometrical conditions are met, then the family of
reflected rays (or their virtual extensions) will have an envelope
in the form of a three-dimensional surface called the caustic
surface. If a "screen" defined by the X_1, X_2-plane is positioned
parallel to the x_1, x_2-plane so that it intersects the caustic
surface, then a cross section can be observed as a bright curve
(the caustic curve) bordering a dark region (the shadow spot) on
the "screen".

If an incident ray strikes the specimen's surface at the point
x_i, then the coordinates of the point at which the virtual
extension of the reflected ray strikes the screen are

$$X_i = x_i - 2z_o(\partial f/\partial x_i) \qquad\qquad (1)$$

where z_o is the distance from the specimen to the screen. It is
assumed that the magnitude of f is small compared to z_o and that
the slope of the reflecting surface is much less than unity.
Equation (1) is the optical mapping of points on the specimen
surface onto points on the screen. A condition for the existence
of a caustic is the non-invertibility of this mapping, and the
determinant of the Jacobian matrix must vanish, that is,

The equation (6) is an expression for r given in the unprimed coordinates is called the caustic curve. Details on the initial ?????
??? and ??? to specify the caustic curve of the screen intensity
in (7) for the size of the caustic curve as compared to the size of the
????? All effects above is of major importance in monitoring the
type of deformation which can be inferred by its detail in the
????? ??? ????.

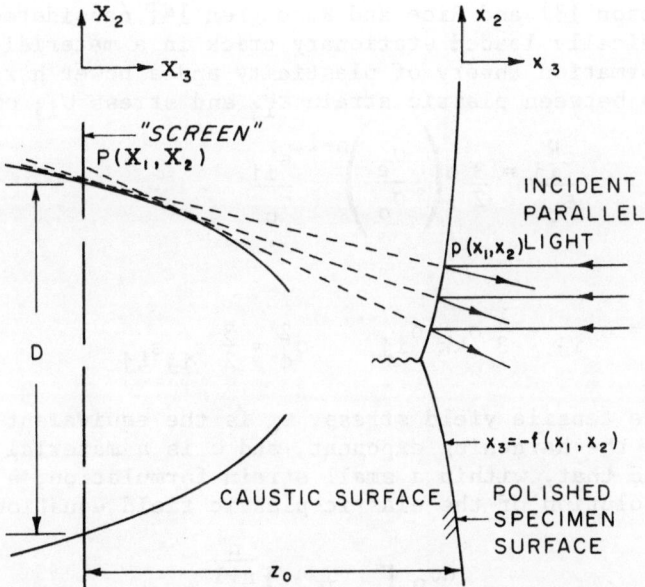

Fig. 1   Schematic Diagram of Optical Setup for Generating Caustics
         by Reflection.

$$\frac{\partial (X_1, X_2)}{\partial (x_1, x_2)} = 0 \tag{2}$$

The condition (2) is an equation for a curve on the specimen surface which is called the _initial curve_. Points on the initial curve map into points on the caustic curve on the screen according to (1). The size of the initial curve compared to the size of the crack tip plastic zone is of major importance in considering the type of information which can be inferred by the method in the presence of crack tip plasticity.

ASYMPTOTIC ELASTIC-PLASTIC CRACK TIP FIELD

     Hutchinson [3] and Rice and Rosengren [4] considered the case of a monotonically loaded stationary crack in a material described by a J_2 deformation theory of plasticity and a power hardening relationship between plastic strain ε_{ij}^p and stress σ_{ij} of the form

$$\frac{\varepsilon_{ij}^p}{\varepsilon_o} = \frac{3}{2} \alpha \left(\frac{\sigma_e}{\sigma_o} \right)^{n-1} \frac{s_{ii}}{\sigma_o} \tag{3}$$

where

$$s_{ij} = \sigma_{ij} - \frac{1}{3} \sigma_{kk} \delta_{ij} \qquad \sigma_e^2 = \frac{3}{2} s_{ij} s_{ij} \tag{4}$$

and σ_o is the tensile yield stress, ε_o is the equivalent yield strain, n is the hardening exponent, and α is a material constant. They observed that, within a small strain formulation, a possible asymptotic solution of the elastic-plastic field equations is

$$\varepsilon_{ij} \rightarrow \frac{\alpha \sigma_o}{E} \left[\frac{JE}{\alpha \sigma_o^2 I_n r} \right]^{\frac{n}{n+1}} E_{ij}(n, \theta)$$

$$\sigma_{ij} \rightarrow \sigma_o \left[\frac{JE}{\alpha \sigma_o^2 I_n r} \right]^{\frac{n}{n+1}} \Sigma_{ij}(n, \theta) \tag{5}$$

as $r \rightarrow 0$. The factors describing angular variation also depend on n. The dimensionless quantity I_n, which is defined in [3], decreases from 5 for n = 1 to 2.57 for n $\rightarrow \infty$ for cases of plane stress. The singular field in (5), which is commonly known as the HRR asymptotic field, includes the value of the J-integral as an intensity factor.

Based on the observation that J is a characterizing parameter
for the crack tip field, it has been suggested that a condition
for onset of crack growth is the attainment of a critical value
of J. This seems reasonable, provided that the one parameter field
prevails over a region large in size compared to the fracture
process zone size [2]. The interest in measuring values of J for
ductile materials stems from the potential usefulness of this
suggested fracture initiation criterion.

RELATIONSHIP BETWEEN J AND CAUSTIC SIZE

Consider a large plate with thickness d of a power law
hardening material containing an edge crack and subjected to boundary loading which produces a plane stress opening mode of deformation. The normal displacement u_3 of the plate is

$$u_3(x_1, x_2) = - f(x_1, x_2) = \frac{d}{2} \varepsilon_{33} \tag{6}$$

where ε_{33} is the total strain in the thickness direction. Near
the crack tip, $\varepsilon_{33}^p \gg \varepsilon_{33}^e$ so that the total strain may be replaced
by the plastic strain in (6). For surface deformation consistent
with the HRR field (5)

$$f(x_1, x_2) = - \frac{\alpha\sigma_o d}{2E} \left[\frac{JE}{\alpha\sigma_o^2 I_n r} \right]^{\frac{n}{n+1}} E_{33}(\theta, n) \tag{7}$$

The dependence of E_{33} on θ has been tabulated in one degree increments for a number of values of n by Shih [5].

The shape of the reflecting surface (7) may now be substituted into the equation for the optical mapping (1), and the
condition for the existence of a caustic (2) may be enforced. If
this is done, then a relationship between the value of J and the
maximum transverse diameter of the caustic curve D is found in the
form

$$J = S_n \frac{\alpha\sigma_o^2}{E} \left[\frac{E}{\alpha\sigma_o z_o d} \right]^{\frac{n+1}{n}} D^{\frac{3n+2}{n}} \tag{8}$$

where S_n is a function of n whose value is essentially 0.072 for
n > 4. Assuming that the bulk material parameters and the
geometrical parameters of the system are known, (8) provides a
means of inferring a value of J from a measurement of a shadow
spot size. The details of the derivation are presented in [6].

It is instructive to examine the full reflected optical field,
rather than just the caustic curve, for the situation at hand. A

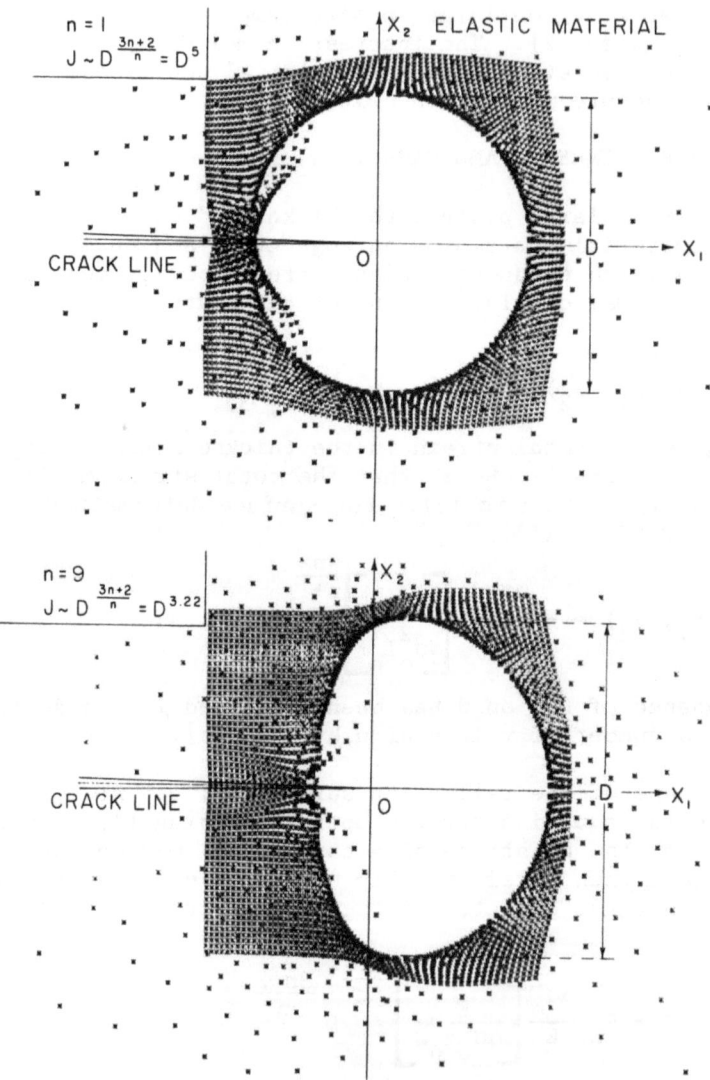

Fig. 2 Simulated Reflected Optical Field for a Rectangular Array
of Light Rays Incident on the Crack Tip Region. Each
Symbol Indicates Where a Reflected Ray Will Strike the
"Screen". Data Shown are for n = 1 and n = 9.

numerical simulation of the optical reflection has been carried
out for this purpose. A closely spaced rectangular array of
parallel light rays was assumed to be incident on the deformed
specimen surface. Due to the deformation, the reflected rays devi-
iate from parallelism. The pattern of reflected rays that would
be observed (photographically) on the screen plane in Fig. 1 is
shown in Fig. 2 for n = 1 and n = 9. The central parts of these
simulated reflected fields contain no rays, and these parts are the
shadow spots. The reflected rays are most dense on the perimeters
of the shadow spots, and these perimeters are the caustic curves.
The numerical scheme by which the reflected optical fields are
constructed is described in [7], where additional results are
presented.

Experiments based on the ideas presented here have been limited
to a small number of tests on a tool steel in a conditions with a
very large strain hardening exponent. Specimens were in the double
cantilever beam configuration with a wedge loading arrangement.
The results of these tests have been reported in [8]. Further
experimental work is required before the potential of the experimental procedure described here can be realized or before all of
the limitations of the procedure can be appreciated.

ACKNOWLEDGMENT

This work has been supported by the Office of Naval Research,
Structural Mechanics Program, and by the NSF Materials Research
Laboratory at Brown University.

The computations were performed on the VAX-11/780 Engineering
Computer Facility at Brown University. This facility was made
possible by grants from the National Science Foundation (Solid
Mechanics Program), the General Electric Foundation, and the
Digital Equipment Corporation.

REFERENCES

1. J. Beinert and J. F. Kalthoff, "Experimental Determination
    of Dynamic Stress Intensity Factors by the Method of Shadow
    Patterns", in Mechanics of Fracture, Vol. VII, edited by G. C.
    Sih, Noordhoof, 1982.
2. J. R. Rice, "Elastic-Plastic Fracture Mechanics", in The
    Mechanics of Fracture, edited by F. Erdogan, AMD Vol. 19, 1976,
    ASME, pp. 23-53.
3. J. W. Hutchinson, "Singular Behavior at the End of a Tensile
    Crack", Journal of the Mechanics and Physics of Solids, Vol.
    16, 1968, pp. 13-31.
4. J. R. Rice and G. F. Rosengren, "Plane Strain Deformation Near
    a Crack Tip in a Power Law Hardening Materials", Journal of the
    Mechanics and Physics of Solids, Vol. 16, 1968, pp. 1-12.

5. C. F. Shih, "Elastic-Plastic Analysis of Combined Mode Crack
    Problems", Ph.D. Thesis, Harvard University, 1973.
6. A. J. Rosakis, "Experimental Determination of the Fracture
    Initiation and Dynamic Crack Propagation Resistance of
    Structural Steels by the Optical Method of Caustics",
    Ph.D. Thesis, Brown University, 1982.
7. C. C. Ma, "Caustic Curves Obtained by Numerically Simulating
    Reflection of Light Rays from the Surface of Plane Stress
    Fracture Specimens", Sc.M. Thesis, Brown University 1982.
8. A. J. Rosakis and L. B. Freund, "Optical Measurement of the
    Plastic Strain Concentration at a Crack Tip in a Ductile
    Steel Plate", Journal of Engineering Materials and Technology,
    Vol. 104, 1982, pp. 115-120.

APPLICATION OF MICROSTATISTICAL FRACTURE MECHANICS

TO DYNAMIC FRACTURE PROBLEMS

Donald A. Shockey, Lynn Seaman and Donald R. Curran
Poulter Laboratory
SRI International
Menlo Park, CA   94025

INTRODUCTION

Thus far in this conference, the dynamic fracture papers have focused on continuum treatments of single crack problems. However, many fracture problems encountered by the Army are not single crack problems and cannot be treated by continuum approaches. In fact, ordnance problems can, and usually do, involve many simultaneously active cracks as well as other simultaneously occurring failure processes. The situation is illustrated in Figure 1, which shows a polished and etched cross section through the impact site of a steel plate that was impacted at 6 km/s by a 6-mm-diameter water-filled polycarbonate sphere. Failure processes in evidence are ductile fracture (voids beneath the crater), brittle fracture (cracks extending inward from the crater profile), adiabatic shear banding (the white etching bands extending inward from the crater profile and also those connecting the ductile voids beneath the crater), the α (bcc) to ε (hcp) to α (bcc) phase transformations (indicated by the hemispherical dark-etching region encompassing the crater), and homogeneous plastic flow (indicated by the curvature in the originally straight and parallel rolling texture). All these five failure processes have occurred more or less simultaneously, and each process influences the other as it occurs.

The complexity of the situation precludes a treatment by classical fracture mechanics concepts. A microstatistical fracture mechanics (MSFM) approach is more appropriate, in which the development of individual microfailure features is treated

274

D. A. SHOCKEY ET AL.

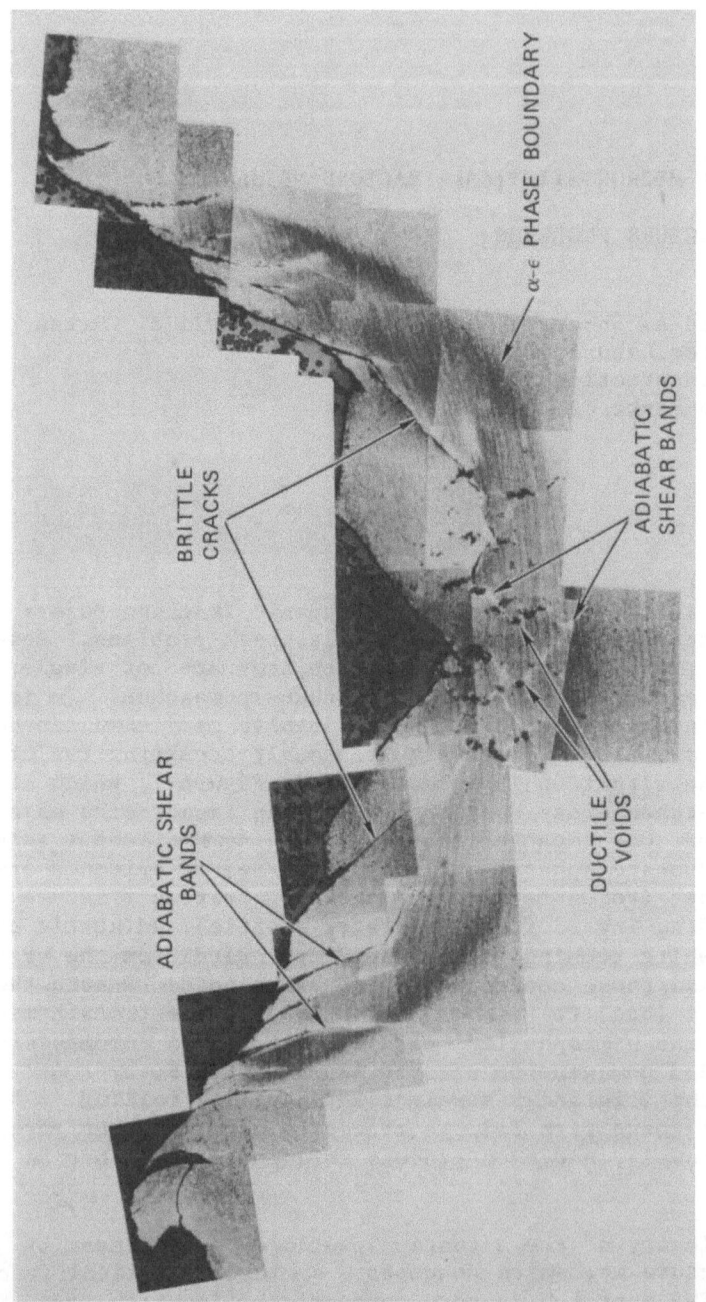

Figure 1 Polished and etched cross section through the impact site of a steel plate that was impacted at 6 km/s by a 6-mm-diameter water-filled polycarbonate sphere.

statistically, and the effect of this development on the state of stress or strain is accounted for continuously.

The first purpose of this paper is to describe this MSFM approach by illustrating its use for the case of shock-induced ductile voids in steel.[1] A microstatistical treatment of shear bands is presented in the following paper. The second purpose is to show that the microstatistical approach merges with the continuum approach. This is done by using the approach and the data generated in the shock wave experiments to compute the behavior of a macrocrack propagating in a DCB specimen.[2]

SHOCK-WAVE-INDUCED DUCTILE FRACTURE

Dynamic ductile fracture can be studied without the complications caused by other accompanying failure processes in simple, well-defined, plate impact experiments. The arrangement shown in Figure 2 is commonly used to produce a short-lived tensile pulse in a specimen under an easily-analyzed, one-dimensional-strain state. Here a flyer plate mounted on a cylindrical projectile is accelerated in a gun barrel by sudden release of pressurized gas and is made to impact a specimen plate at the gun muzzle. Care is taken to ensure simultaneous impact between the two plates at all locations, so that lateral strains are zero until unloading waves from the periphery can propagate into the specimen. To prevent subsequent and uncontrolled impact loads, the specimen plate is decelerated in a chamber filled with energy-absorbing material.

The initial compression waves produced in the colliding plates upon impact reflect at the plate boundaries to produce a tensile pulse in the specimen whose amplitude is specified and controlled by the impact velocity. The pulse duration is controlled by the plate thicknesses. Thus, if the flyer plate is tapered, as shown in Figure 3, the duration of the pulse experienced by the specimen varies with location, and the dependence of ductile fracture damage on time-at-stress can be investigated in a single experiment. If a series of experiments are performed at different impact velocities, the effect of stress can also be studied. The procedure is illustrated here for an armor steel.

Plate Impact Experiments

High hardness armor steel made by Great Lakes Steel Company and designated as XAR30 was selected for investigation. Chemical composition and tensile properties are given in Tables I and II. Specimen disks 3.81 cm in diameter were machined to various thicknesses (see Table III) and mounted in the target plate. The edges were beveled at an 8-degree angle to allow the specimen to

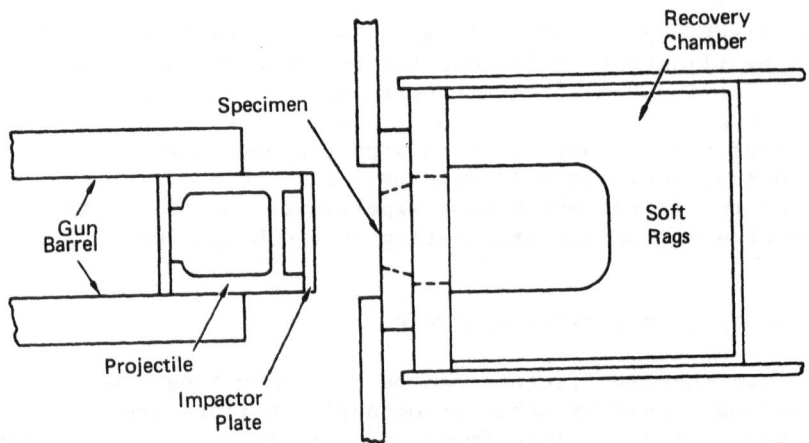

Figure 2   Plate impact arrangement for studies of shock-induced
           fracture in materials under uniaxial strain conditions.

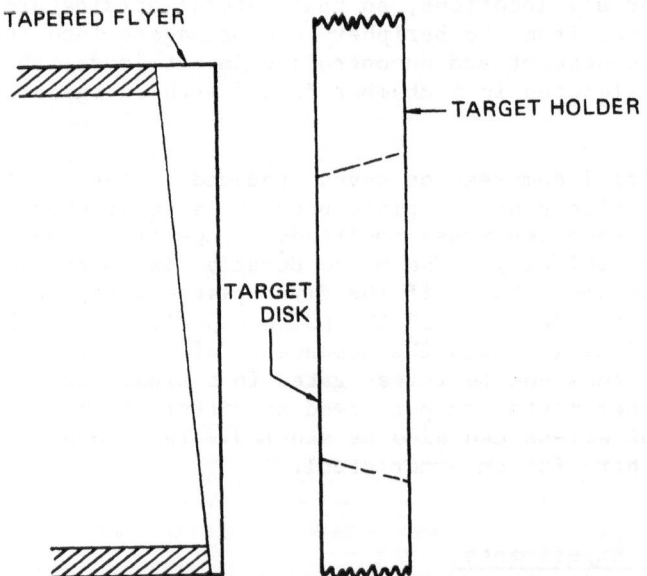

Figure 3   Tapered flyer impact arrangement for
           varying pulse duration in a dynamic
           fracture experiment.

Table I

CHEMICAL COMPOSITION OF XAR30 ARMOR STEEL
(Weight percent)

C	Mn	P	S	Si	Cr	Mo	Zr	B
0.28	0.94	0.010	0.025	0.66	0.59	0.21	0.14	0.0014

Table II

TENSILE PROPERTIES OF XAR30 ARMOR STEEL*

Orientation	0.2% Yield Strength (ksi)	Tensile Strength (ksi)	Elongation (%)	Reduction in Area (%)
Longitudinal	208	262	14.5	49.8
Long transverse	216	253	10.5	39.8
Short transverse*	211	261	10.0	12.5

*Short transverse properties are the average of three tests performed at SRI.
All other values are the average of two tests by Hickey.[3]

Table III

DYNAMIC FRACTURE EXPERIMENTS ON XAR30 ARMOR STEEL

Experiment Number	Specimen Thickness (mm)	Flyer Thickness (mm)	Angle of Taper (deg)	Flyer Velocity (mm/μs)	Remarks
2024-1*	7.62	3.81	5.7°	0.451	Full fracture; clear gage record
2024-2*	10.1	4.44	5.7	0.357	Large continuous fracture; clear gage record
2024-3	5.08	2.24	5.7°	0.226	Numerous microfractures
2024-4	12.1	5.58	6.84°	0.200	Numerous microfractures
2024-5	2.54	1.12	2.3°	0.259	Numerous microfractures

*Instrumented with back-surface ytterbium piezoresistive stress transducers.

release easily from the target plate upon impact and fly into the
catcher tank.

Tapered flyer plates 5.72 cm in diameter and having an average
thickness half that of the specimen were machined from the same
material and mounted on the front of a 15-cm-long aluminum
projectile. The angle of taper varied for different experiments
from 2.3 to 6.8 degrees as shown in Table III. The impacting
surfaces were ground flat and parallel to within 0.01 mm.

The projectile was accelerated down the barrel of the gas gun
upon sudden release of compressed helium, and careful alignment of
the flyer plate and specimen resulted in flat plate impact. The
impact velocity was measured by electrical contacts at the gun
muzzle.

Wave profiles were recorded close to the rear surface of two
of the specimens. Ytterbium piezoresistive stress transducers,
mounted in 9.5-mm-thick blocks of C-7 epoxy that were glued to the
specimens, produced the oscilloscope traces shown in Figure 4.
Both records show the Hugoniot elastic limit (HEL), a flat-topped
loading wave, and a clear fracture signal. The HEL measures the
yield strength of the material under dynamic uniaxial strain
conditions and is useful in specifying the constitutive equation of
the armor. The flat-topped loading wave indicates that impact
planarity was good and that significant attenuation did not occur
before fracture. The second peak in the records, known as the
fracture signal, is caused by a reloading recompression of the
transducers by waves emanating from the internal fracture surfaces
as they form and grow.

Recovered specimens were sectioned to reveal the internal
fracture damage as shown in Figure 5. Specimens 2024-3 and 2024-4
were sectioned to produce a cross section in the direction of
maximum taper, and Specimen 2024-5 was sectioned parallel to the
direction of taper at three locations.

The fracture damage in Specimens 2024-3, 2024-4, and 2024-5
was analyzed quantitatively by counting and measuring the traces of
the microfractures on the polished section surfaces. Measurements
were made by using a large area record reader with which an
operator positions a cross hair on one end of a void trace, pushes
a button to record the coordinates, and repeats the process for the
other end. A simple computer program uses these data to compute
the length, orientation, and position of the trace within the
specimen. The size distributions on the sectioned surfaces are
then converted to actual void size distributions per unit volume by
means of a statistical transformation implemented in the BABS1
computer code. This procedure is described in detail in
Reference 4.

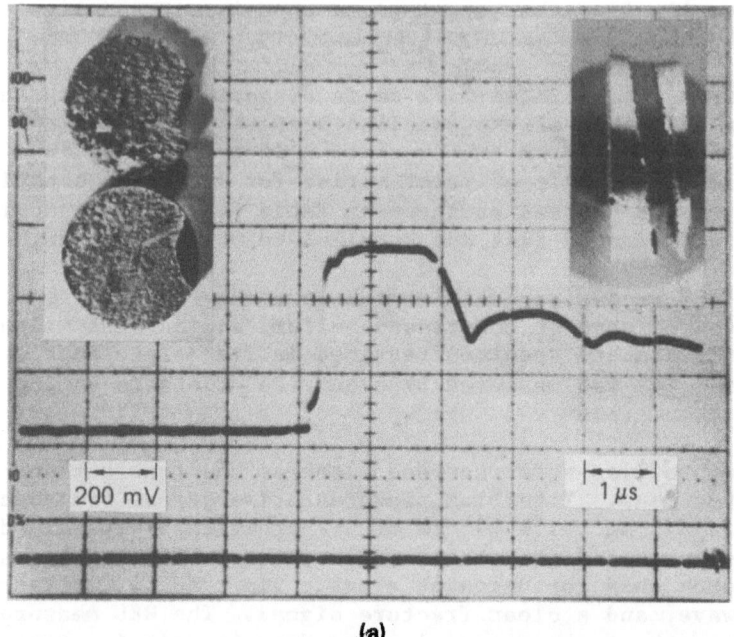

(a)

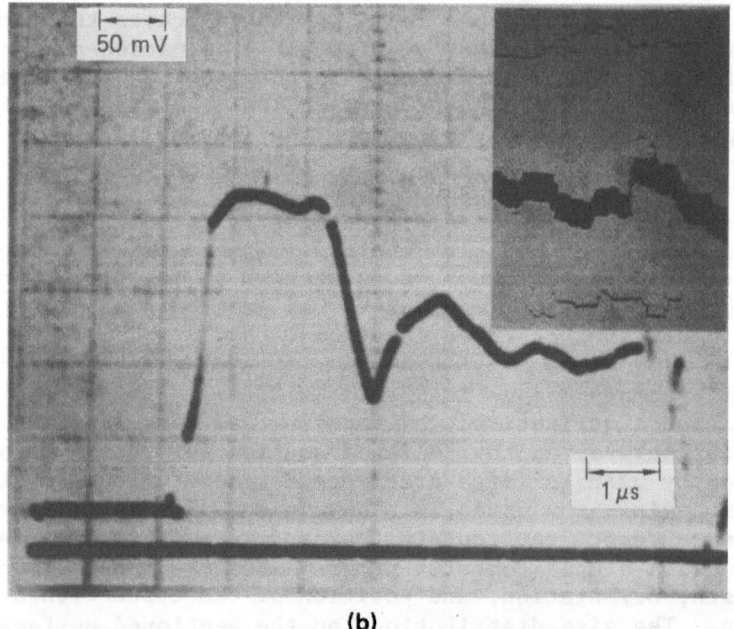

(b)

Figure 4   Voltage-time records from ytterbium stress gages
in PMMA attached to the back surfaces of plate
impact specimens.

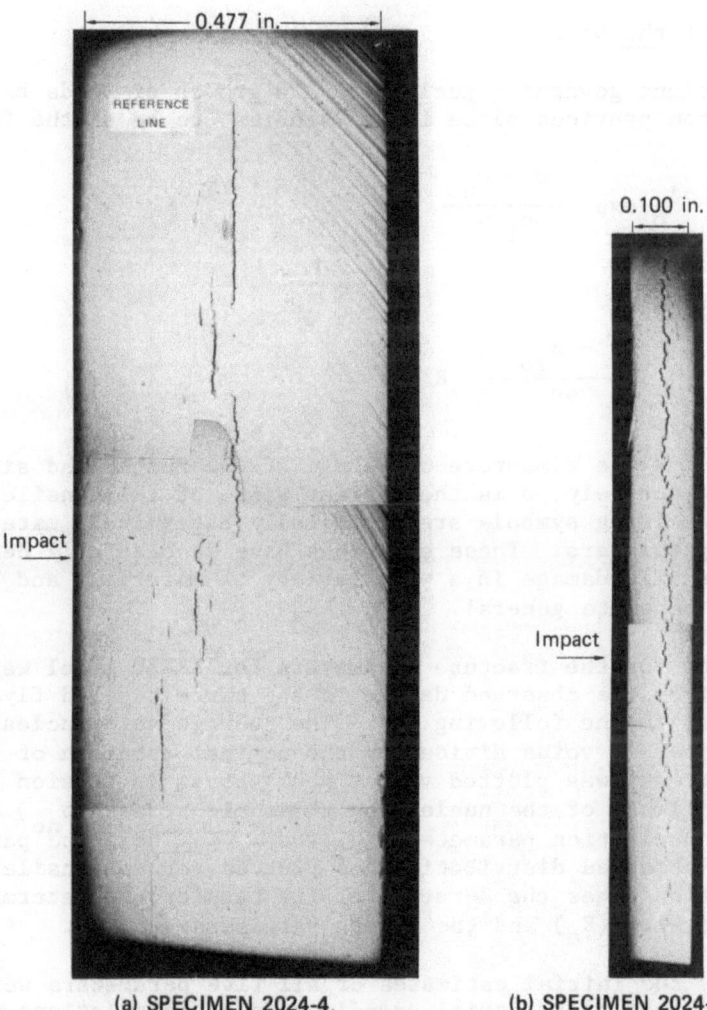

(a) SPECIMEN 2024-4,
SECTION A-B
(Surface is parallel to
taper of flyer)

(b) SPECIMEN 2024-5,
SECTION A-A
(Surface is perpendicular to
taper of flyer)

Figure 5   Polished cross sections showing internal shock-induced
fracture damage.

The results of this quantitative damage analysis are void size distribution curves for various positions on a cross section.   An example is shown in Figure 6.

Analysis of the Data

Equations governing nucleation and growth of voids have been deduced from previous plate impact studies[5] to be of the form

$$\dot{N} = \dot{N}_o \exp \frac{\sigma - \sigma_{no}}{\sigma_1} \tag{1}$$

and

$$\dot{R} = \frac{\sigma - \sigma_{go}}{4\eta} R \tag{2}$$

where $\dot{N}$ and $\dot{R}$ are time-rate-of-change of the number and size of the voids, respectively, σ is the current value of the tensile stress, and the remaining symbols are empirically determined, material-specific parameters. These equations have successfully described dynamic tensile damage in a wide variety of materials and hence appear to be quite general.

Values for the fracture parameters for XAR30 steel were obtained from the observed damage in the three tapered flyer experiments in the following way. The average void nucleation rate (total number of voids divided by the nominal duration of the tensile stress) was plotted versus peak stress in tension to obtain a first estimate of the nucleation threshold stress (σ_{no}) and the other nucleation parameters (σ_1 and $\dot{N}_o$). The shape parameter R_1 of the observed distribution was plotted versus tensile impulse (peak tension times the duration of the tension) to determine the nucleation size (R_o) and the growth rate parameter, η.

After the initial estimates of all five parameters were determined from plots, trial one-dimensional calculations were performed to approximate the impact conditions at several points in the target. These calculations were repeated with different fracture parameters until the computed and measured damage compared satisfactorily. Then a two-dimensional calculation was performed to simulate the entire impact. It was not necessary to modify the parameters further and repeat the two-dimensional simulation. The fracture parameters found for XAR30 armor are listed in Table IV. For comparison, the parameters for Armco iron and aluminum, obtained in earlier work,[4,5] are also included in the table.

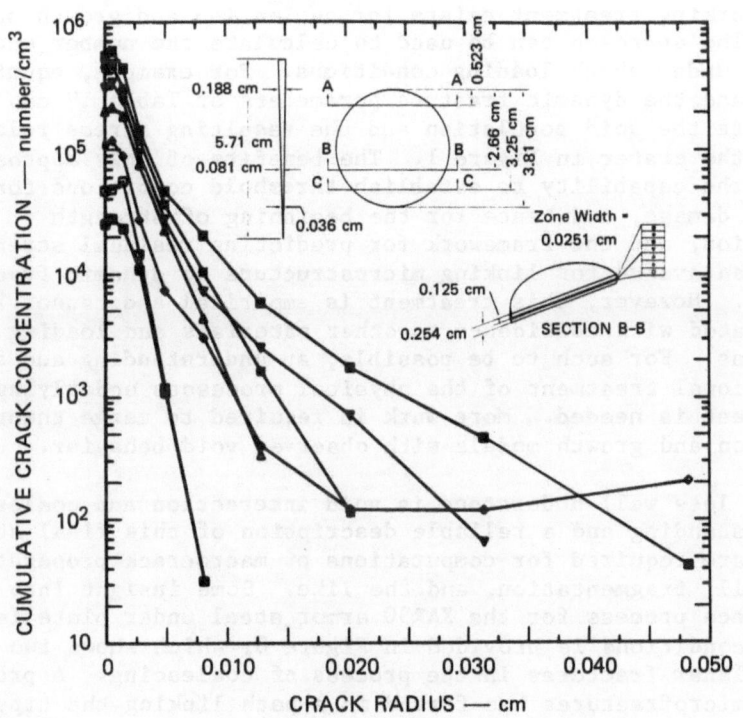

Figure 6   Crack size distributions on section B-B of
specimen 2024-5.

Figure 7 indicates the agreement obtained between measured and computed damage. This figure shows the void-size distribution on the planes of maximum damage on two sections of Specimen 2024-5. The crack size distributions are exponential in the computations and therefore appear as straight lines, so the detail of the curves is not predicted. However, average numbers and sizes of voids are predicted satisfactorily.

Summary of Progress in MSFM and Required Research Efforts

A working treatment exists for nucleation and growth of voids. The approach can be used to calculate the number and sizes of voids under shock loading conditions. For example, equations (1) and (2) and the dynamic fracture parameters of Table IV can be used to compute the void population and the resulting stress relaxation beneath the crater in Figure 1. The benefits of this approach include the capability to establish threshold conditions for fracture damage, and hence for the beginning of strength degradation, and the framework for predicting residual strength, as well as an avenue for linking microstructure to dynamic fracture response. However, this treatment is empirical and cannot be extrapolated with confidence to other materials and loading conditions. For such to be possible, an understanding and a computational treatment of the physical processes underlying void development is needed. More work is required to merge theoretical nucleation and growth models with observed void behavior.

Far less well understood is void interaction and coalescence. An understanding and a reliable description of this final stage of failure are required for computations of macrocrack propagation, full spall, fragmentation, and the like. Some insight into the coalescence process for the XAR30 armor steel under plate impact loading conditions is provided in Figure 8, which shows two parallel but nonplanar fractures in the process of coalescing. A profusion of tiny microfractures has formed in a path linking the tips, suggesting that coalescence is a nucleation and growth process on a smaller scale.

MACROCRACK PROPAGATION

The microstatistical fracture mechanics (MSFM) described above must merge with the continuum fracture mechanics approaches discussed in the previous papers of this session. The classical fracture mechanics approach treats macrocrack propagation as the movement through a homogeneous medium of a continuous crack front separating broken from unbroken material. In reality, however, macrocrack propagation occurs by nucleation, growth, and coalescence of voids in the process zone slightly ahead of the crack front. Figure 9 shows microvoids that have formed at the tip of a notch in

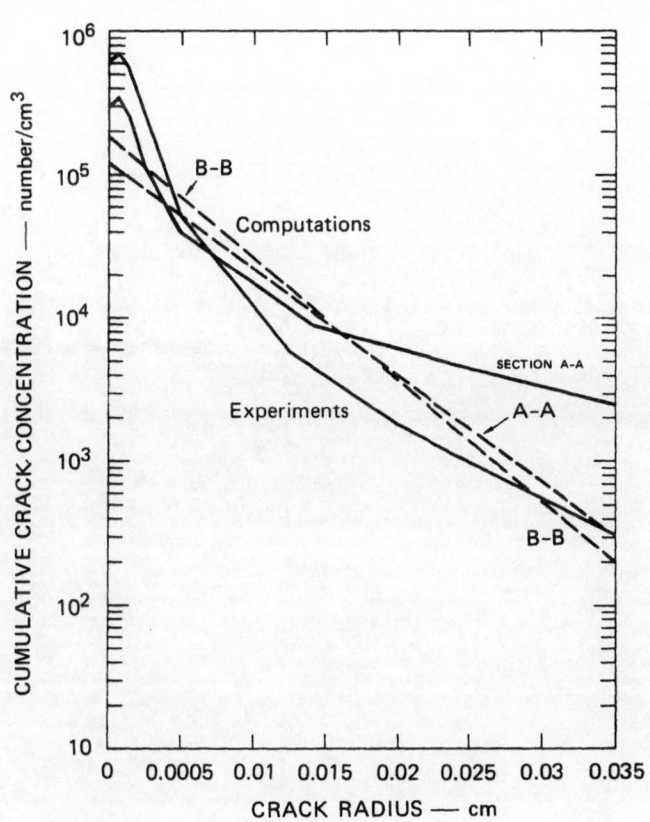

Figure 7   Comparison of measured and computed damage on
planes of maximum damage in tapered flyer impact.
Experiment 2024-5.

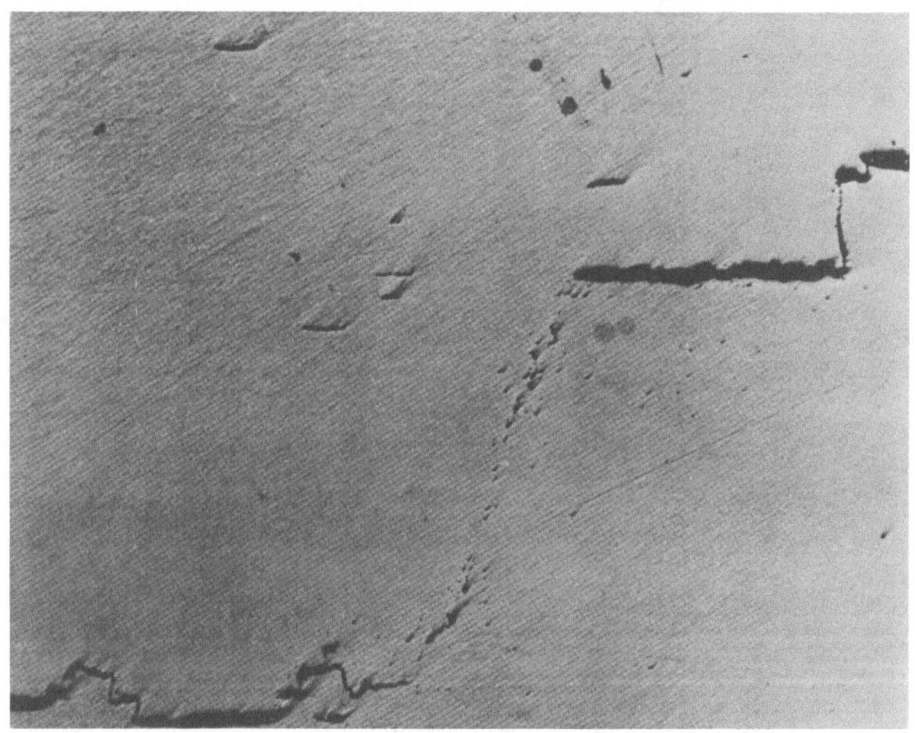

Figure 8   Polished cross section in XAR30 armor steel showing
           incipient coalescence of two planar macrofractures.

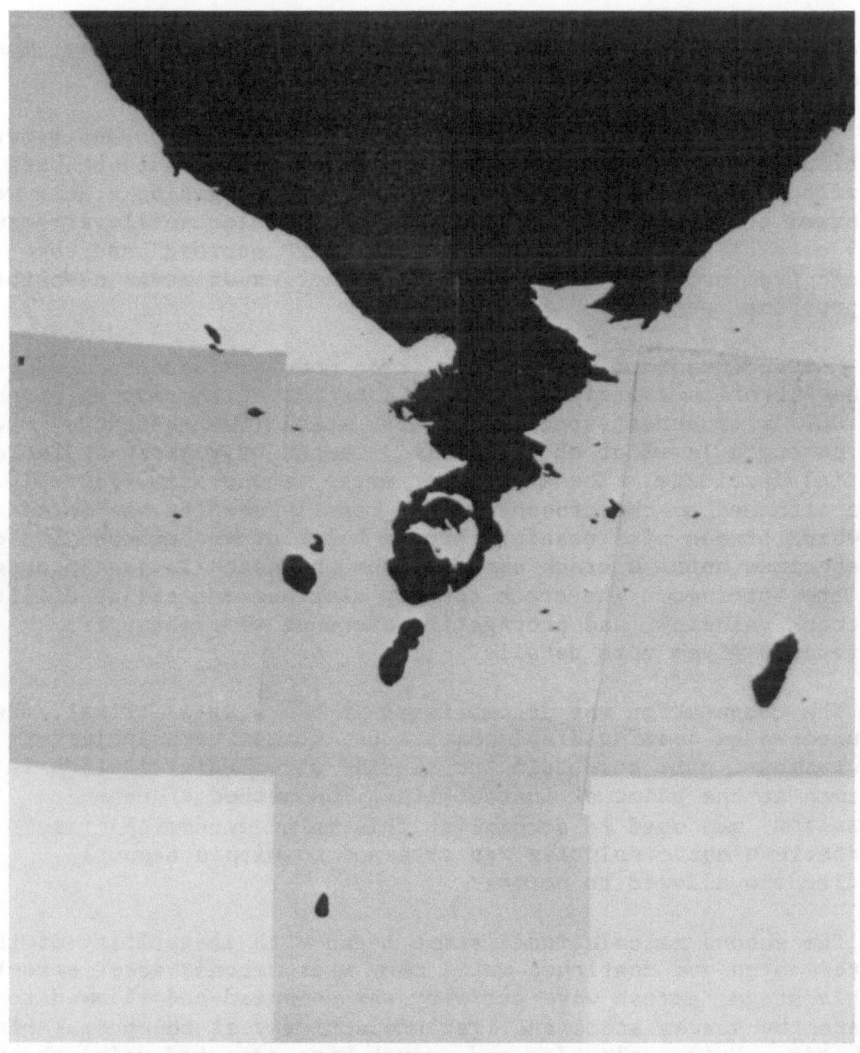

Figure 9   Polished midplane section of a cherry impact specimen
that was loaded to a subcritical deflection showing the
ductile crack emerging from the notch base and isolated
voids near the crack tip.

a charpy specimen. A comparison of the pre-notch-tip damage with
the fracture damage produced by plate impact shows that the damage
morphology is the same and suggests strongly that the fracture
phenomenology is also the same, namely, nucleation, growth, and
coalescence of individual microvoids. Further evidence for crack
growth by pre-crack-tip void activity can be found on fracture
surfaces, which are comprised of the characteristic dimples, the
remnants of severed voids.

To determine whether the microfracture nucleation and growth
functions extracted from plate impact experiments could be used to
describe microfracture activity ahead of a propagating crack, we
performed a computational simulation of a double-cantilever-beam
(DCB) crack arrest experiment using the MSFM approach and the
dynamic fracture data from the plate impact experiments described in
the previous section.

A two-dimensional wave propagation code was used to compute the
stress histories experienced by the material at the moving crack tip
in a DCB crack arrest specimen of 4340 steel (quenched and tempered
to a strength level of about 200 ksi) tested by workers at Battelle
Memorial Institute. The experiment setup is shown in Figure 10. A
wedge attached to the crosshead of a tensile machine was forced
downward between pins passing through holes at the notched end of
the specimen until a crack emerged from the notch tip and propagated
into the specimen. The crack opening displacement at instability,
the crack velocity, and propagation distance were measured.
Reference 6 gives more details.

The computation was accomplished in two stages. First, the
measured wedge opening displacements and forces were applied to the
computational grid to obtain the elastic stress distribution in the
specimen at the point of instability. The method of dynamic
relaxation[7] was used to accomplish this in an economical time. When
the static elastic solution was obtained to within about 1%,
yielding was allowed to occur.

The second calculational stage began with instability of the
blunted notch and continued until many microseconds after arrest.
In this stage, stress wave activity was computed and allowed to
dictate the stress state and fracture activity at the propagating
crack tip. Void nucleation and growth were computed using the model
(Equations 1 and 2) and parameters (Table IV) deduced from the plate
impact work. The strength of the computational cell was gradually
reduced with increasing void volume. When void volume equaled cell
volume, coalescence was said to be complete, cell rupture occurred,
and the crack advanced one cell length. Reference 2 gives details
of the computational procedures.

The crack ran three times farther than calculated and only half
as fast. Better agreement should result if a finer computational

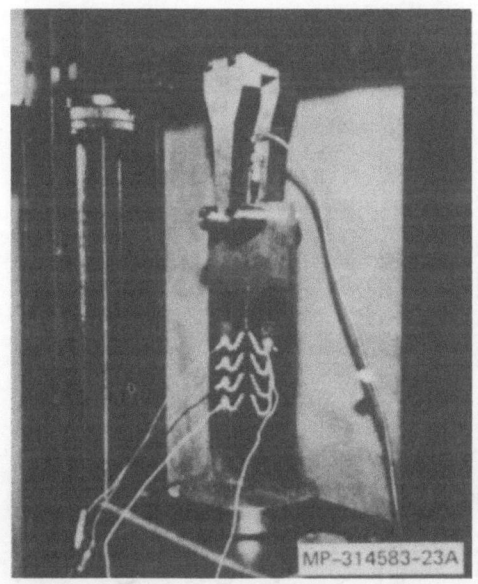

(a) **Wedge loading arrangement for DCB specimen showing the velocity measuring conducting strips.**

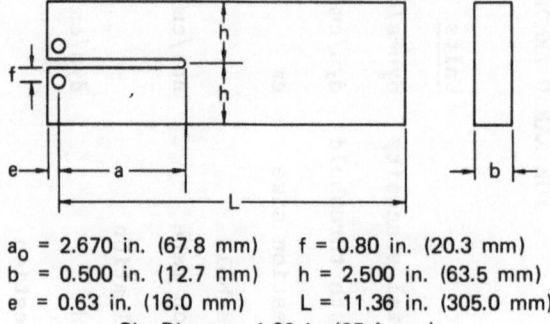

a_0 = 2.670 in. (67.8 mm)   f = 0.80 in. (20.3 mm)
b = 0.500 in. (12.7 mm)   h = 2.500 in. (63.5 mm)
e = 0.63 in. (16.0 mm)   L = 11.36 in. (305.0 mm)
                Pin Diameter 1.00 in (25.4 mm)

(b) **Dimensions of test pieces.**

Figure 10   Crack arrest experiment on a wedge-loaded DCB specimen of 4340 Steel performed at Battelle Memorial Institute.

Table IV

DYNAMIC FRACTURE PARAMETERS
FOR XAR30 ARMOR STEEL, ARMCO IRON, AND 1145 ALUMINUM

Parameter	Units	XAR30	Armco Iron	1145 Al
η, material viscosity	dyn-s/cm^2	4545	417	75
σ_{go}, growth threshold	dyn/cm^2	-1.0×10^8	-2.0×10^8	-4.0×10^9
R_o, nucleation size	cm	4.0×10^{-3}	5.0×10^{-5}	1.0×10^{-4}
$\dot{N}_o$, threshold nucleation rate	no./cm^3/s	4.0×10^8	4.6×10^{12}	3.0×10^9
σ_{no}, nucleation threshold	dyn/cm^2	-2.5×10^{10}	-3.0×10^9	-3.0×10^9
σ_1, nucleation sensitivity	dyn/cm^2	-1.786×10^9	-4.56×10^9	-4.0×10^8

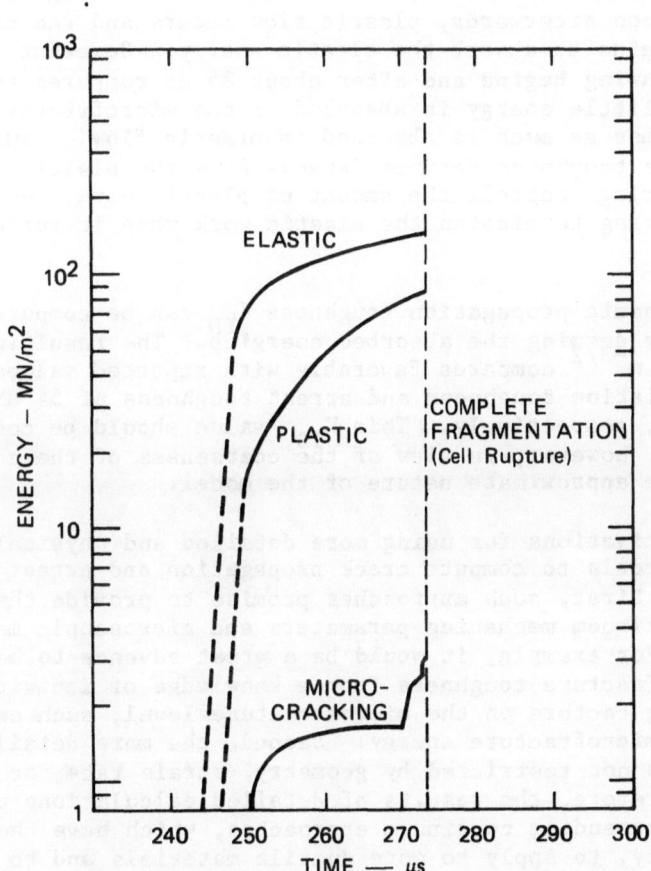

Figure 11   Energy partitioning in a computational cell in
             a DCB specimen as the cell is traversed by the
             crack.

mesh is used; nevertheless, the work shows the feasibility of
computing macrocrack behavior from microevents at the crack tip.

An example of the insight into the fracture process made
possible by the MSFM approach is given in Figure 11, which shows the
variation with time of the elastic strain energy and the energies
absorbed by plastic flow and by microcracking. As the crack tip
approaches and loads a cell, the elastic strain energy rises
sharply. Soon afterwards, plastic flow occurs and the crack tip
material begins to absorb the elastic energy. Somewhat later,
micro-fracturing begins and after about 25 μs ruptures the cell.
Relatively little energy is absorbed in the microfracturing process;
about 20 times as much is absorbed in plastic flow. Thus, although
the fracture toughness derives largely from the plastic work, the
microfracturing controls the amount of plastic work, because
microfracturing terminates the plastic work when it ruptures the
cell.

The dynamic propagation toughness K_{ID} can be computed from
Figure 11 by summing the absorbed energies. The resulting value
$K_{ID} = 61$ MN m$^{-3/2}$ compares favorably with reported values for
static initiation toughness and arrest toughness of 54 MN m$^{-3/2}$ and
68 MN m$^{-3/2}$, respectively. This K_{ID} value should be considered
preliminary, however, in view of the coarseness of the computational
grid and the approximate nature of the model.

The motivations for using more detailed and physically
realistic models to compute crack propagation and arrest are
threefold. First, such approaches promise to provide the link
between continuum mechanics parameters and microscopic material
behavior. For example, it would be a great advance to be able to
derive the fracture toughness from a knowledge of individual
contributing factors on the microstructure level, such as plastic
energy and microfracture energy. Second, the more detailed approach
used here is not restricted by geometry, strain rate, or even stress
state. Therefore, the results of detailed calculations may be
helpful in extending continuum approaches, which have the advantage
of simplicity, to apply to more ductile materials and to smaller
specimens. Third, the role played by microstructure in controlling
toughness can be studied and perhaps understood in a more
quantitative way.

Thus the MSFM merges with continuum treatments much in the way
that quantum mechanics merges with Newtonian mechanics. Both have
their advantages and disadvantages and special areas of
applicability. Continuum treatments have the advantage of
simplicity and low cost, and therefore should be used whenever they
can provide sufficient answers. However, the more complex and
expensive MSFM treatment may be the only viable approach when
detailed information is needed such as size, velocity, and
trajectory distributions of fragments ejected from a plate of armor
steel after impact, Figure 1.

ACKNOWLEDGMENTS

This work was supported by the Army Materials and Mechanics Research Center, Watertown, MA, under Contract No. DAAG46-72-C-0182 and by the Electric Power Research Institute, Palo Alto, CA, under Contract No. RP 499-1.

REFERENCES

1. L. Seaman, and D. A. Shockey, Final Technical Report AMMRC CTR 75-2, Army Materials and Mechanics Research Center, Watertown, Massachusetts (1975).

2. D. A. Shockey, M. Austin, L. Seaman, and D. R. Curran, Final Technical Report to Electric Power Research Institute, Palo Alto, California, Contract RP499-1-1 (1976).

3. C. F. Hickey, Jr., Toughness data for monolithic high-hardness armor steel, AMMRC FTR 72-3, Army Materials and Mechanics Research Center, Watertown, Massachusetts (July 1972).

4. L. Seaman, T. W. Barbee, Jr., and D. R. Curran, Dynamic fracture criteria of homogeneous materials, Technical Report No. AFWL-TR-71-156, Air Force Weapons Laboratory, Kirkland Air Force Base, New Mexico (December 1971).

5. T. W. Barbee, Jr., L. Seaman, R. Crewdson, and D. R. Curran, Dynamic fracture criteria for ductile and brittle metals, J. Matls., 7: (3) 393-401 (1972).

6. G. T. Hahn, R. G. Hoagland, M. F. Kanninen, and A. R. Rosenfield, "Dynamic Crack Propagation," G. C. Sih ed., Noordhoff International Publishing, Leyden, (1972), p. 649.

7. J. R. H. Otter, A. C. Cassell, and R. E. Hobbs, Paper No. 6986, Dynamic relaxation, in: "Proceedings of the Institute of Civil Engineering," (1967).

SCALING OF SHEAR BAND FRACTURE PROCESSES

Lynn Seaman, D. R. Curran and D. A. Shockey

SRI International
Menlo Park, CA 94025

INTRODUCTION

Recently a shear band model has been constructed describing the micro processes of damage (Ref 1). The model considers arrays of shear bands in several size groups and orientations. Under large plastic strains the bands are nucleated, grow, and coalesce to form fragments. With this model we can now estimate the importance of scaling and rate effects. The two effects are intimately related because rate effects are a common source for deviations from replica scaling. Here the model is outlined briefly, with emphasis on the nucleation process which is the main contributor to rate effects. Then a series of calculations are described in which a range of strain rates was applied. The number and size of bands and the strain to failure are examined as a function of strain rate.

This study was undertaken because scaling arises in many practical applications. It is often convenient to perform small-scale tests to predict the effects at full scale. Also, we may want to perform tests at a slower strain rate than that used in the intended application, or we may use one kind of experiment to predict the results of another kind. In all these cases there are changes of scale and changes of strain rate, and we need to know how to correctly account for scale.

The model has only been verified over a limited range of strain rates for one experiment type and therefore should not be considered a fully proven model. However, this scaling study was undertaken to explore what insights might be provided by this model, and to show the potential value of a verified model.

MODEL DESCRIPTION

   The shear band model considered is a microstatistical
fracture model. It treats a statistical distribution of bands
as a probability density at each point in the material. The
bands are nucleated, grow, and coalesce to separate the material
into fragments. Here these features are briefly discussed, with
a more detailed derivation of the nucleation process because
that process contributes the main rate effect.

   The bands are envisioned in the model as small, circular,
planar regions that have lost their cohesive strength. The
bands appear in a range of sizes, such as that shown in Figure
1, and in a range of orientations. They occur first on planes
of maximum shear strain. Bands begin to appear in the model
when the plastic shear strain on any plane exceeds a critical
value. This critical value is that strain at which the increase
in strength caused by work-hardening is just balanced by the
decrease in strength associated with thermal softening. Currently, this calculation of critical strain is made assuming an
adiabatic process. (Actually, at low strain rates an adiabatic
approximation may not be appropriate because of the finite heat
conduction rates.)

   After the threshold strain is reached, bands are nucleated
according to a rate function derived according to the following
considerations. In each orientation of interest in the material, shear bands are assumed to nucleate whenever the decrease
in yield strength associated with work softening exceeds the
increase from work hardening. The thermal heating comes from
the plastic work. Perturbations in the plastic flow caused by
local heterogeneities will then grow and cause localization of
the flow into bands. For this localization to happen, nearly
adiabatic conditions must obtain; that is, the strain rate must
be high enough that most of the heat is retained at the perturbation site. Simple heat flow considerations suggest that
plastic strain rates of 10^3 to 10^4/s would be sufficient to
retain the heat in perturbations as small as 0.1 mm in characteristic size. Hence, in addition to the critical strain for
nucleation obtained from considerations of thermal softening and
work hardening under adiabatic conditions, a critical strain
rate is required to maintain approximately adiabatic
conditions.

   The nucleation rate function is assumed to be composed of
three factors that are dependent on a strain and a strain rate
threshold:

$$\dot{N} = N_o f/T \qquad\qquad (1)$$

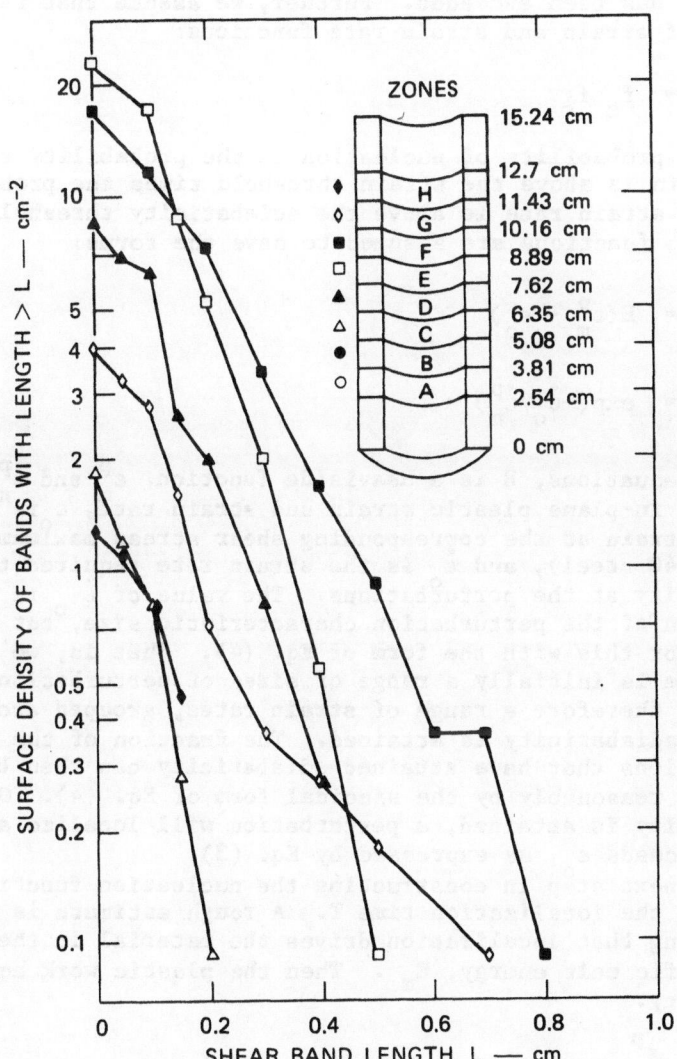

**Figure 1   Shear band distribution for Experiment 2
in 4340 steel, R_c40**

where N_o is the number of potential nucleation sites per unit volume, that is, the number of perturbations in the plastic flow; f is the fraction of those sites that are at the threshold for localization (f is the probability of localization), and T is the time required for localization to occur once the threshold has been exceeded. Further, we assume that f is a product of strain and strain rate functions:

$$f = f_\varepsilon \, f_{\dot{\varepsilon}} \tag{2}$$

Thus, the probability of nucleation is the probability that the strain is above the strain threshold times the probability that the strain rate is above the adiabaticity threshold. These two functions are assumed to have the forms:

$$f_\varepsilon = H(\varepsilon_m^P - \varepsilon_o) \tag{3}$$

$$f_{\dot{\varepsilon}} = \exp(-\dot{\varepsilon}_o / \dot{\varepsilon}_m^P) \tag{4}$$

In these equations, H is a Heaviside function, ε_m^P and $\dot{\varepsilon}_m^P$ are the continuum in-plane plastic strain and strain rate, ε_o is the plastic strain at the corresponding shear stress maximum (about 20% in 4340 steel), and $\dot{\varepsilon}_o$ is the strain rate required to attain adiabaticity at the perturbations. The value of $\dot{\varepsilon}_o$ is actually a function of the perturbation characteristic size, but we account for this with the form of Eq. (4). That is, we assume that there is initially a range of sizes of perturbations in the flow, and therefore a range of strain rates, grouped around $\dot{\varepsilon}_o$, at which adiabaticity is attained. The fraction of the perturbations that have attained adiabaticity can then be expressed reasonably by the sinoidal form of Eq. (4). Once adiabaticity is attained, a perturbation will localize as soon as ε_m^P exceeds ε_o, as expressed by Eq. (3).

The next step in constructing the nucleation function is to determine the localization time T. A rough estimate is obtained by assuming that localization drives the material in the band to the specific melt energy, E_m. Then the plastic work equals the melt energy.

$$\int_{\varepsilon_o}^{\varepsilon^P} \frac{\sigma_y d\varepsilon^P}{\rho} = E_m \tag{5a}$$

or

$$\int_0^T \frac{\sigma_y}{\rho} \dot{\varepsilon}^P \, dt = E_m \tag{5b}$$

where σ_y and $\dot{\varepsilon}_m^P$ are the yield stress and plastic strain rate.
For a constant strain rate, Eq. (5b) is approximately

$$\frac{\sigma_y}{\rho} \; \dot{\varepsilon}_m^P \; T \; = \; E_m \tag{6a}$$

or

$$T \approx (\frac{\rho E_m}{\sigma_y}) \; \frac{1}{\dot{\varepsilon}_m^P} \tag{6b}$$

where σ_y and $\dot{\varepsilon}_m^P$ are chosen to be prelocalization values of the
yield stress and plastic strain rate. Because both the yield
stress and the strain rate in the band will change during the
localization process, Eq. (6) gives only a rough estimate of the
localization time T.

With the above caveat in mind, we combine Eqs. (1) to (4)
and (6) to obtain the nucleation rate expression:

$$\dot{N} = \alpha N_o H(\varepsilon_m^P - \varepsilon_o) \; \dot{\varepsilon}_m^P \; \exp(-\dot{\varepsilon}_o/\dot{\varepsilon}_m^P) \tag{7}$$

where $\alpha = \sigma_y/(\rho E_m)$

and α is about 5 for steels. Thus we have a nucleation rate
function that depends explicitly on the imposed strain rate.

The bands grow under the imposed strain rates according the
following relation, subject to some restrictions:

$$\frac{dR}{dt} = T_1 \; \frac{d\varepsilon^{PS}}{dt} \; R \tag{8}$$

where R is the band radius in plan and T_1 is the growth coefficient. If the expression for dR/dt exceeds the shear velocity
V_s, dR/dt is set equal to V_s . The second restriction is that
the total strain taken by the band cannot exceed the total imposed strain. The strain taken by the band is

$$\Delta\varepsilon \; = \pi b \sum_i N_i(R_{2i}^3 - R_{1i}^3) \tag{9}$$

where $b = B/R$, B is the average slip over the plane of the band,
and R_1 and R_2 are the band sizes before and after growth
respectively. The growth velocity is modified as needed to keep
the strain taken by the bands from exceeding the imposed plastic

shear strain. The integral of Eq. (8) for a constant strain
rate over some time interval is

$$R_2 = R_1 \exp(T_1 \Delta \epsilon^{ps}) \tag{10}$$

where $\Delta \epsilon^{ps}$ is the increment in plastic shear strain. Although
Eq. (8) appears as a rate equation, the growth is actually independent of strain rate unless the shear velocity V or the
strain limit Eq. (9) is reached.

The model provides for a strong interaction between the
nucleation and the growth processes. When plastic shear strain
is applied, that strain can be absorbed by homogeneous plastic
strain, by growth of existing bands, and by nucleation of new
bands. We have assumed that strain would be taken in that mode
in which it is easiest for the material to accomodate the
strain. The amount of strain that can be taken in growth is
limited by the growth law above. If growth does not absorb all
the applied strain, then nucleation can occur. Because we allow
nucleation of bands of finite sizes, the nucleation also absorbs
some strain. If some applied strain still remains, this residual strain is taken homogeneously.

In the model there is the assumption that the bands are
initially isolated, and their nucleation and growth can be
treated as if they were each a single band in an infinite
medium. However, at some time the bands begin to interact,
coalesce, and finally to form fragments. The condition for full
fragmentation is written by considering the fragmented state.
Each fragment has some small number of faces, each formed by
bands. The size of the fragment faces are related to the band
sizes. With these considerations, we have derived a
criterion τ for fragmentation:

$$\tau = \pi \Sigma N_i R_i^3 \tag{11}$$

This parameter τ describes the degree of fragmentation.
When τ reaches one, full fragmentation is declared. Again, this
is a rate-independent process.

SCALING CALCULATIONS

All the calculations described here are point computations
in which a constant strain rate was imposed. The strain rate
was varied to examine the effect of scale and rate on the shear

banding results. The material parameters were determined earlier
by simulations of some fragmenting round experiments on 4340
steel for strain rates of 10^3 to 4×10^3; these parameters were
not varied during this series of calculations. Of principle
concern in the results of these computations are the number of
bands nucleated, the manner in which the damage increases, the
time at which fragmentation occurs, and the amount of strain at
fragmentation.

   The histories of the numbers of shear bands (Fig. 2)
indicate some of the interaction that is occurring between
nucleation and growth. At the very high strain rate, the
history shows a nearly linear rise with time, so nucleation is
continuing throughout and growth is never taking all the
strain. For the intermediate strain rate (8×10^4), the growth
begins to have an important effect early and eventually takes
all the strain, so that no further nucleation occurs. At the
lowest strain rate, there is no effect of growth until near the
time of fragmentation. Then the growth quickly takes all the
strain and nucleation halts. The damage histories in Fig. 3
emphasize the same effects. The damage begins very early in the
high strain rate case because of the large amount of strain
associated with nucleation. However, for the lowest strain
rates, the damage is negligible until late, when growth becomes
important.

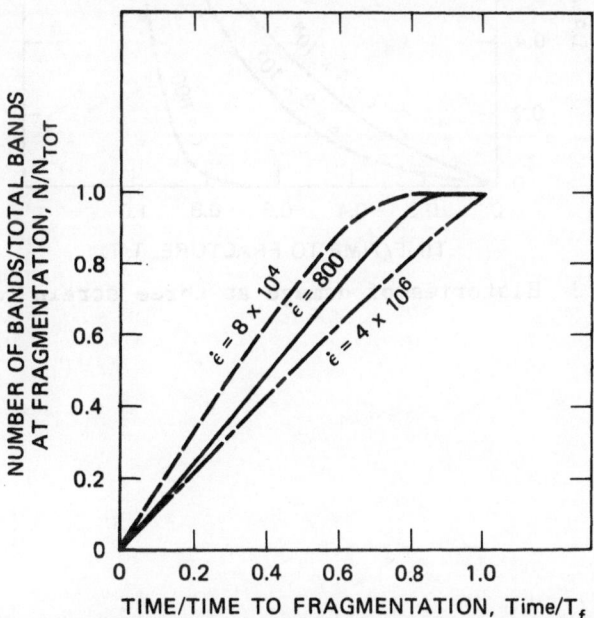

Figure 2   Nondimensionalized histories of the number of
           shear bands for three constant strain rates.

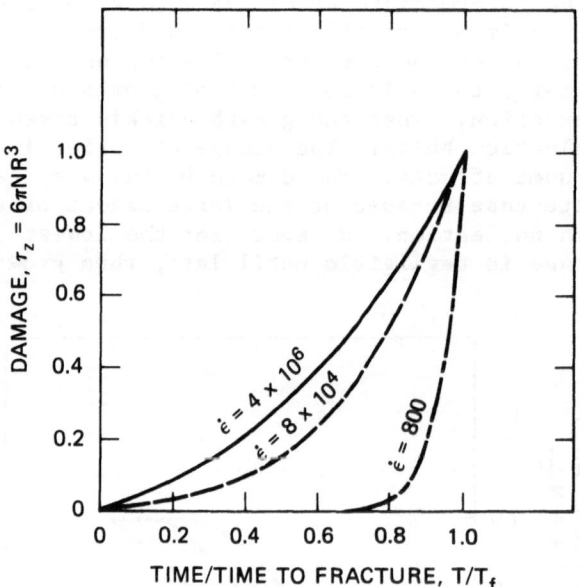

Figure 3   Histories of damage at three strain rates.

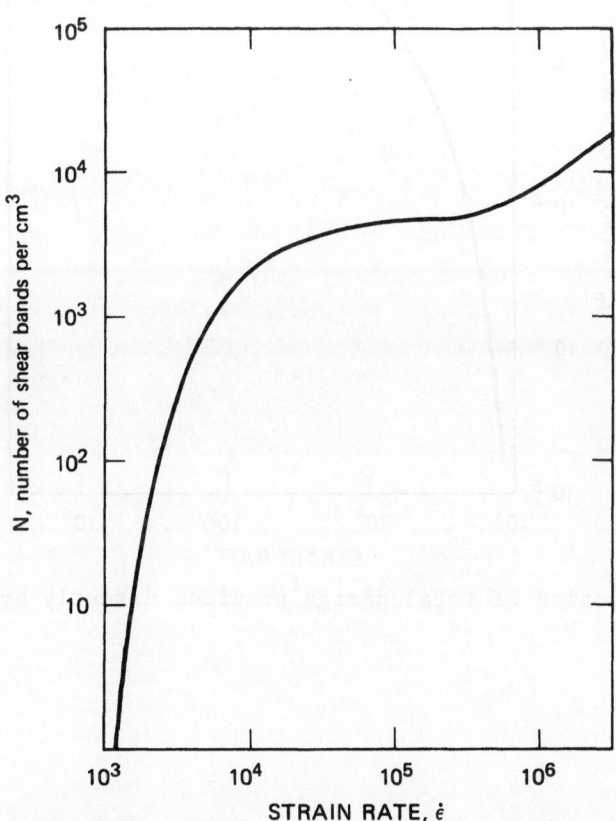

Figure 4   Variation of the number of bands formed with strain
            rate.

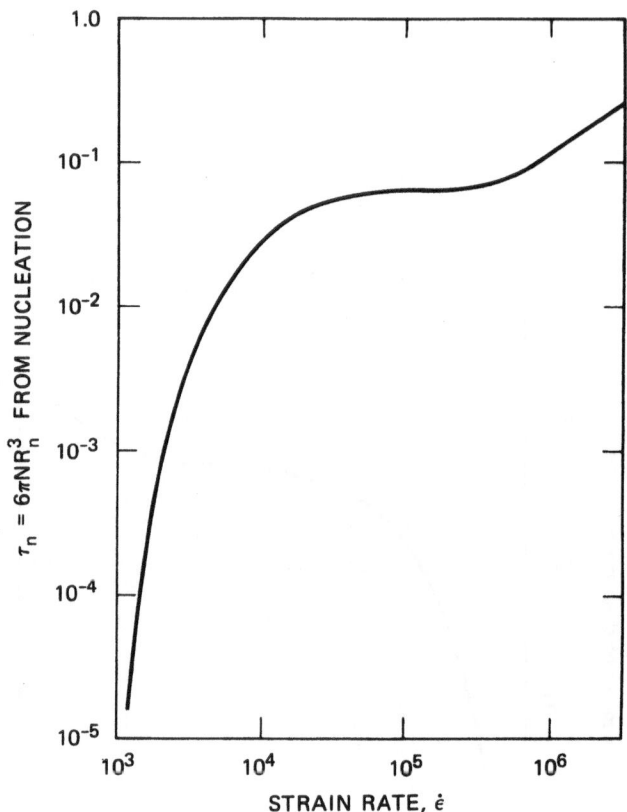

Figure 5   Fraction of total damage provided directly by nucleation.

        The number of shear bands varies dramatically with the
strain rate, as shown in Fig. 4. This graph illustrates the
familiar fact that very few bands are formed at low strain
rates, but a great number at high rates.   A similar relationship
is shown in Fig. 5, where the amount of damage arising just from
nucleation is shown.   This damage gives some indication of the
relative importance of growth and nucleation as a function of
strain rate. At the highest strain rates, we are approaching the
point where damage is nucleation-dominated and fragmentation
occurs with the bands essentially at their nucleation size.

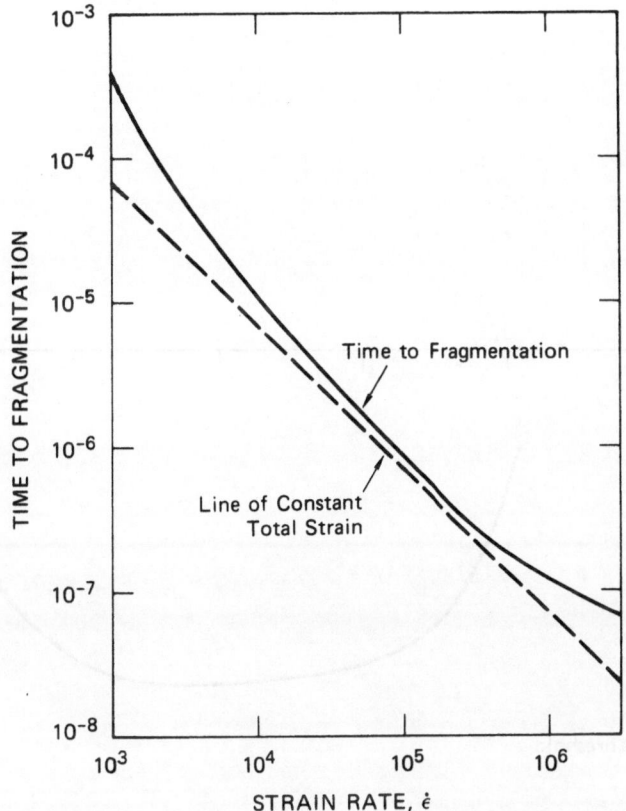

Figure 6   Variation of the time to failure with strain rate.

The time to fragmentation varies strongly with strain rate
as shown in Fig. 6. If the time scaled in the usual way for
geometric scaling, the fragmentation would follow the slope of
the line of constant total strain. From this plot it appears
that the time to fragmentation would scale over the range of
strain rates from about 10^4 to 10^6, but would deviate from
geometric scaling outside this range. Further insight into this
aspect of scaling is given in Fig. 7: strain at full separation
versus strain rate. It appears that larger strains to fracture
are required at both very low strain rates and at very high
rates.

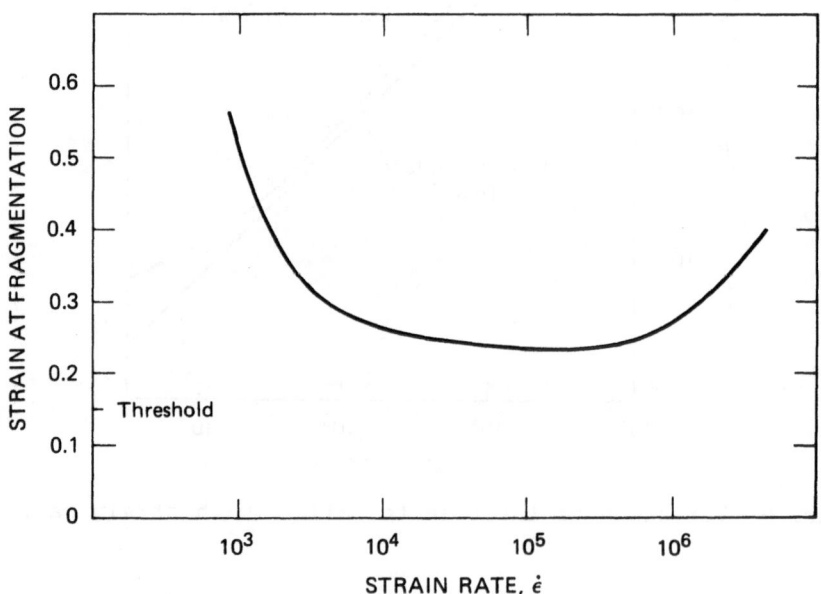

Figure 7   Variation of the strain at fragmentation with strain rate.

With the aid of the foregoing results, let us examine some specific scaling questions. For example, can we use different scales in similar experiments, and can we use the data from one type of experiment to predict the results of another type? The strain rates of common tests are about 10^3 for full-scale fragmenting rounds (larger at early times, but no bands are being formed yet because the strains are still too small), 10^4 for projectile penetrations, and 10^5 for plate impacts. These rate estimates are only nominal because the actual rates vary throughout the material and with time in each test. With these rate estimates, we can see from Fig. 4 that the number of bands will vary by 2 or 3 orders of magnitude between fragmenting rounds and projectile penetrations. Also, we can see that if a small-scale round is used, the numbers of fragments will be significantly different from those in the full-scale round (strain rate is doubled when the scale is cut in half). For rounds of different sizes, the time to fragmentation was found to not scale geometrically (Fig. 6); hence, if the time is an important factor in the study, some correction on time would be required.

The foregoing results point out some of the pitfalls of scaling shear band fracture processes. They also show means whereby we can more certainly plan an experimental program. We can select a set of experimental scales and types that can span the range of strain rates of actual interest. We will need to represent the processes (especially nucleation) with a model that correctly accounts for the effects of strain rate. These results were obtained for a model calibrated to a narrow range of strain rates. Therefore these scaling effects are intended to be indicative of the type of behavior to expect, but not definitive.

Although results were obtained with a sophisticated model, the actual calculations were simple and inexpensive. Only a single cell was considered, and so only a few minutes were required to calculate the response of the cell to the constant strain rate loads for the whole range of strain rates.

REFERENCES

1. D. C. Erlich, L. Seaman, R. D. Caligiuri, and D. R. Curran, Computational Model for Armor Penetration, Annual Report to Ballistic Research Laboratory, Aberdeen Proving Ground, MD 21005 ,and to Army Materials and Mechanics Research Center, Watertown, MA 02172, November 1980.

APPLICATION OF GURSON VOID GROWTH MODEL TO DYNAMIC FRACTURE

Ian M. Fyfe

Department of Aeronautics & Astronautics
University of Washington
Seattle, WA 98195

INTRODUCTION

Strain localization is usually considered to be the precursor
of ductile failure, and, beginning with Considère (1885), a great deal
of effort has been directed towards determining the plastic instability point that precedes such well-known phenomena as necking,
shear band formation and crack growth. The work of Hutchinson and
Neale (1977) is one of the more recent examples of this work. However, at the microscopic level the nucleation, growth and coalescence of voids is generally accepted as an important mechanism in the
ductile failure process. In a recent paper Chu and Needleman (1980)
coupled the work of Hutchinson and Neale with the void growth continuum theory of Gurson (1977) to determine forming-limit diagrams
for biaxial stretched sheets. In dynamic fracture the void growth
concept may be even more important in that failure often occurs with
no well defined strain localization. The spallation of thick plates
under impact conditions is an example. In a study on the inertia
effects on ductile failure Fyfe and Rajendran (1980) found that plastic instability is inhibited by inertia effects to the extent that
classical plastic instability criteria under high strain-rate loading
may of necessity require redefinition.

In this paper stress wave induced ductile failure is examined
within the context of the Gurson model. The objective is to determine if a critical void volume fraction can be used to serve as a
measure to predict when either failure or strain localization may
occur. In a complex process of this kind it is hardly to be expected
that a critical void volume fraction will be reached under all ductile
failure conditions; for example, shear bands may develop as a result

309

of some other triggering mechanism. For this reason a critical void
volume level may be regarded as one of a number of indicators of
impending failure.

THE GURSON MODEL

The macroscopic effect of the nucleation and growth of voids
can be described within the framework of continuum theory, if it is
assumed that the material is a ductile porous one, idealized as an
aggregate consisting of voids and a plastically incompressible ductile matrix. The presence of voids makes the aggregate plastically
compressible. Gurson followed this approach by assuming spherically
symmetric deformations around a spherical cavity. The yield function
which results from this analysis has the following form

$$\Phi = \frac{3J_2}{Y_m^2} + 2f \cosh \left[\frac{\sigma_{11} + \sigma_{22} + \sigma_{33}}{2Y_m} \right] - f^2 - 1 = 0 \qquad (1)$$

where Y_m is the flow stress of the matrix material, f is the current
volume fraction of voids, σ_{ij} are the principle values of the Cauchy
stress acting on an element of the void matrix aggregate, and J_2 is
the second invariant of the deviatoric stress tensor. The equivalent
plastic strain rate of the matrix $\dot{\varepsilon}_m^p$ is related to the flow stress Y_m
by expressing the plastic strain as the difference between the total
and the elastic strains, so that

$$\dot{\varepsilon}_m^p = \dot{\varepsilon}_m - \dot{\varepsilon}_m^e = \left(\frac{1}{E_t} - \frac{1}{E} \right) Y_m \qquad (2)$$

where E and E_t are the Young's and tangent modulus respectively of
the incompressible matrix.

Introducing two new variables Y_m and f requires two additional
equations to describe such materials and they are obtained by
assuming that the equivalent plastic work in the matrix is equal to
the equivalent plastic work in the aggregate, resulting in the expression

$$\sigma_{ij}\dot{\varepsilon}_{ij}^p = (1 - f)Y_m\dot{\varepsilon}_m^p \qquad (3)$$

In addition, by requiring that the void volume fraction increases
due to the growth of existing voids and the nucleation of new voids,
by cracking or decohesion of inclusions or second phase particles we
have the second equation

$$\dot{f} = \dot{f}_{growth} + \dot{f}_{nucleation} \qquad (4)$$

As the matrix material is taken to be plastically incompressible, the increment due to growth is given by

$$\dot{f}_{growth} = (1 - f)(\dot{\varepsilon}_{11} + \dot{\varepsilon}_{22} + \dot{\varepsilon}_{33}) \qquad (5)$$

The nucleation portion of (4) is taken to be of the relatively simple
form as proposed by Needleman and Rice (1978)

$$\dot{f}_{nucleation} = A \, \dot{Y}_m + B \, \dot{\sigma}_{kk}/3 \qquad (6)$$

in which nucleation depends on the increment of the effective stress
Y_m and the hydrostatic tension of $\sigma_{kk}/3$. The two parameters A and B
in (6) are chosen so that void nucleation follows a normal distribution. If the nucleation of cavities are correlated exclusively in
terms of the equivalent plastic strains, then

$$A = \left(\frac{1}{E_t} - \frac{1}{E}\right)\frac{f_n}{s\sqrt{2\pi}} \, e^{-\frac{1}{2}\left(\frac{\varepsilon_m^P - \varepsilon_n}{s}\right)^2}, \; B = 0 \qquad (7)$$

where f_n is the volume fraction of void nucleating particles, ε_n is
the mean strain for nucleation, and s is the corresponding standard
deviation. For materials with stress controlled nucleation the
equivalent parameters are

$$A = B = \frac{f_n}{s\sqrt{2\pi}} \, e^{-\frac{1}{2}\left(\frac{(Y_m + \sigma_{kk}/3) - \sigma_n}{s}\right)^2} \qquad (8)$$

where in this case σ_n is the mean nucleation stress. The plastic
deformation rates of the voided aggregate are then required to
satisfy

$$\dot{\varepsilon}_{ij}^P = \Lambda \, \frac{\partial \Phi}{\partial \sigma_{ij}} \qquad (9)$$

where Λ is determined in the usual fashion from the consistency
condition. However as was pointed out by Needleman and Rice the use
of equation (8) results in difficulties, in that the plastic strain
increment is no longer normal to the yield surface.

As materials subjected to impact and high strain-rate loading
are usually strain-rate dependent, equation (9) may not be applicable and some other constitutive model for the aggregate may be more
suitable. The only requirement is that it contain a plastic potential which can be replaced by equation (1). As most constitutive
models use the von Mises' yield function, either directly or
indirectly, this need not be too severe a restriction.

As an example of the above replacement, if the Perzyna visco-plastic model (Fyfe, 1975) is used, the plastic strain rate for an incompressible material is given by

$$\varepsilon^P_{ij} = \dot{\gamma}\left[\exp\{\alpha(\sqrt{J_2} - K^*)/K_o\} - 1\right]s_{ij}/\sqrt{J_2} \tag{10}$$

while if the Gurson yield function is used, this equation has the form

$$\dot{\varepsilon}^P_{ij} = \dot{\gamma}\left[\exp\{\alpha(\sqrt{\psi} - K^*_m)/K_{mo}\} - 1\right](s_{ij} + \frac{fK_m}{3}\sinh \xi)/\sqrt{\psi} \tag{11}$$

where $\psi = J_2 + 2fK^2_m \cosh\xi - f^2K^2_m$,

$$\xi = \left[\frac{\sigma_{11} + \sigma_{22} + \sigma_{33}}{2Y_m}\right] \quad , \quad K_m = Y_m/\sqrt{3}$$

and $\dot{\gamma}$ and α are material constants, s_{ij} are components of the deviatoric stress, and K^*_m and K_{mo} are respectively the static strain hardening and the initial yield strength of the matrix in simple shear.

An interesting feature of the above theory is that if f and $\dot{f}$ are zero, equation (1) reduces to the von Mises' yield function, equation (3) is the plastic incompressibility requirement of classical plasticity and the constitutive equation (11) is identical to (10). It thus can be readily seen that the Gurson model can be considered as an extension of the equations normally used to determine the plastic response. For this reason it is possible to incorporate void growth into most computer programs in a fairly direct manner.

INCIPIENT SPALLATION

Spallation due to impact loading is, for many materials, manifest as the growth and coalescence of voids into a well defined failure surface, and although failure of this type may be rather specialized, it does provide a suitable vehicle to examine the critical void volume concept.

A computer program which had been assembled to examine the response of a plate to uniaxial strain impact was adapted to include the above theory. This program, along with some experimental data reported by Butcher and Young (1971), was used to determine void volume fraction levels. The experimental conditions were such that they produced, in 6061-T6 aluminum, incipient spallation - defined as the damage level which could just be discerned using 50X magnification. The input conditions and sample dimensions of the experiment are given in the following table.

Table 1 6061-T6 Aluminum Plate Impact Incipient Spall Experiments

Experiment No.	Projectile Thickness (mm)	Target Thickness (mm)	Velocity (m/sec)
1	0.793	3.048	208
2	1.562	6.332	186
3	2.921	12.852	167
4	4.572	15.723	131

In order to determine the constants which describe the distribution function associated with the void nucleation of equation (7)
the effective plastic strain was calculated as a function of time
at the spallation plane for the given experimental conditions. The
results are presented in Figure 1. As the target in experiment 4
sustained damage at the lowest strain values, the distribution constants were taken to allow maximum nucleation over this range of
strain, and ε_n and s were given the values 0.008 and 0.002 respectively.

The value of f_n, and hence the critical void volume fraction f_c,
is very much dependent on what level of damage is considered to be

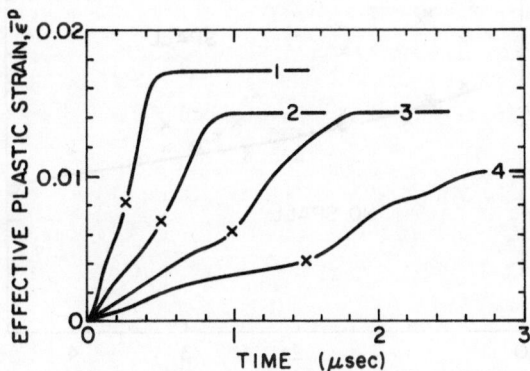

Figure 1.   The effective plastic history at the spallation plane
calculated for the experimental additions as given in Table 1.
- X - The onset of tensile mean stress.

critical. Kreer (1971) in a study of spallation in 6061-T6 suggested
that the initial void volume level should be f_c = 0.017. Another
approach, by Rajendran and Fyfe (1982), used the plastic instability
criterion of Hutchinson and Neale. In this study, the void volume
fraction was calculated for the failure of thin rings (failure in
this case being defined as the point of plastic instability) and it
was determined that f_c = 0.0121 was an appropriate value. As both
values were equally likely, and relatively close, the latter was
used. The value of f_n was adjusted in the analysis of experiment 4
until f_c was reached. This adjustment gave a value of f_c = 0.0124.

With the material constants thus tentatively established, the
other experimental configurations were analyzed, and the results are
shown in Figure 2, together with the experimental data from numerous
other studies, as assembled by Kreer. The agreement with the Butcher
and Young experiment for a flyer-plate thickness of 4.57 mm is, of
course, not surprising as this experiment was used to obtain the
unknown constants of the void nucleation distribution. However, the
experimental observation that higher impact velocities are required
for thinner flyer plates (i.e. shorter duration pulses) is clearly
followed.

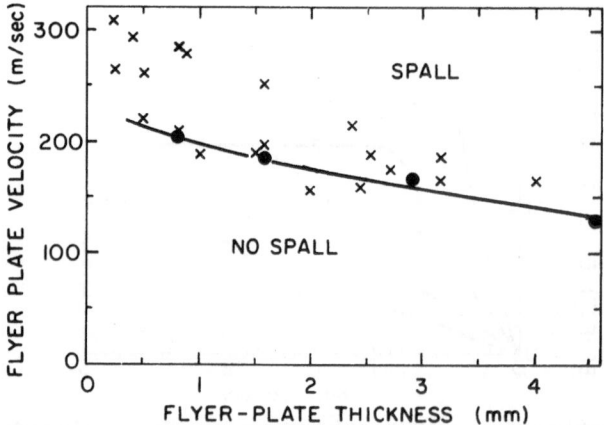

Figure 2. Incipient spallation predicted by the Gurson void growth
model in 6061-T6 aluminum.

The above analysis was repeated for the nucleation model as
defined by equation (8). It was found that the higher mean stresses
which coincide with the higher impact velocities resulted in greater
void nucleation and growth. Thus, incipient spallation was predicted
at significantly lower impact velocities than experimental evidence
would indicate when values for the thinner flyer plate tests were
calculated.

Using the constants determined in the above analysis the failure
of dynamically loaded annular cylindrical specimens was examined.
The loading initiated at the inner boundary produced a compressive
symmetric stress wave which propagated through the thickness of the
cylinder. With impulsive loading it is possible to induce spallation
when the wave reflects from the stress free outer surface. The techniques to produce this type of failure, and the loading pulse
required, are described in a paper by Schmidt and Fyfe (1973). An
example of this type of failure is shown in Figure 3, in which the
presence and initial coalescence of spherical voids at the spallation
plane can be seen.

A computer program which had been developed for the plane strain
deformations associated with this kind of loading was adapted for the
Gurson model. The void volume fraction was calculated throughout the
cylinder, whose outer diameter was such that the incipient spall
level was reached. These results are given in Figure 4. In this
case it was found that the void volume fraction reached a maximum
quite close to the experimentally observed spallation plane, and the

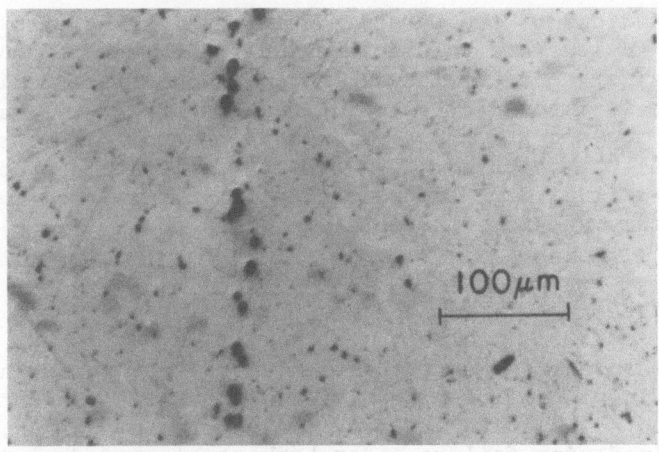

Figure 3. Spallation in a hollow cylinder of 6061-T6 aluminum,
I.D. 7.5 mm, O.D. 19 mm.

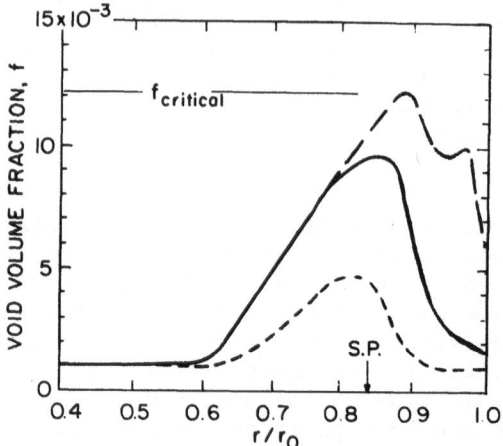

Figure 4. Void volume fraction distribution in a dynamically loaded thick-walled cylinder. (S.P. - spall radius, ------ 1.6 μsec., ———— 2.2 μsec., — — 3.0 μsec.)

time to reach the maximum level was also compatible with the experiments.

The limitations of the Gurson model requires that its application be restricted to low void volume levels, if the assumption of the single void is not to be violated. An additional restriction on the above analysis is imposed by assuming that the E_t of the matrix and the aggregate are one and the same, and so no appreciable softening of the material has occurred. The first restriction has been addressed by Tvergaard (1981), while the second is only imposed for computational expediency and is compatible with this study.

THIN RING FAILURE

The low mean stress and the presence of shear bands in impact induced thin ring expansion makes this configuration an interesting one in the study of ductile failure. As mentioned earlier, Rajendran and Fyfe examined ring failure under both static and dynamic conditions using the Gurson model. In the static case the long-wavelength

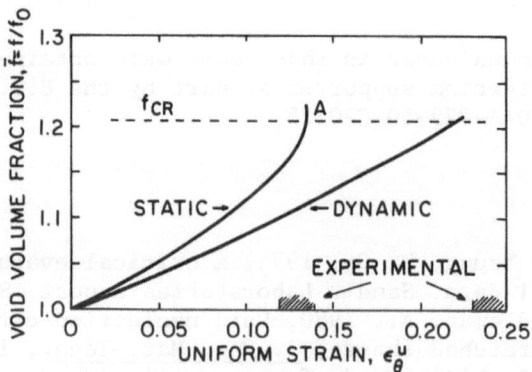

Figure 5. Void volume growth in expanding thin rings. A – point of plastic instability.

plastic instability criterion of Hutchinson and Neale was used to determine the hoop strain at failure. The void volume fraction that was reached at this point was then fixed as the critical value. The dynamic experiments were used to determine if a critical void volume fraction determined in this fashion was strain-rate independent. Results are shown in Figure 5 where it can be seen that strain rate does not appear to have an appreciable effect. In this case it was assumed that f_n was zero and that all the voids nucleated immediately plastic flow occurred. The initial value f_o was 0.01 for the results shown.

CONCLUSIONS

The results of this study show that the nucleation process is a key element in void growth failure analysis, and further work is required in this area. As is to be expected, the Gurson model is particularly suited to the high mean stress failure conditions of spallation and the value of f_c should be the highest in this case. In the ring configuration localized plastic flow may occur before any significant hole growth, in which event some instability criterion is required. If the critical void volume fraction is a function of the stress state, application of the Gurson model to notched tensile tests, crack formation and growth are exciting possible applications.

ACKNOWLEDGMENT

   The results communicated in this paper were obtained in the
course of an investigation supported in part by the U.S. Army Research
Office under Grant DAA-G29-80-C-0128.

REFERENCES

Butcher, B. M. and Young, E. G., 1971, A critical evaluation of
      6061-T6 spall data, Sandia Laboratories Report, SC-DR-710173.
Chu, C. C. and Needleman, A., 1980, Void nucleation effects in
      biaxially stretched sheets, J. Eng. Mat. Tech., 102:249.
Considère, A., 1885, L'Emploi du fer et de l'acier dans les
      constructions, Ann. Ponts Chaussee, 9:574.
Fyfe, I. M., 1975, The applicability of elastic/viscoplastic theory
      in stress wave propagation, J. Appl. Mech., 42:141.
Fyfe, I. M. and Rajendran, A. M., 1980, Dynamic pre-strain and
      inertia effects on the fracture of metals, J. Mech. Phys.
      Solids, 28:17
Gurson, A. L., 1977, Continuum theory of ductile rupture by void
      nucleation and growth: Part 1 - Yield criterion and flow
      rules for porous ductile materials, J. Eng. Mat. Tech., 99:2
Hutchinson, J. W. and Neale, K. W., 1977, Influence of strain-rate
      sensitivity on necking under uniaxial tension, Acta Met.,
      25:839.
Kreer, J. R., 1971, Dynamic fracture in 6061-T6 aluminum, Tech.
      Report AFWL-TR-70-180, Air Force Systems Command.
Needleman, A. and Rice, J. R., 1978, Limits to ductility set by
      plastic flow localization, in: "Mechanics of Sheet Metal
      Forming," D. P. Koistinen and N. M. Wang, eds., Plenum Press,
      New York.
Rajendran, A. M. and Fyfe, I. M., 1982, Inertia effects on the
      ductile failure of thin rings, J. Appl. Mech., 49:31
Schmidt, R. M. and Fyfe, I. M., 1973, An examination of dynamic
      fracture under biaxial strain conditions, Exp. Mech., 13:163
Tvergaard, V., 1981, Ductile fracture by cavity nucleation between
      larger voids, Report 210, Danish Center for Appl. Math. and
      Mech.

CONTENTS (VOL. II)

WORK IN PROGRESS

SOME PRELIMINARY VIEWS OF PLASMA INTERACTION (ELECTRO-MAGNETIC LAUNCH SYSTEMS)

Dr. A. Buckingham
Lawrence Livermore National Laboratory

ADIABATIC PLASTIC INSTABILITY IN ARMOR STEELS

Drs. G. Olson, M. Azrin and N. Tsangarakis
Army Materials and Mechanics Research Center

EXPANDING RING TESTS TO DETERMINE DYNAMIC FLOW STRESS AND FRACTURE

Dr. S. J. Bless
University of Dayton Research Institute

MODELING OF DYNAMIC BEHAVIOR OF ARMCO IRON

Dr. W. Cook
Air Force Armament Laboratory

SPALL IN MILD STEEL

Dr. S. J. Bless, University of Dayton Research Institute and
Dr. G. Spitole, Air Force Armament Laboratory

321